Springer Tracts in Modern Physics
Volume 190

Available online at
SpringerLink.com

Starting with Volume 165, Springer Tracts in Modern Physics is part of the [SpringerLink] service. For all customers with standing orders for Springer Tracts in Modern Physics we offer the full text in electronic form via [SpringerLink] free of charge. Please contact your librarian who can receive a password for free access to the full articles by registration at:

www.springerlink.com

If you do not have a standing order you can nevertheless browse online through the table of contents of the volumes and the abstracts of each article and perform a full text search.

There you will also find more information about the series.

Springer
Berlin
Heidelberg
New York
Hong Kong
London
Milan
Paris
Tokyo

Physics and Astronomy

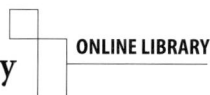

ONLINE LIBRARY

http://www.springer.de

Springer Tracts in Modern Physics

Springer Tracts in Modern Physics provides comprehensive and critical reviews of topics of current interest in physics. The following fields are emphasized: elementary particle physics, solid-state physics, complex systems, and fundamental astrophysics.

Suitable reviews of other fields can also be accepted. The editors encourage prospective authors to correspond with them in advance of submitting an article. For reviews of topics belonging to the above mentioned fields, they should address the responsible editor, otherwise the managing editor. See also www.springer.de

Managing Editor

Gerhard Höhler

Institut für Theoretische Teilchenphysik
Universität Karlsruhe
Postfach 69 80
76128 Karlsruhe, Germany
Phone: +49 (7 21) 6 08 33 75
Fax: +49 (7 21) 37 07 26
Email: gerhard.hoehler@physik.uni-karlsruhe.de
www-ttp.physik.uni-karlsruhe.de/

Elementary Particle Physics, Editors

Johann H. Kühn

Institut für Theoretische Teilchenphysik
Universität Karlsruhe
Postfach 69 80
76128 Karlsruhe, Germany
Phone: +49 (7 21) 6 08 33 72
Fax: +49 (7 21) 37 07 26
Email: johann.kuehn@physik.uni-karlsruhe.de
www-ttp.physik.uni-karlsruhe.de/~jk

Thomas Müller

Institut für Experimentelle Kernphysik
Fakultät für Physik
Universität Karlsruhe
Postfach 69 80
76128 Karlsruhe, Germany
Phone: +49 (7 21) 6 08 35 24
Fax: +49 (7 21) 6 07 26 21
Email: thomas.muller@physik.uni-karlsruhe.de
www-ekp.physik.uni-karlsruhe.de

Fundamental Astrophysics, Editor

Joachim Trümper

Max-Planck-Institut für Extraterrestrische Physik
Postfach 16 03
85740 Garching, Germany
Phone: +49 (89) 32 99 35 59
Fax: +49 (89) 32 99 35 69
Email: jtrumper@mpe-garching.mpg.de
www.mpe-garching.mpg.de/index.html

Solid-State Physics, Editors

Andrei Ruckenstein
Editor for The Americas

Department of Physics and Astronomy
Rutgers, The State University of New Jersey
136 Frelinghuysen Road
Piscataway, NJ 08854-8019, USA
Phone: +1 (732) 445 43 29
Fax: +1 (732) 445-43 43
Email: andreir@physics.rutgers.edu
www.physics.rutgers.edu/people/pips/
Ruckenstein.html

Peter Wölfle

Institut für Theorie der Kondensierten Materie
Universität Karlsruhe
Postfach 69 80
76128 Karlsruhe, Germany
Phone: +49 (7 21) 6 08 35 90
Fax: +49 (7 21) 69 81 50
Email: woelfle@tkm.physik.uni-karlsruhe.de
www-tkm.physik.uni-karlsruhe.de

Complex Systems, Editor

Frank Steiner

Abteilung Theoretische Physik
Universität Ulm
Albert-Einstein-Allee 11
89069 Ulm, Germany
Phone: +49 (7 31) 5 02 29 10
Fax: +49 (7 31) 5 02 29 24
Email: frank.steiner@physik.uni-ulm.de
www.physik.uni-ulm.de/theo/qc/group.html

Guido Altarelli   Klaus Winter (Eds.)

Neutrino Mass

With 118 Figures and 10 Tables

 Springer

Professor Guido Altarelli

CERN-TH
1211 Geneva 23, Switzerland
E-mail: guido.altarelli@cern.ch

Professor Klaus Winter

CERN-TH
1211 Geneva 23, Switzerland
E-mail: klaus.winter@cern.ch

Cataloging-in-Publication Data applied for

A catalog record for this book is available from the Library of Congress.

Bibliographic information published by Die Deutsche Bibliothek

Die Deutsche Bibliothek lists this publication in the Deutsche Nationalbibliografie; detailed bibliographic data is available in the Internet at dnb.ddb.de.

Physics and Astronomy Classification Scheme (PACS): 12.15, 13.10, 14.60

ISSN print edition: 0081-3869
ISSN electronic edition: 1615-0430   ⁄
ISBN 3-540-40328-0 Springer-Verlag Berlin Heidelberg New York

Springer-Verlag Berlin Heidelberg New York
a member of BertelsmannSpringer Science+Business Media GmbH

www.springer.de

© Springer-Verlag Berlin Heidelberg 2003
Printed in Germany

Typesetting: Authors and LE-TeX GbR, Leipzig using a Springer LaTeX macro package
Production: LE-TeX Jelonek, Schmidt & Vöckler GbR, Leipzig
Cover concept: eStudio Calamar Steinen
Cover production: *design & production* GmbH, Heidelberg

Printed on acid-free paper   SPIN: 10842454   56/3141/YL   5 4 3 2 1 0

Preface

Neutrino physics began with the direct experimental detection of neutrinos and continued with the discovery of different neutrino flavors and the study of their interactions, first in charged-current weak processes and then via the newly discovered neutral-current weak reactions. In the last few decades, the study of neutrino interactions, now incorporated into the standard electro-weak theory, has reached a level of maturity [K. Winter (ed.), *Neutrino Physics*, 1st edn. (Cambridge University Press, Cambridge 1991); 2nd edn. (2000)]. At present the main focus of neutrino physics is on the study of neutrino oscillations, which determine the mass-squared splittings and mixings of neutrinos, and on experiments aimed at a determination of the absolute scale of neutrino masses. The result that neutrino masses are very small suggests that neutrinos, the only neutral fundamental fermions, are Majorana particles with small masses inversely proportional to the large scale of lepton-number-nonconserving interactions.

The investigation of neutrino oscillations started with the pioneering work of Davis (2002 Nobel laureate) and his group using a chlorine detector in the Homestake mine in the USA. In 1968 this experiment made the first report of a deficit in the detected rate of neutrinos from the sun with respect to the expectation based on the theory of solar energy generation. Pontecorvo attributed the solar neutrino deficit to neutrino flavor oscillations. Since then, the investigation of solar neutrino oscillations with radiochemical and counter experiments has continued; the main experimental results have been obtained by Homestake, GALLEX, SAGE, (Super-)Kamiokande, SNO, and KamLAND. Today there is growing evidence that the solar neutrino deficit is indeed due to the disappearence of ν_e's, which oscillate mainly into ν_μ's and ν_τ's. The mass-squared difference that corresponds to solar oscillations is about 7×10^{-5} eV^2 and the mixing angle is large.

The early studies of neutrino oscillations with reactor- and accelerator-generated neutrino beams led to no evidence for oscillations (but in a domain of sensitivity to mass-squared differences well above that indicated by the solar neutrino deficit). The experiments CHORUS and NOMAD at CERN have recently been completed, further improving the existing limits on oscillations. At various times, several experiments have made positive claims which were later excluded by other experiments. At present, there is a claim by the Los Alamos accelerator experiment LSND of oscillations with a mass-squared difference of the order of $1\,\mathrm{eV}^2$, which has not been confirmed, although it has not been completely excluded, by the KARMEN collaboration. The LSND signal region will be explored precisely by the MiniBooNE experiment now starting to take data.

By studying neutrinos generated in the earth's atmosphere by cosmic rays, clear evidence for neutrino oscillations, with a mass-squared difference around $3 \times 10^{-3}\,\mathrm{eV}^2$ and a near-maximal mixing angle, has been accumulated over the years, and the conclusion was finally established by Super-Kamiokande. The Super-Kamiokande experiment in Japan uses an underground 50 kiloton water Cherenkov detector and has systematically explored the up–down asymmetry of atmospheric neutrino fluxes over distances of the order of the earth's diameter. The dominant oscillation for atmospheric neutrinos is $\nu_\mu \to \nu_\tau$ while the oscillation $\nu_\mu \to \nu_e$ at the same frequency is strongly suppressed, as directly shown by CHOOZ, a reactor experiment in France. The K2K experiment uses the Super-Kamiokande detector and 1 GeV neutrinos produced by an accelerator 250 km away to check the disappearance of ν_μ's in an independent setup that deals only with neutrinos produced and detected on the earth.

Neutrino oscillations determine differences between mass-squared values, and the actual scale of neutrino masses remains to be fixed experimentally. In particular, the scale of neutrino masses is important for cosmology as neutrinos are candidates for hot dark matter: nearly degenerate neutrinos with a common mass around 1–2 eV would significantly contribute to Ω_m, the matter density in the universe in units of the critical density. The detection of $0\nu\beta\beta$ decay would be extremely important for the determination of the overall scale of neutrino masses, the confirmation of their Majorana nature, and the experimental clarification of the ordering of levels in the associated spectrum. For neutrino masses in a range remarkably consistent with present oscillation data, the decay of heavy right-handed neutrinos with lepton number nonconservation can provide a viable and attractive model of baryogenesis through leptogenesis.

This book is devoted to a review of the theory and experiments on neutrino oscillations, masses, and mixings. The formalism of neutrino masses and of neutrino oscillations in vacuum and in matter is described in the article by Kayser in Chap. 1. The experimental bounds on the direct determination of neutrino masses are reviewed by Weinheimer in Chap. 2. The astrophysical and cosmological issues of importance for neutrino physics are described by Kainulainen and Olive in Chap. 3. The experimental results on neutrino oscillations are summarized by Geisser in Chap. 4. In the article by Fogli and Lisi in Chap. 5 the theoretical interpretation of neutrino oscillation data in terms of mass-squared differences and mixing angles is reviewed. Theoretical models of neutrino masses and mixings are discussed in Chap. 6 by Altarelli and Feruglio. Finally, the physics potential of future long-baseline experiments is presented in Chap. 7 by Lindner.

Geneva, June 2003

Guido Altarelli
Klaus Winter

Contents

4 Experimental Results on Neutrino Oscillations

5 Theoretical Interpretation
of Current Neutrino Oscillation Data

6 Theoretical Models of Neutrino Masses and Mixings

7 The Physics Potential of Future Long-Baseline
Neutrino Oscillation Experiments

List of Contributors

Guido Altarelli
CERN-TH
1211 Geneva 23, Switzerland
guido.altarelli@cern.ch

Ferruccio Feruglio
Dipartimento di Fisica,
Università di Padova and INFN
Padova, Italy
feruglio@pd.infn.it

Gianluigi Fogli
Dipartimento di Fisica and
Istituto Nazionale di Fisica Nucleare
Sezione di Bari
Via Amendola 173
70126 Bari, Italy
gianluigi.fogli@ba.infn.it

Achim Geiser
Deutsches Elektronen-Synchrotron
DESY
Hamburg, Germany
Achim.Geiser@desy.de

Kimmo Kainulainen
Theory Division, CERN
1211 Geneva 23, Switzerland
and
NORDITA
Blegdamsvej 17
2100 Copenhagen Ø, Denmark
kimmo.kainulainen@cern.ch

Boris Kayser
Theoretical Physics Department
Fermi National
Accelarator Laboratory
P.O. Box 500
Batavia, IL 60510, USA
boris@fnal.gov

Manfred Lindner
Physik Department T30d,
Technische Universität München
James-Franck-Straße 1
85747 Garching, Germany
lindner@ph.tum.de

Eligio Lisi
Dipartimento di Fisica and
Istituto Nazionale di Fisica Nucleare
Sezione di Bari
Via Amendola 173
70126 Bari, Italy
eligio.lisi@ba.infn.it

Keith A. Olive
Theoretical Physics Institute
School of Physics and Astronomy
University of Minnesota
Minneapolis, MN 55455, USA
olive@umphys.spa.umn.edu

Christian Weinheimer
Helmholtz-Institut
für Strahlen- und Kernphysik
Rheinische
Friedrich-Wilhelms-Universität
53115 Bonn, Germany
weinheimer@iskp.uni-bonn.de

Klaus Winter
Humboldt University, Berlin
and
CERN-TH
1211 Geneva 23, Switzerland
klaus.winter@cern.ch

1 Neutrino Mass, Mixing, and Flavor Change

Boris Kayser

1.1 Neutrino Masses and Mixing, and the Seesaw

The evidence that neutrinos change from one flavor to another is compelling [1]. Barring exotic possibilities, neutrino flavor change implies neutrino mass and mixing. Thus, neutrinos almost certainly have nonzero masses and mix.

That neutrinos have masses means that there is a spectrum of three or more neutrino mass eigenstates, $\nu_1, \nu_2, \nu_3, \ldots$ That neutrinos mix means that the neutrino state coupled by the charged-current weak interaction to the W boson and a specific charged lepton (such as the electron) is none of the neutrino mass eigenstates but, rather, is a mixture of them. Consider, for example, the leptonic W decay $W^+ \rightarrow \ell_\alpha^+ + \nu_\alpha$, yielding the specific charged lepton ℓ_α. Here, the "flavor" α of the lepton can be e, μ or τ, and ℓ_e is the electron, ℓ_μ the muon, and ℓ_τ the τ. In the W decay, the produced neutrino state $|\nu_\alpha\rangle$, referred to as the neutrino of flavor α, is the superposition

$$|\nu_\alpha\rangle = \sum_i U_{\alpha i}^* |\nu_i\rangle \qquad (1.1)$$

of the mass eigenstates $|\nu_i\rangle$. Here, U is a matrix known as the leptonic mixing matrix.[1]

Through our studies of neutrinos, we hope to eventually discover what physics lies behind their masses and mixing.[2] This underlying physics may contain neutrino mass terms of two different kinds: Dirac and Majorana. As depicted in Fig. 1.1, a Dirac mass term turns a neutrino into a neutrino or an antineutrino into an antineutrino, while a Majorana mass term converts a neutrino into an antineutrino or vice versa. Thus, Dirac mass terms conserve the lepton number L that distinguishes leptons from antileptons, while Majorana mass terms do not. The quantum number L is also conserved by the Standard Model (SM) couplings of neutrinos to other particles. Thus, if we assume that the interactions between neutrinos and other particles are well described by these SM couplings – a very plausible assumption in view of the

---

[1] Increasingly, U is referred to as the MNS matrix or the MNSP matrix, to honor the pioneering contributions of Maki, Nakagawa, Sakata, and Pontecorvo.

[2] For a discussion of this physics, see Chap. 6.

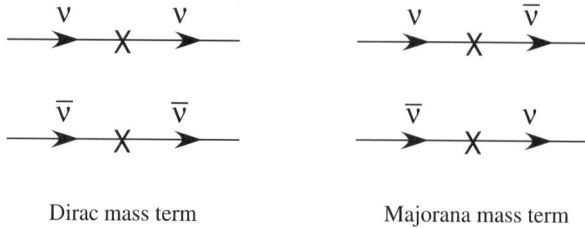

Dirac mass term   Majorana mass term

Fig. 1.1. The effects of Dirac and Majorana mass terms. The action of the mass terms is represented by the symbol X

great success of the SM – then any L nonconservation that we might oberve in neutrino experiments would have to arise from Majorana mass terms, not from interactions.

A Dirac mass term may be constructed out of a chirally left-handed neutrino field ν_L^0 and a chirally right-handed one ν_R^0:[3]

$$\mathcal{L}_D = -m_D \overline{\nu_R^0} \nu_L^0 + \text{h.c.} ,\qquad (1.2)$$

where 'h.c.' denotes the Hermitian conjugate. A Majorana mass term may be constructed out of ν_L^0 alone, in which case we have the "left-handed Majorana mass"

$$\mathcal{L}_{m_L} = -\frac{m_L}{2} \overline{(\nu_L^0)^c} \nu_L^0 + \text{h.c.} ,\qquad (1.3)$$

or out of ν_R^0 alone, in which case we have the "right-handed Majorana mass"

$$\mathcal{L}_{m_R} = -\frac{m_R}{2} \overline{(\nu_R^0)^c} \nu_R^0 + \text{h.c.}\qquad (1.4)$$

In these expressions, m_D, m_L, and m_R are mass parameters, and for any field ψ, ψ^c is the corresponding charge-conjugate field. In terms of ψ, $\psi^c = C\overline{\psi}^{\mathrm{T}}$, where C is the charge conjugation matrix, and T denotes transposition.

(In writing both mass and interaction terms, we use the superscript zero to denote underlying fields out of which a model is constructed. Fields without a superscript zero correspond to physical particles of definite mass.)

Table 1.1 indicates the effects of the various fields in the mass terms on neutrinos and antineutrinos. From this table, we see that each type of mass term does indeed induce the transitions ascribed to it in Fig. 1.1.

---

[3] A chirally left-handed (right-handed) field $\psi_{L(R)}$ is one that obeys the relations $P_{L(R)}\psi_{L(R)} = \psi_{L(R)}$ and $P_{R(L)}\psi_{L(R)} = 0$. Here, $P_L = (1/2)(1 - \gamma_5)$ and $P_R = (1/2)(1 + \gamma_5)$ are the left-handed and right-handed chiral projection operators, respectively. When the quanta of the field are massless, ψ_L annihilates particles of left-handed helicity, and creates antiparticles of right-handed helicity; ψ_R annihilates particles of right-handed helicity, and creates antiparticles of left-handed helicity. When the quanta are not massless, there are corrections of order mass/energy to these rules.

Table 1.1. The effects of the fields. "A" signifies that the given particle is annihilated by the field, and "C" that the particle is created

Field	Effect on ν	Effect on $\bar{\nu}$
$\nu^0_{L,R}$	A	C
$\overline{\nu^0_{L,R}}$	C	A
$(\nu^0_{L,R})^c$	C	A
$\overline{(\nu^0_{L,R})^c}$	A	C

An electrically charged fermion such as a quark cannot have a Majorana mass term, because such a term would convert it into an antiquark, in violation of electric charge conservation. However, for the electrically neutral neutrinos, Majorana mass terms are not only allowed but rather likely, given that the neutrinos are now known to be particles with mass.[4] To see why Majorana mass terms are likely, suppose first that some neutrino is described by the SM. In the original version of that model, the neutrino would be massless. It would be described by the left-handed (LH) field that we have called ν^0_L, and the model would contain no right-handed (RH) counterpart to ν^0_L. Let us suppose that we now try to extend the SM to accommodate a nonzero mass for this neutrino in the same way that the SM already accommodates nonzero masses for the quarks and charged leptons. The latter masses, of course, are all of Dirac type, and arise from Yukawa couplings of the form

$$-f_q\varphi\overline{(q^0_L)}q^0_R + \text{h.c.} \tag{1.5}$$

Here, q^0 is some quark, φ is the neutral Higgs field, and f_q is a coupling constant. When φ develops a vacuum expectation value $\langle\varphi\rangle_0$, the coupling (1.5) yields a term

$$-f_q\langle\varphi\rangle_0\overline{(q^0_L)}q^0_R + \text{h.c.} \tag{1.6}$$

This is a Dirac mass term of the form of (1.2) for the quark q^0, and $f_q\langle\varphi\rangle_0$ is the mass. Extending the SM to include a mass for our neutrino that parallels the masses of the quarks is a simple matter of adding to the model a RH neutrino field ν^0_R and a Yukawa coupling $-f_\nu\varphi\overline{(\nu^0_L)}\nu^0_R +$ h.c., where f_ν is a suitable coupling constant. When φ develops its vacuum expectation value, this coupling will yield a Dirac mass term

$$\mathcal{L}_D = -f_\nu\langle\varphi\rangle_0\overline{(\nu^0_L)}\nu^0_R + \text{h.c.} \tag{1.7}$$

for this neutrino. This term imparts to the neutrino a mass $m_\nu = f_\nu\langle\varphi\rangle_0$. Now suppose, for example, that we would like m_ν to be of order 0.05 eV, the neutrino mass scale suggested by the observed atmospheric neutrino oscillations. Since $\langle\varphi\rangle_0 = 174\,\text{GeV}$, the coupling f_ν must then be of order 10^{-13}.

---

[4] The author first heard the argument that follows from Belen Gavela.

Such an infinitesimal coupling constant may not be out of the question, but it certainly strikes one as unlikely to be the ultimate explanation of neutrino mass.

In addition, to generate a Dirac mass for the neutrino, we were obliged to introduce the RH neutrino field ν_R^0. In the SM, right-handed fermion fields are weak-isospin singlets. Hence, so are their charge conjugates. Thus, once the field ν_R^0 exists, there is nothing in the SM to prevent the occurrence of a right-handed Majorana mass term like that in (1.4): such a term violates neither the conservation of weak isospin nor that of electric charge. Consequently, if nature contains a Dirac neutrino mass term, then it is highly likely that it contains a Majorana mass term as well. And, needless to say, if nature does not contain a Dirac neutrino mass term, then it certainly contains a Majorana mass term, which would then be the only source of neutrino mass.

Suppose that a neutrino has a Dirac mass, as the quarks and charged leptons do, and also a right-handed Majorana mass like that in (1.4), as suggested by the previous argument. Its total mass term $\mathcal{L}_{m_\nu}$ is then

$$
\begin{aligned}
\mathcal{L}_{m_\nu} &= -m_D \overline{\nu_R^0} \nu_L^0 - \frac{m_R}{2} \overline{(\nu_R^0)^c} \, \nu_R^0 + \text{h.c.} \\
&= -\frac{1}{2} \left[\overline{(\nu_L^0)^c}, \overline{\nu_R^0} \right] \begin{bmatrix} 0 & m_D \\ m_D & m_R \end{bmatrix} \begin{bmatrix} \nu_L^0 \\ (\nu_R^0)^c \end{bmatrix} + \text{h.c.}
\end{aligned}
\tag{1.8}
$$

Here, we have used the identity $\overline{(\nu_L^0)^c} m_D (\nu_R^0)^c = \overline{\nu_R^0} m_D \nu_L^0$. The matrix

$$
\mathcal{M}_\nu = \begin{bmatrix} 0 & m_D \\ m_D & m_R \end{bmatrix}
\tag{1.9}
$$

appearing in $\mathcal{L}_{m_\nu}$ is referred to as the neutrino mass matrix.

It is natural to suppose that the Dirac mass m_D of our neutrino is of the same order of magnitude as the Dirac masses of the quarks and charged leptons, since in the SM all of these Dirac masses arise from couplings to the same Higgs field. Of course, the Dirac masses of the quarks and charged leptons are their total masses, so we expect m_D to be of the same order of magnitude as the mass of a typical quark or charged lepton. Furthermore, since nothing in the SM requires the right-handed Majorana mass m_R to be small, we expect that this mass will be *large*: $m_R \gg m_D$.

The mass matrix $\mathcal{M}_\nu$ can be diagonalized by the transformation

$$
Z^T \mathcal{M}_\nu Z = \mathcal{D}_\nu ,
\tag{1.10}
$$

where

$$
\mathcal{D}_\nu = \begin{bmatrix} m_1 & 0 \\ 0 & m_2 \end{bmatrix}
\tag{1.11}
$$

is a diagonal matrix whose diagonal elements are the positive-definite eigenvalues of $\mathcal{M}_\nu$,[5] Z is a unitary matrix, and T denotes transposition. To first

---

[5] Any negative eigenvalue can be converted into a positive one by a suitable modification of Z.

order in the small parameter $\rho \equiv m_D/m_R$,

$$Z = \begin{bmatrix} 1 & \rho \\ -\rho & 1 \end{bmatrix} \begin{bmatrix} i & 0 \\ 0 & 1 \end{bmatrix} . \tag{1.12}$$

Using this Z in (1.10), we find that to order ρ^2,

$$\mathcal{D}_\nu = \begin{bmatrix} m_D^2/m_R & 0 \\ 0 & m_R \end{bmatrix} . \tag{1.13}$$

Thus, the mass eigenvalues are $m_1 \simeq m_D^2/m_R$ and $m_2 \simeq m_R$.

To recast $\mathcal{L}_{m_\nu}$ in terms of mass eigenfields, we define the two-component column vector ν_L by

$$\nu_L \equiv Z^{-1} \begin{bmatrix} \nu_L^0 \\ (\nu_R^0)^c \end{bmatrix} . \tag{1.14}$$

(The column-vector field ν_L is chirally left-handed, since the charge conjugate of a field with a given chirality always has the opposite chirality.) We then define the two-component field ν, with components ν_1 and ν_2, by

$$\nu \equiv \nu_L + (\nu_L)^c \equiv \begin{bmatrix} \nu_1 \\ \nu_2 \end{bmatrix} . \tag{1.15}$$

Using the fact that scalar covariant combinations of fermion fields can connect only fields of opposite chirality, it is easy to show that the $\mathcal{L}_{m_\nu}$ of (1.8) may be rewritten as

$$\mathcal{L}_{m_\nu} = -\sum_{i=1}^{2} \frac{m_i}{2} \overline{\nu_i}\nu_i . \tag{1.16}$$

We recognize the ith term of this expression as the usual mass term for a neutrino ν_i. The mass of that neutrino appears to be $m_i/2$, but we shall see shortly that it is actually m_i.

From the definition in (1.15), we see that $\nu_i = \nu_{Li} + \nu_{Li}^c$ goes into itself under charge conjugation. A neutrino whose field has this property is identical to its antiparticle [2], and is known as a Majorana neutrino. Thus, the eigenstates of the combined Dirac–Majorana mass term $\mathcal{L}_{m_\nu}$ of (1.8) are Majorana neutrinos.

Fermions that are distinct from their antiparticles are known as Dirac particles. The mass term for a Dirac fermion f of mass m_f is $-m_f \overline{f}f$. But the mass term for a Majorana neutrino ν of mass m_ν is $-(1/2)m_\nu \overline{\nu}\nu$. To see why there is this extra factor of $1/2$, we note that if ν has a mass term $-k\overline{\nu}\nu$ in the Lagrangian density (where k is some constant), then the mass of ν is

$$\langle \nu \text{ at rest} \mid \int d^3x \, k\overline{\nu}\nu \mid \nu \text{ at rest} \rangle . \tag{1.17}$$

If ν is a Majorana particle, this matrix element is twice as large as it would be if ν were a Dirac particle. To see why, suppose first that ν is a Dirac particle

($\bar{v} \neq v$). Then the field v can absorb a neutrino or create an antineutrino. The field $\bar{v}$ can absorb an antineutrino or create a neutrino. Thus, in the matrix element (1.17), it is the field v that absorbs the initial neutrino, and the field $\bar{v}$ that creates the final neutrino. Now suppose that v is a Majorana particle. Then the fields v and $\bar{v}$ still do just what they did in the Dirac case, except that now there is no difference between the "antineutrino" and the neutrino. The field v can either absorb or create this neutrino, and so can the field $\bar{v}$. Thus the matrix element (1.17) has two terms. In the first, the field v absorbs the initial neutrino and the field $\bar{v}$ creates the final one. In the second, the field $\bar{v}$ absorbs the initial neutrino and the field v creates the final one. It is straightforward to show that these two terms are equal, and that each of them is equal to the single term present in the Dirac case. Hence, for a given k, the matrix element (1.17) is twice as large in the Majorana case as in the Dirac case. As is well known, in the Dirac case the matrix element is just equal to k, so that in the Majorana case k is half the mass.

With m_D of the order of the mass of a typical quark or charged lepton, and $m_R \gg m_D$, the mass of v_1,

$$m_1 \cong m_D^2/m_R , \qquad (1.18)$$

can be very small. Thus, if we identify v_1 as one of the light neutrinos, we have an elegant explanation of why it is so light. This explanation, in which physical neutrino masses are small because the RH Majorana mass m_R is large, is known as the seesaw mechanism, and (1.18) is referred to as the seesaw relation [3]. The mass m_R is assumed to reflect some high mass scale where new physics responsible for neutrino mass resides. Interestingly, if m_R is just a little below the grand unification scale – say $m_R \sim 10^{15}$ GeV – and $m_D \sim m_{top} \simeq 175$ GeV, then from (1.18) we obtain $m_1 \sim 3 \times 10^{-2}$ eV. This is in the range of the neutrino mass suggested by the experiments on atmospheric neutrino oscillations [4].

The reader will have noticed that under our assumptions about m_D and m_R, the mass of v_2,

$$m_2 \cong m_R , \qquad (1.19)$$

is far from small. The eigenstate v_2 cannot be one of the light neutrinos, but is a hypothetical very heavy, neutral lepton. Such neutral leptons figure prominently in attempts to explain the baryon–antibaryon asymmetry of the universe in terms of leptogenesis.

The seesaw mechanism, based as it is on the $\mathcal{L}_{m_v}$ of (1.8), predicts that the light neutrinos such as v_1, as well as the hypothetical heavy neutral leptons such as v_2, are Majorana particles. The light-neutrino aspect of this prediction is one of the factors driving a major effort [5] to look for neutrinoless double beta decay. This is the L-violating reaction Nucl $\rightarrow$ Nucl$'$ + 2e$^-$, in which one nucleus decays to another plus two electrons. Observation of this reaction at any nonzero level would show that the light neutrinos were indeed Majorana particles [6].

So far, we have analyzed the simplified case in which there is only one light neutrino and one heavy neutral lepton. In the real world, there are three leptonic generations, with a light neutrino in each one, and the particles in different generations mix. It is quite easy to extend our analysis to accommodate this situation.[6]

In the SM, there are left-handed weak-eigenstate charged leptons $\ell^0_{L\alpha}$, where α = e, μ, or τ. Each $\ell^0_{L\alpha}$ couples to an LH weak-eigenstate neutrino $\nu^0_{L\alpha}$ via the charged-current weak interaction

$$\mathcal{L}_W = -\frac{g}{\sqrt{2}} W^-_\rho \sum_{\alpha=e,\mu,\tau} \overline{\ell^0_{L\alpha}} \gamma^\rho \nu^0_{L\alpha} + \text{h.c.} \tag{1.20}$$

Here, W is the charged weak boson, and g is the semiweak coupling constant. To allow for neutrino masses, we add to the model the RH fields $\nu^0_{R\alpha}$, where α = e, μ, or τ. Then, in analogy with (1.8), we introduce the neutrino mass term

$$\mathcal{L}_{m_\nu} = -\frac{1}{2} [\overline{(\nu^0_L)^c}, \overline{\nu^0_R}] \begin{bmatrix} 0 & m^T_D \\ m_D & m_R \end{bmatrix} \begin{bmatrix} \nu^0_L \\ (\nu^0_R)^c \end{bmatrix} + \text{h.c.} \tag{1.21}$$

Here, ν^0_L is the column vector

$$\nu^0_L \equiv \begin{bmatrix} \nu^0_{Le} \\ \nu^0_{L\mu} \\ \nu^0_{L\tau} \end{bmatrix}, \tag{1.22}$$

and similarly for ν^0_R. The quantities m_D and m_R are now 3×3 matrices. In writing (1.21), we have used the fact that, for a given α and β, $\overline{(\nu^0_{L\alpha})^c}(m^T_D)_{\alpha\beta}(\nu^0_{R\beta})^c = \overline{(\nu^0_{R\beta})}(m_D)_{\beta\alpha}(\nu^0_{L\alpha})$. Thus, when one sums over α and β, the contributions of the submatrices m_D and m^T_D to $\mathcal{L}_{m_\nu}$ are identical, and add up to conventional Dirac mass terms without the factor of 1/2 in front of $\mathcal{L}_{m_\nu}$. Since $\overline{(\nu^0_{R\alpha})}(\nu^0_{R\beta})^c = \overline{(\nu^0_{R\beta})}(\nu^0_{R\alpha})^c$, the matrix m_R may be taken to be symmetric. Thus, the 6×6 mass matrix

$$\mathcal{M}_\nu = \begin{bmatrix} 0 & m^T_D \\ m_D & m_R \end{bmatrix} \tag{1.23}$$

is symmetric. Such a matrix may be diagonalized by the transformation in (1.10), but where Z is now a 6×6 unitary matrix and $\mathcal{D}_\nu$ is a 6×6 diagonal matrix whose diagonal elements m_i, $i = 1, \ldots, 6$, are the positive-definite eigenvalues of the $\mathcal{M}_\nu$ defined in (1.23).

To reexpress $\mathcal{L}_{m_\nu}$ in terms of mass-eigenstate neutrinos, we introduce the column vector ν_L via (1.14) as before. Of course, in that relation ν^0_L and $(\nu^0_R)^c$ each now have three components, Z is a 6×6 matrix, and ν_L has

---

[6] For simplicity, we shall not include the further very interesting possibility that there are light sterile neutrinos in addition to the three light active ones.

six components. We then introduce the field v via a six-component version of (1.15):

$$\mathsf{v} \equiv \mathsf{v_L} + (\mathsf{v_L})^c \equiv \begin{bmatrix} \mathsf{v}_1 \\ \mathsf{v}_2 \\ \vdots \\ \mathsf{v}_6 \end{bmatrix} . \tag{1.24}$$

It is then easily shown, as before, that the mass term $\mathcal{L}_{m_\mathsf{v}}$ of (1.21) may be rewritten as

$$\mathcal{L}_{m_\mathsf{v}} = -\sum_{i=1}^{6} \frac{m_i}{2} \overline{\mathsf{v}_i} \mathsf{v}_i . \tag{1.25}$$

Thus, the v_i are the neutrinos of definite mass, the mass of v_i being m_i. From (1.24), we see that each v_i is a Majorana neutrino.

To complete the treatment of the leptonic sector, we introduce for the charged leptons a (Dirac) mass term $\mathcal{L}_{m_\ell}$ given by [7]

$$\mathcal{L}_{m_\ell} = -\overline{\ell_\mathrm{R}^0} \mathcal{M}_\ell \ell_\mathrm{L}^0 + \text{h.c.} \tag{1.26}$$

Here,

$$\ell_\mathrm{L}^0 \equiv \begin{bmatrix} \ell_{\mathrm{L}e}^0 \\ \ell_{\mathrm{L}\mu}^0 \\ \ell_{\mathrm{L}\tau}^0 \end{bmatrix} \tag{1.27}$$

is a column vector whose αth component is the LH weak-eigenstate charged-lepton field $\ell_{\mathrm{L}\alpha}^0$. The quantity ℓ_R^0 is an analogous column vector whose αth component is the RH weak-isospin singlet charged-lepton field $\ell_{\mathrm{R}\alpha}^0$. Finally, $\mathcal{M}_\ell$ is the 3×3 charged-lepton mass matrix. This matrix may be diagonalized by the transformation [7]

$$A_\mathrm{R}^\dagger \mathcal{M}_\ell A_\mathrm{L} = \mathcal{D}_\ell , \tag{1.28}$$

where $A_{\mathrm{L,R}}$ are two distinct 3×3 unitary matrices, and

$$\mathcal{D}_\ell = \begin{pmatrix} m_\mathrm{e} & 0 & 0 \\ 0 & m_\mu & 0 \\ 0 & 0 & m_\tau \end{pmatrix} \tag{1.29}$$

is the diagonal matrix whose diagonal elements are the charged-lepton masses.

If we define the three-component column vectors $\ell_{\mathrm{L,R}}$ via

$$\ell_{\mathrm{L,R}}^0 = A_{\mathrm{L,R}} \, \ell_{\mathrm{L,R}} \tag{1.30}$$

and then introduce the vector

$$\ell \equiv \ell_\mathrm{L} + \ell_\mathrm{R} , \tag{1.31}$$

we quickly find that

$$\mathcal{L}_{m_\ell} = -\overline{\ell} \mathcal{D}_\ell \ell = -\sum_{\alpha=\mathrm{e},\mu,\tau} \overline{\ell_\alpha} m_\alpha \ell_\alpha . \tag{1.32}$$

Thus, the components ℓ_α of the vector ℓ are the charged leptons of definite mass: e, μ, and τ.

To recast the SM weak interaction $\mathcal{L}_{\mathrm{W}}$, (1.20), in terms of mass eigenstates, it is convenient to write the 6×6 matrix Z in the form

$$ Z = \begin{bmatrix} V & Y \\ X & W \end{bmatrix} , \tag{1.33} $$

in which V, W, X, and Y are 3×3 submatrices. If the Dirac mass matrix m_{D} is much smaller than the Majorana mass matrix m_{R}, then X and Y are much smaller than V and W for the same reason as the off-diagonal elements of the 2×2 version of Z, (1.12), are small. Similarly, from (1.13) we may conclude that in the three-generation, six-neutrino case, the first three neutrinos, $\nu_{1,2,3}$, are light, but the second three, $\nu_{4,5,6}$, are very heavy. To emphasize this, we shall call the first three neutrinos $\nu_{1,2,3}^{\mathrm{Light}}$ and the second three $\nu_{1,2,3}^{\mathrm{Heavy}}$, From experimental searches for heavy neutral leptons, we know that there are none with masses below 80 GeV [8]. Thus, in neutrino experiments at energies less than this (and even at much higher energies if the heavy neutrinos are at the TeV or even the grand unification scale), it is only the light neutrinos that play a significant role. Now, from the 6×6 analogue of (1.14) and from (1.33),

$$ \nu_{L\alpha}^0 = \sum_{i=1}^{3} \left(V_{\alpha i} \nu_{Li}^{\mathrm{Light}} + Y_{\alpha i} \nu_{Li}^{\mathrm{Heavy}} \right) \cong \sum_{i=1}^{3} V_{\alpha i} \nu_{Li}^{\mathrm{Light}} , \tag{1.34} $$

where we have used $Y \ll V$ in the second step. From (1.34) and (1.30), we may rewrite the weak interaction (1.20) as

$$ \mathcal{L}_{\mathrm{W}} \cong -\frac{g}{\sqrt{2}} W_\rho^- \overline{\ell_{\mathrm{L}}} \gamma^\rho U \nu_{\mathrm{L}}^{\mathrm{Light}} - \frac{g}{\sqrt{2}} W_\rho^+ \overline{\nu_{\mathrm{L}}^{\mathrm{Light}}} \gamma^\rho U^\dagger \ell_{\mathrm{L}} . \tag{1.35} $$

Here,

$$ \nu_{\mathrm{L}}^{\mathrm{Light}} \equiv \begin{bmatrix} \nu_{\mathrm{L}1}^{\mathrm{Light}} \\ \nu_{\mathrm{L}2}^{\mathrm{Light}} \\ \nu_{\mathrm{L}3}^{\mathrm{Light}} \end{bmatrix} \tag{1.36} $$

is a column vector whose ith component is the left-handed projection of the field of the ith light-neutrino mass eigenstate, and

$$ U \equiv A_{\mathrm{L}}^\dagger V \tag{1.37} $$

is the "leptonic mixing matrix". This is the same matrix as the one called U in (1.1). However, we are now assuming that there are only three light neutrinos, so that U is 3×3, and we are relating U to the matrices A_{L} and V that take part in the diagonalization of the underlying charged-lepton and neutrino mass matrices.

Equation (1.35) expresses the charged-current weak interaction in terms of charged leptons and neutrinos of definite mass. Since the matrix Z is unitary, and X and Y are much smaller than V and W, the matrix V is to a good approximation unitary itself. From the unitarity of A_{L} and (1.37), this means that the leptonic mixing matrix U is approximately unitary as well.

It is not hard to count the number of independent parameters necessary to fully determine U [9]. This matrix has nine entries, each of which may have a real and an imaginary part, leading to a total of 18 parameters. On these parameters, unitarity imposes nine constraints. First of all, each of the three columns of U must be a vector of unit length. Secondly, any two columns must be orthogonal to each other. This requirement leads to three orthogonality conditions, each of which has a real and an imaginary part, for a total of six constraints. With the nine unitarity constraints taken into account, nine parameters are left.

With the charged-lepton and neutrino indices indicated explicitly, (1.35) for the weak interaction reads

$$\mathcal{L}_{\mathrm{W}} = -\frac{g}{\sqrt{2}} \mathrm{W}_\rho^- \sum_{\substack{\alpha=e,\mu,\tau \\ i=1,2,3}} \overline{\ell_{\mathrm{L}\alpha}} \gamma^\rho U_{\alpha i} \nu_{\mathrm{L}i}^{\mathrm{Light}} + \mathrm{h.c.} \tag{1.38}$$

From (1.38), we see that, apart from the overall strength factor $g/\sqrt{2}$, $U_{\alpha i}$ is essentially just the amplitude $\langle \ell_\alpha^- | H_{\mathrm{W}} | \nu_i^{\mathrm{Light}} \rangle$ for the transition $\nu_i^{\mathrm{Light}} \to \ell_\alpha^-$ via emission or absorption of a W boson, caused by the action of the weak Hamiltonian H_{W} corresponding to $\mathcal{L}_{\mathrm{W}}$. Now, we are always free to redefine what we mean by the state $\langle \ell_\alpha^- |$ by multiplying it by a phase factor: $\langle \ell_\alpha^- | \to e^{i\varphi_\alpha} \langle \ell_\alpha^- |$. Obviously, this phase redefinition causes the $U_{\alpha i}$, for all i, to undergo the change $U_{\alpha i} \to e^{i\varphi_\alpha} U_{\alpha i}$. Thus, phase redefinition of the three charged leptons can be used to remove three phase parameters from U, leaving a matrix that contains $9 - 3 = 6$ parameters. One might think that additional phase parameters could be removed by phase redefinition of the neutrinos. If the neutrinos are Dirac particles, this is true. But if, as we are assuming, they are Majorana particles, then one can show that phases removed from U by phase-redefining the neutrinos simply show up somewhere else, and still have the same physical effects as they do when they are located in U [9, 10]. Thus, we shall leave them in U, which consequently retains six parameters. These may be chosen to be mixing angles, which would be present even if U were real, and complex phase factors. To see how many of the six parameters are mixing angles and how many are complex phase factors, we assume for a moment that the latter are turned off (set to unity), so that U is real. It then contains nine real entries. On these entries, unitarity imposes six constraints: each column of U must be a vector of unit length and each pair of columns must be orthogonal. Thus, when the complex phase factors are turned off, U contains $9 - 6 = 3$ independent parameters – the mixing angles. Since the complex U, with the complex phase factors turned on, contains a total of six parameters, three of these must be complex phase factors.

A common parameterization of U in terms of mixing angles and phases is [11, 12]

$$U = \begin{bmatrix} 1 & 0 & 0 \\ 0 & c_{23} & s_{23} \\ 0 & -s_{23} & c_{23} \end{bmatrix} \begin{bmatrix} c_{13} & 0 & s_{13}\,\mathrm{e}^{-\mathrm{i}\delta} \\ 0 & 1 & 0 \\ -s_{13}\,\mathrm{e}^{\mathrm{i}\delta} & 0 & c_{13} \end{bmatrix} \begin{bmatrix} c_{12} & s_{12} & 0 \\ -s_{12} & c_{12} & 0 \\ 0 & 0 & 1 \end{bmatrix} \begin{bmatrix} \mathrm{e}^{\mathrm{i}\alpha_1/2} & 0 & 0 \\ 0 & \mathrm{e}^{\mathrm{i}\alpha_2/2} & 0 \\ 0 & 0 & 1 \end{bmatrix}$$

$$= \begin{bmatrix} c_{12}c_{13}\,\mathrm{e}^{\mathrm{i}\alpha_1/2} & s_{12}c_{13}\,\mathrm{e}^{\mathrm{i}\alpha_2/2} & s_{13}\,\mathrm{e}^{-\mathrm{i}\delta} \\ (-s_{12}c_{23}-c_{12}s_{23}s_{13}\,\mathrm{e}^{\mathrm{i}\delta})\,\mathrm{e}^{\mathrm{i}\alpha_1/2} & (c_{12}c_{23}-s_{12}s_{23}s_{13}\,\mathrm{e}^{\mathrm{i}\delta})\,\mathrm{e}^{\mathrm{i}\alpha_2/2} & s_{23}c_{13} \\ (s_{12}s_{23}-c_{12}c_{23}s_{13}\,\mathrm{e}^{\mathrm{i}\delta})\,\mathrm{e}^{\mathrm{i}\alpha_1/2} & (-c_{12}s_{23}-s_{12}c_{23}s_{13}\,\mathrm{e}^{\mathrm{i}\delta})\,\mathrm{e}^{\mathrm{i}\alpha_2/2} & c_{23}c_{13} \end{bmatrix} .$$

$$(1.39)$$

Here, $c_{ij} \equiv \cos\theta_{ij}$ and $s_{ij} \equiv \sin\theta_{ij}$, where θ_{12}, θ_{13}, and θ_{23} are the three mixing angles, and δ, α_1, and α_2 are the three phases. The phase δ, referred to as the Dirac phase, is the leptonic analogue of the single phase that may be found in the 3×3 quark mixing matrix. The phases α_1 and α_2, known as Majorana phases, are the extra physically significant phases that U may contain when the neutrino mass eigenstates are Majorana particles. As may be seen in (1.39), the phase α_1 is common to all elements of the first column of U. Thus, this phase could be removed from U by phase-redefining the neutrino ν_1^{Light}. Similarly, α_2 could be removed by redefining ν_2^{Light}. However, as we have mentioned, when neutrinos are Majorana particles, phases removed from U by phase-redefining neutrinos simply reappear elsewhere, and continue to have the same physical consequences as they had when located in U [2].

At the origin of coordinates, $x^\mu = 0$, the weak interaction $\mathcal{L}_{\mathrm{W}}$ of (1.35) transforms under CP as

$$(\mathrm{CP})\mathcal{L}_{\mathrm{W}}(\mathrm{CP})^{-1} = -\frac{g}{\sqrt{2}}\mathrm{W}_\rho^- \overline{\ell_{\mathrm{L}}}\gamma^\rho U^* \nu_{\mathrm{L}}^{\mathrm{Light}} - \frac{g}{\sqrt{2}}\mathrm{W}_\rho^+ \overline{\nu_{\mathrm{L}}^{\mathrm{Light}}}\gamma^\rho U^{\mathrm{T}}\ell_{\mathrm{L}} . \quad (1.40)$$

In writing this expression, we have taken some arbitrary phase factors that in principle could be present to be unity. Comparing the CP mirror image of $\mathcal{L}_{\mathrm{W}}$ in (1.40) with $\mathcal{L}_{\mathrm{W}}$ itself, (1.35), we see that if $U^* \neq U$ – that is, if U contains any of the phases δ, α_1, and α_2, so that it is not real – then the weak interaction is not CP invariant. In our discussions of neutrino oscillation and double beta decay, we shall see examples of the CP-violating physical effects that these phases can produce.

1.2 What is a Majorana Neutrino?

As we have seen, the seesaw mechanism predicts that neutrinos are Majorana particles. We have also seen that, quite apart from the specific details of the seesaw mechanism, it is rather likely that nature contains Majorana neutrino mass terms. From the procedure we followed to diagonalize the combined

Majorana and Dirac mass term of (1.8) (see (1.8)–(1.16) and the accompanying discussion), it is clear that when Majorana mass terms are present, the neutrino mass eigenstates are Majorana particles. Thus, it is rather likely that neutrinos are indeed Majorana particles. Since the behavior of Majorana neutrinos can – at first – be rather puzzling, it is worth trying to clarify the nature of these particles.

A Majorana neutrino mass eigenstate ν_i is a particle whose field goes into itself under charge conjugation. Thus, the neutrino is identical to its antiparticle: $\nu_i = \overline{\nu}_i$. Now, in descriptions of neutrino processes, it is sometimes *assumed* that there is a conserved lepton number L, where L(negatively charged lepton) $= L$(neutrino) $= -L$(positively charged lepton) $= -L$(antineutrino) $= 1$. Particles are then identified as neutrinos or antineutrinos in accordance with the process through which they are produced. For example, if the production process is $\pi^+ \to \mu^+ + \nu_\mu$, the outgoing neutral particle is identified as a neutrino, not an antineutrino, because $L(\pi^+) = 0$ and $L(\mu^+) = -1$, and it is being assumed that L is conserved. Similarly, if the production process is $\pi^- \to \mu^- + \overline{\nu}_\mu$, the outgoing neutral particle is identified as an antineutrino. Now, we know that, when interacting in a detector, the "neutrino" produced in π^+ decay will create a μ^-, while the "antineutrino" produced in π^- decay will create a μ^+. (For simplicity, we are disregarding mixing and neutrino oscillation here.) This behavior appears to suggest that "neutrinos" and "antineutrinos" are different particles, and that L is indeed conserved. But there is another, equally viable interpretation of this behavior. We know from measurements of the muon polarization in pion decays that the "ν_μ" produced in $\pi^+ \to \mu^+ + \nu_\mu$ has LH (negative) helicity, while the "$\overline{\nu}_\mu$" produced in $\pi^- \to \mu^- + \overline{\nu}_\mu$ has RH (positive) helicity. Let us now assume that nature contains Majorana mass terms, so that lepton number L is not conserved, and neutrinos are Majorana particles. For simplicity, we also continue to neglect mixing, so that ν_μ is a mass eigenstate. Then, *for a given helicity h*, ν_μ and $\overline{\nu}_\mu$ are the same particle. Nevertheless, the neutral particles produced in π^+ and π^- decay still differ from each other, because they have opposite helicity. Under the assumption that they are Majorana neutrinos, helicity is the *only* difference between them. But helicity is a sufficient difference to explain why the neutral particle originating from π^+ decay will yield a μ^- when it interacts, while the particle originating from π^- decay will yield a μ^+. After all, the weak interaction is maximally parity-violating, so it is not surprising at all that oppositely polarized particles interact differently. Indeed, it is easily verified that in the charged-current weak interaction of (1.35), the first term completely dominates for an incoming Majorana neutrino with negative helicity, while the second term completely dominates for the same incoming neutrino when its helicity is positive. As we can see, the first term will create a negatively charged lepton, but the second term will create a positively charged one. Thus, what happens when a neutrino interacts can be understood without invoking a conserved lepton number. It can be explained by assuming that neutrinos are Majorana particles, and simply noting that by reversing the helicity of a Majorana neutrino, we can reverse the charge of the lepton that this neutrino creates when it interacts. In this picture, the role played by the "neutrino" when L conservation is assumed is played by the LH helicity state of the Majorana neutrino, and the role played by the "antineutrino" is played by the RH helicity state of the same particle.

Correlating the charge of a produced lepton with the helicity of the Majorana neutrino that produces it leads to a puzzle. Suppose a Majorana neutrino, as seen by observer (a), is moving to the right with LH helicity, as shown in Fig. 1.2a. As seen by observer (b), who is moving to the right faster than the neutrino, the latter is moving to the left. However, its spin is still pointing to the left, just as it was for observer (a). Thus, as seen by observer (b), the neutrino has RH helicity, as shown in Fig. 1.2b. Now, suppose our neutrino interacts with a target that is at rest in the frame of observer (a) (frame (a)) and creates a μ^-, a lepton with the charge expected in view of the neutrino's helicity. But, as seen by observer (b), this same neutrino has RH helicity. Does this mean that, as seen from the frame of observer (b) (frame (b)), the neutrino's interaction with the target produces a μ^+, rather than a μ^-? Clearly, it had better not mean that. Lorentz-transforming the μ^- from frame (a) to frame (b) certainly does not change its electric charge. Consequently, the collision of the neutrino with the target at rest in frame (a) yields a μ^-, regardless of whether the collision is viewed from frame (a) or frame (b). But how can this be the case, given that, when an incident neutrino has RH helicity, as ours does in frame (b), the weak interaction of (1.35) appears to strongly favor the production of a positively charged lepton over that of a negatively charged one?

The solution to this puzzle is that any collision between a neutrino and a target depends on *two* weak currents: the leptonic current in (1.35) *and* a current for the target. Each of these currents is a Lorentz four-vector, and can look very different in different frames. But the amplitude for the collision is the scalar product of the two currents, and a scalar product of two four-vectors is Lorentz invariant. Thus, the result of the collision, and in particular the charge of the produced lepton, will be the same as seen by all observers.

To illustrate this point, let us consider a collision between a Majorana muon neutrino ν_μ and a spinless target N (a spinless nucleus, for example). We neglect mixing, so that ν_μ is a mass eigenstate. We assume that ν_μ has LH

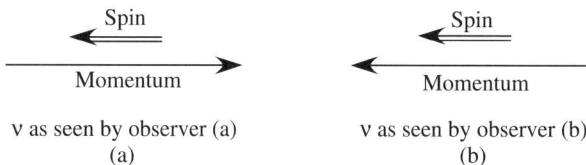

Fig. 1.2. A neutrino as seen from two different frames of reference

helicity in the rest frame of N, and that the collision produces an outgoing muon and a spinless nuclear recoil N'. Given the helicity of the ν_μ, we expect the probability for the muon to be negative to far outweigh that for it to be positive, but we shall allow both possibilities, and compute the amplitudes for the two reactions $\nu_\mu + N \to \mu^\mp + N'_\pm$, where N'_+ is a nuclear recoil whose charge is one unit greater that that of N, and N'_- is a nuclear recoil whose charge is one unit less. To keep the illustrative calculation simple, we take the matrix element of the weak current of the nuclear target, J^ρ_N, to have the form

$$\langle N'_\pm(k')|J^\rho_N|N(k)\rangle = c(k+k')^\rho . \tag{1.41}$$

Here, k and k' are, respectively, the four-momenta of N and $N'_\pm$, and c is a constant whose value we assume to be the same for N'_+ and N'_-. We consider the case of forward scattering, in which, in the rest frame of N, the μ and the N' both leave the collision with momenta parallel to the momentum of the incident ν_μ. The reaction, as seen in this frame, is depicted in Fig. 1.3a. We assume that in this frame all particles, save the initial nucleus, are highly relativistic, and that, to a sufficiently good approximation, the initial nucleus and the nuclear recoil have the same mass. For this case, we find by explicit calculation in the rest frame of N, using the leptonic current of (1.35) and the nuclear current of (1.41), that

$$\frac{A[\nu_\mu(\text{LH}) + N \to \mu^+ + N'_-]}{A[\nu_\mu(\text{LH}) + N \to \mu^- + N'_+]} \cong \frac{1}{2}\frac{m_\nu}{E_\nu}\frac{m_\mu}{E_\mu}\frac{E_{N'}}{m_{N'}} . \tag{1.42}$$

Here, $A[\ldots]$ is the amplitude for the process in the brackets, and E_ν, E_μ, and $E_{N'}$ are, respectively, the energies of the neutrino, muon, and nuclear recoil, whose masses are, respectively, m_ν, m_μ, and $m_{N'}$. As expected, μ^- production dominates over μ^+ production because of the small value of m_ν/E_ν, and in the limit where $m_\nu/E_\nu \to 0$, this dominance is total.

Next, we calculate the amplitudes for $\mu^+N'_-$ and $\mu^-N'_+$ production in a frame where all particles are highly relativistic, and all of them, including the ν_μ and N, move in the direction opposite to that of the ν_μ in the rest

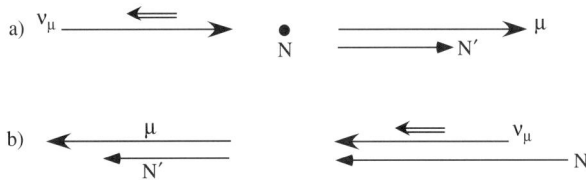

Fig. 1.3. (a) The forward reaction $\nu_\mu + N \to \mu + N'$ as seen in the rest frame of N, where the ν_μ has LH helicity. The *single lines* show momenta, and the *double line* shows the ν_μ spin. **(b)** The same reaction as seen in a frame in which all particles move at highly relativistic speeds in a direction opposite to that of the ν_μ in the rest frame of N

frame of N. The view from this frame, in which the ν_μ has RH helicity, is shown in Fig. 1.3b. By explicit calculation in this frame, we find that

$$\frac{A[\nu_\mu(\text{RH}) + \text{N} \to \mu^+ + \text{N}'_-]}{A[\nu_\mu(\text{RH}) + \text{N} \to \mu^- + \text{N}'_+]}$$

$$= \frac{E_\nu^* + m_\nu + |\boldsymbol{p}_\nu^*|}{E_\nu^* + m_\nu - |\boldsymbol{p}_\nu^*|} \frac{E_\mu^* + m_\mu + |\boldsymbol{p}_\mu^*|}{E_\mu^* + m_\mu - |\boldsymbol{p}_\mu^*|} \frac{(E_\text{N}^* + E_{\text{N}'}^*) - (|\boldsymbol{k}^*| + |\boldsymbol{k}'^*|)}{(E_\text{N}^* + E_{\text{N}'}^*) + (|\boldsymbol{k}^*| + |\boldsymbol{k}'^*|)} \ .$$

$$(1.43)$$

Here, A once again denotes an amplitude, and the helicity of the ν_μ that appears here is the helicity seen in the new frame. The quantities E_ν^* and $|\boldsymbol{p}_\nu^*|$ are, respectively, the energy and momentum of the neutrino in this frame, and similarly for E_μ^* and $|\boldsymbol{p}_\mu^*|$, E_N^* and $|\boldsymbol{k}^*|$, and $E_{\text{N}'}^*$ and $|\boldsymbol{k}'^*|$.

It is tedious but straightforward to reexpress the right-hand side of (1.43) in terms of quantities in the rest frame of N. When one does this, one finds that the ratio of the amplitudes in (1.43) is exactly the same as the ratio of the amplitudes in (1.42). That is, the relative rates at which $\mu^+\text{N}'_-$ and $\mu^-\text{N}'_+$ are produced are exactly the same in both of the frames we have considered, as demanded by Lorentz invariance. In particular, μ^- production dominates over μ^+ production in both frames, despite the fact that in one of the frames the incoming neutral lepton is right-handed.

1.3 Neutrino Flavor Change

There is now a strong conviction that neutrinos do have nonzero masses and mix. As indicated at the start of this chapter, this conviction is based on the compelling evidence that neutrinos can change from one flavor to another. In this section, we shall briefly review the physics of neutrino flavor change, and see why this phenomenon implies neutrino masses and mixing.

Neutrino flavor change in vacuo is the process in which a neutrino is created together with a charged lepton ℓ_α of flavor α, then travels a macroscopic distance L in vacuum, and finally interacts with a target to produce a second charged lepton ℓ_β whose flavor β is different from that of the first charged lepton. That is, in the course of traveling from source to target, the neutrino morphs from a ν_α to a ν_β. The process, commonly referred to as $\nu_\alpha \to \nu_\beta$ oscillation, is depicted in the upper diagram of Fig. 1.4. As shown in the lower diagram of Fig. 1.4, the intermediate neutrino can be any of the (light) mass eigenstates ν_i, and the amplitude for the oscillation is the coherent sum of the contributions of the various mass eigenstates. (From this point on, we use the simplified notation ν_i, without a superscript "Light", to mean a light-neutrino mass eigenstate.) The contribution of a given ν_i is a product of three factors. First, from (1.1) or (1.38), the amplitude for the created ν_α to be the mass eigenstate ν_i is $U_{\alpha i}^*$. Secondly, the amplitude for this ν_i to travel

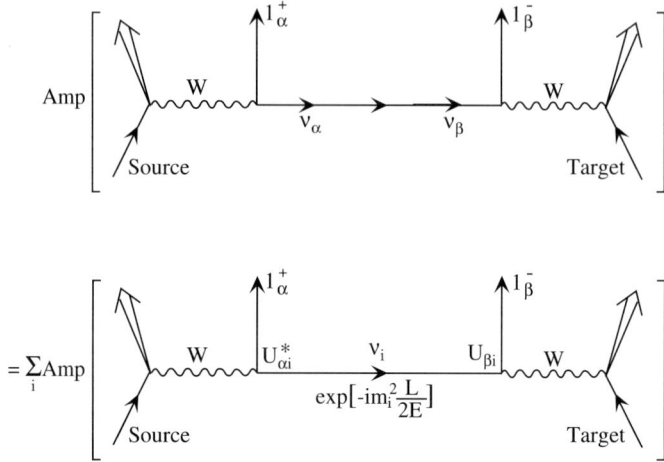

Fig. 1.4. Neutrino flavor change in vacuo. As shown in the *upper diagram*, the neutrino is created together with a charged lepton ℓ_α^+ by a source. After traveling a distance L, it interacts with a target and produces a second charged lepton ℓ_β^-. As shown in the *lower diagram*, the amplitude for this process is a sum over the contributions of all the neutrino mass eigenstates ν_i

a distance L if the neutrino energy is E is $\exp(-im_i^2 L/2E)$ [11].[7] Thirdly, the amplitude for ν_i, having arrived at the target, to produce the particular charged lepton ℓ_β^- is, from (1.38), $U_{\beta i}$. Thus, the amplitude $\mathrm{Amp}[\nu_\alpha \to \nu_\beta]$ for $\nu_\alpha \to \nu_\beta$ oscillation is given by

$$\mathrm{Amp}[\nu_\alpha \to \nu_\beta] = \sum_i U_{\alpha i}^* e^{-im_i^2(L/2E)} U_{\beta i} , \qquad (1.44)$$

where the sum runs over all the light mass eigenstates. Squaring this relation and using the (at least approximate) unitarity of the mixing matrix U, we find that the probability $P(\nu_\alpha \to \nu_\beta)$ for $\nu_\alpha \to \nu_\beta$ oscillation is given by [11]

$$
\begin{aligned}
P(\nu_\alpha \to \nu_\beta) &= |\mathrm{Amp}[\nu_\alpha \to \nu_\beta]|^2 \\
&= \delta_{\alpha\beta} - 4 \sum_{i>j} \mathrm{Re}\, (U_{\alpha i}^* U_{\beta i} U_{\alpha j} U_{\beta j}^*) \sin^2[\Delta m_{ij}^2 (L/4E)] \\
&\quad + 2 \sum_{i>j} \mathrm{Im}\, (U_{\alpha i}^* U_{\beta i} U_{\alpha j} U_{\beta j}^*) \sin[\Delta m_{ij}^2 (L/2E)] , \qquad (1.45)
\end{aligned}
$$

where $\Delta m_{ij}^2 \equiv m_i^2 - m_j^2$. This expression for $P(\nu_\alpha \to \nu_\beta)$ is valid for an arbitrary number of neutrino mass eigenstates, and holds whether β is different from α or not. However, we see that if all the neutrino masses, and

---

[7] Some of the treatment in the present section follows the treatment in this reference.

consequently all the splittings Δm_{ij}^2, vanish, then $P(\nu_\alpha \to \nu_\beta) = \delta_{\alpha\beta}$. Thus, the oscillation in vacuo of ν_α into a *different* flavor ν_β implies neutrino mass. From (1.44), we see that this change of flavor also implies neutrino mixing: in the absence of mixing, the matrix U is diagonal, so that $\text{Amp}[\nu_\alpha \to \nu_\beta]$ vanishes if $\beta \neq \alpha$. Finally, from (1.45), we see that the probability for neutrino oscillation really does oscillate as a function of L/E, justifying the name "oscillation".

Assuming that CPT invariance holds,

$$P(\overline{\nu_\alpha} \to \overline{\nu_\beta}) = P(\nu_\beta \to \nu_\alpha) . \tag{1.46}$$

However, from (1.45), we see that

$$P(\nu_\beta \to \nu_\alpha; U) = P(\nu_\alpha \to \nu_\beta; U^*) . \tag{1.47}$$

Thus,

$$P(\overline{\nu_\alpha} \to \overline{\nu_\beta}; U) = P(\nu_\alpha \to \nu_\beta; U^*) . \tag{1.48}$$

That is, the probability for $\overline{\nu_\alpha} \to \overline{\nu_\beta}$ is the same as for $\nu_\alpha \to \nu_\beta$, except that U is replaced by U^*. But this means that if U is not real, then $P(\overline{\nu_\alpha} \to \overline{\nu_\beta})$ differs from $P(\nu_\alpha \to \nu_\beta)$ by a reversal of the last term of (1.45). This difference is a violation of CP invariance, which would require $\nu_\alpha \to \nu_\beta$ and $\overline{\nu_\alpha} \to \overline{\nu_\beta}$ to have equal probability.

Neutrino oscillation depends on the interference of different contributions to an amplitude (see (1.44)), so it is a quintessentially quantum mechanical phenomenon. It raises a number of subtle questions, some of which have been addressed by treatments based on wave packets [13]. However, it has also been shown that, for a number of the observations of oscillation that have been made in practice, a wave packet treatment is not necessary [14]. Sophisticated analyses of oscillation continue to yield new insights [15]. However, they lead to the same oscillation probability as we have obtained here.

If neutrinos pass through enough matter between their source and a target detector, then their coherent forward scattering from particles in this matter can significantly modify their oscillation pattern. This is true even if, as in the Standard Model, forward scattering of neutrinos from other particles does not by itself change neutrino flavor. The flavor change in matter that grows out of an interplay between flavor-nonchanging neutrino–matter interactions and neutrino mass and mixing is known as the Mikheyev–Smirnov–Wolfenstein (MSW) effect [16].

To treat a neutrino in matter, it is convenient to describe its state by a column vector in flavor space,

$$\begin{bmatrix} a_e(t) \\ a_\mu(t) \\ a_\tau(t) \end{bmatrix} , \tag{1.49}$$

where $a_e(t)$ is the amplitude for the neutrino to be a ν_e at time t, and similarly for the other flavors. The time evolution of the neutrino state is then described by a Schrödinger equation in which the Hamiltonian $\mathcal{H}$ is a 3×3 matrix that acts on this column vector [17]. To illustrate, we shall make the simplifying assumption that we are dealing with an effectively "two-neutrino" problem, in which only ν_e and ν_μ, and two corresponding mass eigenstates ν_1 and ν_2, matter. In this case the neutrino is described by a two-component column vector,

$$\begin{bmatrix} a_e(t) \\ a_\mu(t) \end{bmatrix} , \tag{1.50}$$

and $\mathcal{H}$ is 2×2. If our neutrino is traveling in vacuo, the mixing is described by the vacuum mixing matrix

$$U_V = \begin{array}{c} \\ e \\ \mu \end{array} \begin{array}{cc} 1 & 2 \\ \begin{bmatrix} \cos\theta_V & \sin\theta_V \\ -\sin\theta_V & \cos\theta_V \end{bmatrix} \end{array} , \tag{1.51}$$

in which θ_V is the mixing angle in vacuo, and the symbols above and to the left of the matrix label the columns and rows. It is easy to show that, apart from an irrelevant multiple of the identity, $\mathcal{H}$ in vacuo is then [17]

$$\mathcal{H}_V = \frac{\Delta m_V^2}{4E} \begin{bmatrix} -\cos 2\theta_V & \sin 2\theta_V \\ \sin 2\theta_V & \cos 2\theta_V \end{bmatrix} . \tag{1.52}$$

Here, $\Delta m_V^2 \equiv m_2^2 - m_1^2$ is the (mass)^2 splitting in vacuo, and E is the neutrino energy. One can straightforwardly show that this $\mathcal{H}_V$ predicts that the probability $P_V(\nu_e \rightarrow \nu_\mu)$ for $\nu_e \rightarrow \nu_\mu$ oscillation in vacuo is given by

$$P_V(\nu_e \rightarrow \nu_\mu) = \sin^2 2\theta_V \sin^2 \left(\Delta m_V^2 \frac{L}{4E} \right) . \tag{1.53}$$

This is the well-known formula for two-neutrino oscillation in vacuo. This formula follows also from (1.45) for the special case of two neutrinos, if we take $\alpha = e$, $\beta = \mu$, $i = 2$, $j = 1$, $\Delta m_{21}^2 \equiv \Delta m_V^2$, and U to be the matrix U_V of (1.51).

In matter, W-exchange-induced coherent forward scattering of ν_e from ambient electrons adds an interaction energy V to the ν_e–ν_e element of $\mathcal{H}$. (The ν_μ–ν_μ element is not affected, because the reaction $\nu_\mu e \rightarrow \nu_\mu e$ cannot be induced by W exchange.) Obviously, V must be proportional to G_F, the Fermi constant, and to N_e, the number of electrons per unit volume. Indeed, it can be shown that [18]

$$V = \sqrt{2}\, G_F N_e . \tag{1.54}$$

In addition, Z-exchange-mediated scattering from ambient particles adds a further interaction energy to all diagonal elements of $\mathcal{H}$. However, since the coupling of Z to neutrinos is flavor-independent, this further addition to $\mathcal{H}$ is a multiple of the identity matrix, and no such addition has any effect on neutrino flavor oscillation [17]. Thus, we may safely omit the Z-exchange-induced energy. The 2×2 Hamiltonian in matter is then

$$\mathcal{H} = \frac{\Delta m_V^2}{4E} \begin{bmatrix} -\cos 2\theta_V & \sin 2\theta_V \\ \sin 2\theta_V & \cos 2\theta_V \end{bmatrix} + \begin{bmatrix} V & 0 \\ 0 & 0 \end{bmatrix}. \qquad (1.55)$$

Harmlessly adding to this $\mathcal{H}$ the multiple $-V/2$ of the identity, we may rewrite it as

$$\mathcal{H} = \frac{\Delta m_M^2}{4E} \begin{bmatrix} -\cos 2\theta_M & \sin 2\theta_M \\ \sin 2\theta_M & \cos 2\theta_M \end{bmatrix}. \qquad (1.56)$$

Here,

$$\Delta m_M^2 = \Delta m_V^2 \sqrt{\sin^2 2\theta_V + (\cos 2\theta_V - x)^2} \qquad (1.57)$$

is the effective mass splitting in matter, and

$$\sin^2 2\theta_M = \frac{\sin^2 2\theta_V}{\sin^2 2\theta_V + (\cos 2\theta_V - x)^2} \qquad (1.58)$$

is the effective mixing angle in matter. In these expressions,

$$x \equiv \frac{V}{\Delta m_V^2 / 2E} \qquad (1.59)$$

is a dimensionless measure of the relative importance of the interaction with matter in the behavior of the neutrino.

If a neutrino travels through matter of constant density, then $\mathcal{H}$ (1.56) is a position-independent constant. As we can see, it is exactly the same as the vacuum Hamiltonian (1.52), except that the vacuum mass splitting and mixing angle are replaced by their values in matter. As a result, the oscillation probability is given by the usual formula, (1.53), but with the mass splitting and mixing angle replaced by their values in matter. The latter values can differ markedly from their vacuum counterparts. A striking example is the case where the vacuum mixing angle θ_V is very small, but $x \cong \cos 2\theta_V$. Then, as we see from (1.58), $\sin^2 2\theta_M \cong 1$. Interaction with matter has promoted a very small mixing angle into a maximal one.

One important example of neutrino propagation in matter is the journey of solar neutrinos, which are created as electron neutrinos in the center of the sun, outward through solar material. Of course, the electron density encountered by these neutrinos is not a constant, so $\mathcal{H}$ depends on the distance r from the center of the sun. Nevertheless, under certain conditions, the propagation of the neutrinos is adiabatic. That is, the electron density $N_e(r)$ varies slowly enough that one may solve the Schrödinger equation for neutrino propagation for one r at a time, and then patch together the solutions. This is true, in particular, for the large-mixing-angle (LMA) version of the MSW picture of what happens to the solar neutrinos, which is the most favored explanation of their observed behavior.

In the LMA MSW scenario, $\Delta m_V^2 \sim 5 \times 10^{-5}$ eV^2 [19]. For the most closely scrutinized solar neutrinos, the ones from ^{8}B decay, typical energies E are 6–7 MeV. For these neutrinos, $\Delta m_V^2/4E \sim 0.2 \times 10^{-5}$ eV^2/MeV. Now, at $r \simeq 0$, where the solar neutrinos are born, the electron density $N_e \simeq 6 \times 10^{25}$ /cm^3 [20]. This value yields the magnitude $V \sim 0.75 \times 10^{-5}$ eV^2/MeV for the interaction energy V at $r \simeq 0$. Consequently, where the neutrinos are born, the interaction (second) term of the Hamiltonian $\mathcal{H}$ of (1.55) dominates over the vacuum (first) term, at least to some extent. As a result, $\mathcal{H}$ is approximately diagonal at $r \simeq 0$. This means that at birth, a ^{8}B neutrino is not only a ν_e but also, approximately, in an eigenstate of $\mathcal{H}$. Since $V > 0$, the neutrino is in the heavier of the two eigenstates. The neutrino then propagates outward adiabatically. This means that it continues to be in an eigenstate of $\mathcal{H}$ – an r-dependent eigenstate that changes slowly as $\mathcal{H}$ changes. It will then emerge from the sun as one of the two eigenstates of the zero-density (vacuum) Hamiltonian. That is, our neutrino leaves the sun as one of the mass eigenstates of $\mathcal{H}_V$. Since, as one may readily verify, the eigenlevels of $\mathcal{H}$ (1.55) never cross, and the neutrino started in the heavier eigenlevel at $r \simeq 0$, it will leave the sun as the heavier of the two mass eigenstates of $\mathcal{H}_V$. If we define $\Delta m_V^2 = m_2^2 - m_1^2$ to be positive, then this is the eigenstate called ν_2. Being an eigenstate of the vacuum Hamiltonian, this state will propagate without mixing all the way to the surface of the earth. Now, from (1.51), ν_2 has the flavor composition

$$|\nu_2\rangle = |\nu_e\rangle \sin\theta_V + |\nu_\mu\rangle \cos\theta_V . \tag{1.60}$$

The probability that a ^{8}B solar neutrino still has the flavor ν_e with which it was born when it arrives at the earth is just the ν_e fraction of this state, $\sin^2\theta_V$.

When information from atmospheric neutrino oscillation observations is taken into account, one learns that the "other flavor" with which solar electron neutrinos mix is not ν_μ but a 50–50 mixture of ν_μ and μ_τ. However, if one simply understands "ν_μ" in our analysis of the solar neutrinos to be a shorthand for this 50–50 mixture, then our analysis remains valid.

Like oscillation in vacuo, neutrino flavor change in matter requires neutrino masses and mixing. If either Δm_V^2 or θ_V vanishes, then the Hamiltonian in matter (1.55) is diagonal. Thus, a neutrino born with a given flavor will retain that flavor forever.

Flavor change has been reported for atmospheric neutrinos, solar neutrinos, and the accelerator neutrinos studied by the Liquid Scintillator Neutrino Detector (LSND) experiment. Each of these three reported flavor changes calls for a splitting Δm^2 that is of an order of magnitude different from the orders of magnitude called for by the other two. Obviously, these three very different splittings cannot all be accommodated if there are only three neutrino mass eigenstates, since there are then only three splittings Δm_{ij}^2, and they obviously satisfy

$$\Delta m_{32}^2 + \Delta m_{21}^2 + \Delta m_{13}^2 = 0 . \tag{1.61}$$

Thus, if all three reported flavor changes prove to be genuine, then nature must contain at least four neutrino mass eigenstates ν_i. Now, three linear combinations of these ν_i, namely ν_e, ν_μ, and ν_τ, couple to the W boson and one of the three charged leptons. If there are exactly four ν_i, then there is a fourth linear combination of them, ν_s, orthogonal to ν_e, ν_μ, and ν_τ, which has no charged-lepton partner, and hence cannot couple to the W. Moreover, since the decays $Z \to \nu_\alpha \bar{\nu}_\alpha$ of the Z into neutrino pairs are found to produce only three distinct neutrino flavors [21], the fourth neutrino ν_s evidently does not couple to the Z either. Thus, ν_s does not have any of the Standard Model weak couplings. Such a neutrino is called "sterile". Obviously, it is quite unlike the "active" neutrinos, ν_e, ν_μ, and ν_τ. Consequently, it will be very interesting to see whether all three of the reported neutrino flavor changes are confirmed, so that nature must contain a sterile neutrino.

1.4 Double Beta Decay

Given the theoretical expectation that neutrinos are Majorana particles, it would obviously be desirable to confirm experimentally that this is indeed the case. As mentioned in Sect. 1.1, the observation of neutrinoless double beta decay, the reaction Nucl → Nucl' + 2e⁻, would provide the sought-for confirmation [6].

If neutrinoless double beta decay (often referred to as 0νββ) does occur, it is quite likely to be dominated by a mechanism in which the parent nucleus emits a pair of virtual W⁻ bosons, turning into the daughter nucleus, and then the W⁻ bosons exchange one or another of the light neutrino mass eigenstates ν_i to create the outgoing electrons. The heart of this mechanism is the second step, W⁻W⁻ → e⁻e⁻ via Majorana neutrino exchange. The diagram for this step is shown in Fig. 1.5. There, the Standard Model weak interaction is assumed to act at each vertex. If neutrinos and antineutrinos differ, this interaction creates the exchanged particle as an antineutrino, but can absorb

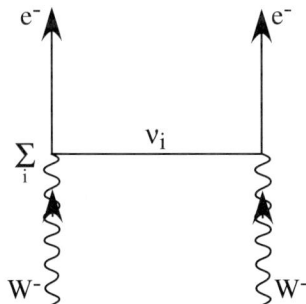

Fig. 1.5. The process at the heart of neutrinoless double beta decay. The exchanged particle can be any of the light neutrinos ν_i

it only as a neutrino. Thus, the diagram is forbidden unless neutrinos and antineutrinos do *not* differ – the Majorana case.

As indicated in Fig. 1.5, the amplitude for $W^-W^- \to e^-e^-$ is the coherent sum of the contributions of all the light neutrino mass eigenstates ν_i. From (1.38), the contribution of ν_i involves the current $\overline{\ell_{Le}}\gamma^\rho U_{ei}\nu_{Li}$, acting at both vertices. Thus, this contribution is proportional to U_{ei}^2. It is also proportional to m_i. The latter factor may be understood by recalling that the exchanged ν_i is produced as an "antineutrino", which in the Majorana case simply means that it has the helicity normally associated with an antineutrino. That is, it is right-handed, except for a small left-handed piece with an amplitude of order m_i/E, E being its energy. It is only this left-handed piece that the LH weak current acting to absorb the ν_i can accommodate without further suppression. Thus, the amplitude for $0\nu\beta\beta$ is proportional to a factor $m_{\beta\beta}$ given by

$$m_{\beta\beta} = \left| \sum_i m_i U_{ei}^2 \right| . \tag{1.62}$$

This $m_{\beta\beta}$ is referred to as the effective neutrino mass for double beta decay.

While neutrino oscillation has provided us with the evidence that neutrino masses are nonzero, this process cannot determine the masses m_i of the individual neutrino mass eigenstates. Rather, oscillation can only determine the (mass)^2 *splittings* Δm_{ij}^2, as (1.45) for $P(\nu_\alpha \to \nu_\beta)$ makes very evident. One approach to gaining some information about the m_i, and thereby some knowledge of the absolute scale of the neutrino masses, is to look for kinematic effects of neutrino mass in the leptonic tritium decays $^3H \to {}^3He + e^- + \overline{\nu}_i$ (see Chap. 2). Another approach is to look for $0\nu\beta\beta$, since a knowledge of $m_{\beta\beta}$ (1.62) would clearly provide at least some information about the scale of the masses m_i.

The effective mass $m_{\beta\beta}$ could, in principle, also provide some information about the CP-violating phases in the matrix U of (1.39). From (1.44), we see that only the Dirac phase δ in this matrix can influence neutrino oscillation. Any Majorana phase, such as α_1, is common to an entire column of U. Thus, this phase cancels out of the oscillation amplitude, in which the ν_i contribution, as we see in (1.44), is proportional to $U_{\alpha i}^* U_{\beta i}$. On the other hand, a Majorana phase, say in the ith column of U, would not cancel out of $m_{\beta\beta}$, since, as (1.62) shows, the ν_i contribution to $m_{\beta\beta}$ is proportional to U_{ei}^2, rather than $U_{ei}^* U_{ei}$. Thus, if we know the masses m_i and the mixing angles in U, and we also know $m_{\beta\beta}$ with sufficient precision, we can in principle learn something about the Majorana phases, or at least demonstrate that they are present. Whether this would be feasible in practice is being explored [22].

1.5 Conclusion

Neutrino flavor change, either in vacuo or in matter, implies neutrino mass and mixing. Thus, the very strong evidence for flavor change makes a compelling case that neutrinos have nonzero masses. Owing to the possibility – unique to neutrinos – of Majorana mass terms, the physics underlying neutrino mass may be quite different from that underlying the masses of the quarks and charged leptons. In addition, if Majorana mass terms are present, the neutrinos are Majorana particles, making them quite different from the other fundamental fermions.

Progress in understanding the world of neutrinos has been quite striking in recent years. However, we are still only beginning to uncover the secrets of this world. Exciting years lie ahead.

References

1. Q. Ahmad et al. (SNO Collaboration), Phys. Rev. Lett. **89**, 011301 (2002); C.K. Jung, C. McGrew, T. Kajita, and T. Mann, Annu. Rev. Nucl. Part. Sci. **51**, 451 (2001).
2. See e.g. B. Kayser, F. Gibrat-Debu, and F. Perrier, *The Physics of Massive Neutrinos* (World Scientific, Singapore, 1989).
3. M. Gell-Mann, P. Ramond, and R. Slansky, in *Supergravity*, eds. D. Freedman and P. van Nieuwenhuizen (North-Holland, Amsterdam, 1979), p. 315; T. Yanagida, in *Proceedings of the Workshop on Unified Theory and Baryon Number in the Universe*, eds. O. Sawada and A. Sugamoto (KEK, Tsukuba, Japan, 1979); R. Mohapatra and G. Senjanovic, Phys. Rev. Lett. **44**, 912 (1980); Phys. Rev. D **23**, 165 (1981).
4. C.K. Jung, C. McGrew, T. Kajita, and T. Mann, Annu. Rev. Nucl. Part. Sci. **51**, 451 (2001).
5. S. Elliott and P. Vogel: arXiv:hep-ph/0202264.
6. J. Schechter and J. Valle, Phys. Rev. D **25**, 2951 (1982); E. Takasugi, Phys. Lett. **149B**, 372 (1984).
7. P. Langacker, Phys. Rep. **72**, 185 (1981).
8. K. Hagiwara et al. (Particle Data Group), Phys. Rev. D **66**, 010001 (2002).
9. See e.g. B. Kayser, in *CP Violation*, ed. C. Jarlskog (World Scientific, Singapore, 1989), p. 334.
10. S. Bilenky, J. Hosek, and S.Petcov, Phys. Lett. **B94**, 495 (1980); J. Schechter and J. Valle, Phys. Rev. D **22**, 2227 (1980); M. Doi, T. Kotani, H. Nishiura, K. Okuda, and E. Takasugi, Phys. Lett. **B102**, 323 (1981).
11. B. Kayser, "Neutrino physics as explored by flavor change", in K. Hagiwara et al. (Particle Data Group), Phys. Rev. D **66**, 010001 (2002), p. 392.
12. G. Branco, L. Lavoura, and J. Silva, *CP Violation* (Oxford University Press, Oxford, 1999).
13. B. Kayser, Phys. Rev. D **24**, 110 (1981); F. Boehm and P. Vogel, *Physics of Massive Neutrinos* (Cambridge University Press, Cambridge, 1987), p. 87; C. Giunti, C. Kim, and U. Lee, Phys. Rev. D **44**, 3635 (1991); J. Rich, Phys.

Rev. D **48**, 4318 (1993); H. Lipkin, Phys. Lett. B **348**, 604 (1995); W. Grimus and P. Stockinger, Phys. Rev. D **54**, 3414 (1996); T. Goldman: arXiv: hep-ph/9604357; Y. Grossman and H. Lipkin, Phys. Rev. D **55**, 2760 (1997); W. Grimus, S. Mohanty, and P. Stockinger, in *Proceedings of the 17th International Workshop on Weak Interactions and Neutrinos*, eds. C. Dominguez and R. Viollier (World Scientific, Singapore, 2000), p. 355.

14. L. Stodolsky, Phys. Rev. D **58**, 036006 (1998).
15. C. Giunti, arXiv:hep-ph/0202063; M. Beuthe, arxiv:hep-ph/0109119; Phys. Rev. D **66**, 013003 (2002); and references therein.
16. L. Wolfenstein, Phys. Rev. D **17**, 2369 (1978); S. Mikheyev and A. Smirnov, Sov. J. Nucl. Phys. **42**, 913 (1986); Sov. Phys. JETP **64**, 4 (1986); Nuovo Cim. **9C**, 17 (1986).
17. See e.g. B. Kayser, in *Flavor Physics for the Millennium, TASI 2000*, ed. J. Rosner (World Scientific, Singapore, 2001), p. 625.
18. See e.g. F. Boehm and P. Vogel, *Physics of Massive Neutrinos* (Cambridge University Press, Cambridge, 1987).
19. Q. Ahmad et al. (The SNO Collaboration), Phys. Rev. Lett. **89**, 011302 (2002).
20. J. Bahcall, *Neutrino Astrophysics* (Cambridge University Press, Cambridge, 1989).
21. LEP Collaborations and LEP Electroweak Working Group, as reported by J. Drees at the 20th International Symposium on Lepton and Photon Interactions at High Energy, Rome, July 2001.
22. V. Barger, S. Glashow, P. Langacker, and D. Marfatia, Phys. Lett. B **540**, 247 (2002); S. Pascoli, S. Petcov, and W. Rodejohann, arXiv:hep-ph/0209059.

2 Laboratory Limits on Neutrino Masses

Christian Weinheimer

The recent neutrino oscillation experiments have obtained nonzero differences of squared neutrino masses and therefore proven that neutrinos are massive particles. The values of the neutrino masses have to be determined in a different way. There are two classes of laboratory experiments, both of which have yielded up to now only upper limits on neutrino masses. The direct mass experiments investigate the kinematics of weak decays, obtaining information about the neutrino mass without further requirements. Here, the tritium β decay experiments give the most stringent results. The search for neutrinoless double β decay is also very sensitive to the neutrino mass states. However, this search is complementary to direct neutrino mass experiments, since it requires neutrinos to be identical to their antiparticles and probes a linear combination of neutrino masses including complex phases.

This chapter is structured as follows. After an introduction in Sect. 2.1, the two approaches are discussed together with the current experimental results in Sects. 2.2 and 2.3, followed by consideration of the outlook for future activities in Sect. 2.4.

2.1 Introduction

Recent experimental results from atmospheric and solar neutrinos give strong evidence that neutrinos oscillate from one flavor state into another (see Chap. 4). Therefore, a neutrino of one specific flavor eigenstate ν_α is a nontrivial superposition of neutrino mass states ν_i. This is described by a unitary mixing matrix $U_{\alpha i}$:

$$\nu_\alpha = \sum_i U_{\alpha i}\nu_i \; . \tag{2.1}$$

Future oscillation experiments will determine the elements $U_{\alpha i}$ with great precision.

However, ν-oscillation experiments do not yield the values of the neutrino masses; in the case of pure vacuum oscillation they are only sensitive to differences between squared neutrino masses $\Delta m_{ij}^2 = |m^2(\nu_i) - m^2(\nu_j)|$. The values Δm_{ij}^2 obtained from oscillation experiments only give lower limits on neutrino masses

$$\max\left(m(\nu_i), m(\nu_j)\right) \geq \sqrt{\Delta m_{ij}^2}\,. \tag{2.2}$$

On the other hand, if the absolute value of one mass eigenstate ν_i is known, all other neutrino masses can be reconstructed with the help of the differences between the squared neutrino masses, if the signs of the different $m^2(\nu_i) - m^2(\nu_j)$ values are known (see Chap. 1).

Information about the neutrino masses can also be deduced from cosmology (see Chap. 3). Independent information about neutrino masses comes from the measurement of the arrival time distribution of neutrinos emitted in a supernova explosion; the present limit from SN1987a is $m(\nu_e) < 23\,\text{eV}$ [1] (see Sect. 2.2 for remarks on $m(\nu_e)$).

Direct information about neutrino masses can be obtained by laboratory experiments using two different approaches: the investigation of the decay kinematics of weak decays and the search for neutrinoless double β decay. The two methods give complementary information about the neutrino masses $m(\nu_i)$.

As a result of assuming CPT invariance, we do not distinguish here between the masses of neutrinos $m(\nu_i)$ and of the corresponding antineutrinos $m(\overline{\nu}_i)$.

2.2 Decay Kinematics of Weak Decays

The investigations of the kinematics of weak decays are based on measurements of the charged decay products. Using energy and momentum conservation, the missing neutrino mass can be reconstructed from the kinematics of the charged particles. The part of the phase space which is most sensitive to the neutrino mass is that which corresponds to the emission of a nonrelativistic massive neutrino. Therefore, decays that release charged particles with a small free kinetic energy are preferred.

In principle, a kinematic neutrino mass measurement yields information about the different mass eigenstates $m(\nu_i)$, since it performs a projection onto energy and mass. Usually, however, the different neutrino mass eigenstates cannot be resolved by the experiment. Therefore, an average over neutrino mass eigenstates is obtained, which is specific to the flavor of the weak decay considered, and hence is labeled $m(\nu_e)$, $m(\nu_\mu)$, or $m(\nu_\tau)$.[1] This fact will be discussed in more detail for the case of the muon neutrino ν_μ.

---

[1] This average value is not a unique quantity but depends also on how the experiment is analyzed, in particular, whether the analysis is done under the assumption of a single neutrino mass state. Considering, however, the small differences of squared masses Δm_{ij}^2 obtained by neutrino oscillation measurements and comparing those differences with the experimental resolution of the present kinematic measurements, this question appears to be of a rather academic nature.

2.2.1 $m(\nu_\mu)$

The muon neutrino mass $m(\nu_\mu)$ has been investigated in the two-body decay of a pion at rest:

$$\pi^+ \to \mu^+ + \nu_\mu \quad \text{or} \quad \pi^- \to \mu^- + \bar{\nu}_\mu \ . \tag{2.3}$$

Energy and momentum conservation result in a sharp muon momentum $p(\mu)$, from which the mass of the muon neutrino $m(\nu_\mu)$ could be obtained as

$$m^2(\nu_\mu) = m^2(\pi) + m^2(\mu) - 2m(\pi)\sqrt{m^2(\mu) + p^2(\mu)} \ . \tag{2.4}$$

Equation (2.4) applies only if the muon neutrino ν_μ is a well-defined mass eigenstate, which is not so in the case of neutrino mixing. Hence, if the muon momentum $p(\mu)$ for pion decay at rest could be measured with sufficient precision, one would detect three different values $p_i^2(\mu)$ with relative fractions $|U_{\mu i}^2|$, corresponding to the three mass eigenstates $m(\nu_i)$ contributing to the muon neutrino ν_μ. Up to now, however, no direct neutrino mass measurement has discriminated between different neutrino masses or has established a signal of any nonzero neutrino mass.

Therefore, in the left-hand side of (2.4), we define a mean squared average of the mass eigenstates of the muon neutrino as

$$m^2(\nu_\mu) = \sum_i |U_{\mu i}^2| m^2(\nu_i) \ . \tag{2.5}$$

To deduce $m^2(\nu_\mu)$ from (2.5), three quantities have to be measured with very high precision: the muon mass $m(\mu) = 105.6583568(52)\,\text{MeV}$ [1], the pion mass $m(\pi) = 139.570180(350)\,\text{MeV}$ [1], and the muon momentum obtained from pion decay at rest $p(\mu) = 29.791998(110)\,\text{MeV}$, which was measured in a dedicated experiment at the Paul Scherrer Institute, Zürich [2]. Putting these values into (2.4), one obtains [2]

$$m^2(\nu_\mu) = -0.016 \pm 0.023\,\text{MeV}^2 \ , \tag{2.6}$$

from which an upper limit on the muon neutrino mass can be derived [1],

$$m(\nu_\mu) < 190\,\text{keV} \quad (90\%\ \text{C.L.}) \ . \tag{2.7}$$

2.2.2 $m(\nu_\tau)$

The most sensitive information about $m(\nu_\tau)$ comes from the investigation of τ pairs produced at electron–positron colliders decaying into multiple pions. Owing to the large mass of the τ, decays into five and six pions give the highest sensitivity because they restrict the available phase space of the ν_τ. However, the corresponding branching ratios are rather small.

The quantity looked at is the invariant mass of the multiple pions M_π. Although M_π does not have a direct physical meaning, the mass of the tau neutrino $m(\nu_\tau)$ restricts M_π owing to energy and momentum conservation. In the rest frame of the decaying τ lepton, M_π^2 is expressed by

$$M_\pi^2 = \left(\sum_j E_j(\pi), \sum_j \boldsymbol{p}_j(\pi) \right)^2 = \left(m(\tau) - E(\nu_\tau), -\boldsymbol{p}(\nu_\tau) \right)^2 \quad (2.8)$$

$$\leq \left(m(\tau) - m(\nu_\tau) \right)^2 . \quad (2.9)$$

The most sensitive investigation has been performed by the ALEPH experiment at LEP. Its two-dimensional analysis in the $M_\pi - \sum_j E_{j,\mathrm{lab}}(\pi)$ plane restricts $m(\nu_\tau)$ as follows [3]:

$$m(\nu_\tau) < 18.2 \,\mathrm{MeV} \quad (95\% \ \mathrm{C.L.}) . \quad (2.10)$$

A further improvement based on data from B-factories can be expected, with an estimated sensitivity limit of 3 MeV.

Using the most recent results on atmospheric and solar neutrino oscillations (see Chap. 4), the neutrino mixing matrix $U_{\alpha i}$ and the squared mass differences Δm_{ij}^2 suggest that the averages $m^2(\nu_\mu)$ and $m^2(\nu_\tau)$ (see (2.5)) are rather close, owing to the strong ν_μ–ν_τ mixing and the very small difference Δm_{23}^2. Therefore, $m(\nu_\tau)$ is already constrained by the limit on the muon neutrino mass (2.7).

2.2.3 $m(\nu_e)$

Attempts are being made to determine the mass of the electron neutrino by investigation of the electron energy spectrum (β spectrum) of a nuclear β decay [4–6]. In a β^- decay

$$(Z, A) \rightarrow (Z+1, A)^+ + e^- + \bar{\nu}_e , \quad (2.11)$$

the available energy is shared between the β electron and the electron antineutrino, because the recoiling nucleus receives practically no kinetic energy owing to its much heavier mass. The phase space region of nonrelativistic neutrinos, where the highest sensitivity to the neutrino mass is achieved, corresponds to the very upper end of the β spectrum. To maximize this part, a β emitter with a very low endpoint energy E_0 is required. This requirement is fulfilled by ^{187}Re and tritium (T or ^{3}H), which have the two lowest endpoint energies, of 2.6 keV and 18.6 keV, respectively.

Although tritium has a higher endpoint energy than has ^{187}Re, its use has several advantages:

– Tritium decays by a super-allowed[2] transition into its mirror nucleus ^{3}He, resulting in a half-life of 12.3 years, compared with the primordial half-life of the forbidden transition of ^{187}Re of 5×10^{10} yr. The short half-life yields a high specific activity and minimizes the inelastic processes of β electrons within the tritium source.
– Owing to the super-allowed decay, the transition matrix element does not depend on the electron energy: the β spectrum is determined entirely by the available phase space.
– Tritium has the simplest atomic shell, minimizing the corrections that are necessary because of the electronic final states or inelastic scattering in the β source.

These arguments clearly favor tritium for the standard setup, which consists of a β source connected to a β spectrometer (sometimes called a "passive source" setup). The advantage of the lower ^{187}Re endpoint energy can only be exploited if the β source and the spectrometer are the same object (sometimes called an "active source" setup); this situation is realized in the case of a cryogenic bolometer, for instance.

β spectrum. According Fermi's Golden Rule, the transition rate for a β decay where an electron with kinetic energy between E and $E + \Delta E$ is emitted is given by the transition matrix element M and the density of final states $\rho(E)$:

$$\frac{\mathrm{d}^2 N}{\mathrm{d}t\,\mathrm{d}E} = \frac{2\pi}{\hbar}|M^2|\rho(E) \ . \tag{2.12}$$

Let us first calculate the density of the final states. The number of different states $\mathrm{d}n$ in a volume V with momenta between p and $p + \mathrm{d}p$, or with energies in the corresponding interval around the total energy E_{tot}, is

$$\mathrm{d}n = \frac{4\pi p^2\,\mathrm{d}p V}{h^3} = \frac{4\pi p E_{\mathrm{tot}}\,\mathrm{d}E_{\mathrm{tot}} V}{h^3} \ , \tag{2.13}$$

This gives a state density per unit energy interval of

$$\frac{\mathrm{d}n}{\mathrm{d}E_{\mathrm{tot}}} = \frac{4\pi p E_{\mathrm{tot}} V}{h^3} = \frac{p E_{\mathrm{tot}} V}{2\pi^2 \hbar^3} \ . \tag{2.14}$$

Since the mass of the nucleus is – especially in our case – much larger than the energies of the two emitted leptons, we can use the following simplification: the nucleus takes up no energy but balances all momenta. Therefore,

---

[2] An allowed or super-allowed transition is not hampered by the conservation of angular momentum. The outgoing lepton pair couples to give a total spin of either 0 or 1. Therefore, the angular momentum of the nucleus changes by $\Delta I_{\mathrm{had}} = 0, \pm 1$ without changing parity.

we can count the densities of states of the electron and of the neutrino independently:[3]

$$\rho(E) = \frac{\mathrm{d}n_e}{\mathrm{d}E}\frac{\mathrm{d}n_\nu}{\mathrm{d}E} = \frac{\mathrm{d}n_e}{\mathrm{d}E_e}\frac{\mathrm{d}n_\nu}{\mathrm{d}E_\nu} = \frac{p_e E_e p_\nu E_\nu V^2}{4\pi^4\hbar^6} . \tag{2.15}$$

We have used in (2.15) the fact that the differential of the kinetic energy $\mathrm{d}E$ is identical to the differentials of the total energies $\mathrm{d}E_e$ and $\mathrm{d}E_\nu$.

Using $E_\nu = E_0 - E$, we can write all energies and momenta in terms of the kinetic energy of the electron E and its mass m, yielding

$$\rho(E) = \frac{p_e(E + m)\sqrt{(E_0 - E)^2 - m^2(\nu_e)}(E_0 - E)V^2}{4\pi^4\hbar^6} . \tag{2.16}$$

The transition matrix element M can be divided into a leptonic part M_{lep} and a hadronic part M_{had}. Usually the coupling is written separately and expressed in terms of the Fermi coupling constant G_F and the Cabibbo angle Θ_C:

$$M = G_F \cos\Theta_C M_{\mathrm{lep}} M_{\mathrm{had}} . \tag{2.17}$$

For an allowed or super-allowed decay such as that of tritium, the leptonic part $|M_{\mathrm{lep}}^2|$ results essentially in the probability for the two leptons to be found at the nucleus, which is $1/V$ for the neutrino and $(1/V)F(E, Z + 1)$ for the electron, yielding

$$|M_{\mathrm{lep}}^2| = \frac{1}{V^2}F(E, Z + 1) . \tag{2.18}$$

The Fermi function $F(E, Z + 1)$ accounts for the final electromagnetic interaction of the emitted β electron with the daughter nucleus. The Fermi function is approximately given by [5]

$$F(E, Z + 1) = \frac{2\pi\eta}{1 - \exp(-2\pi\eta)} , \tag{2.19}$$

where the Sommerfeld parameter $\eta = \alpha(Z + 1)/\beta$ with fine structure constant α and relativistic velocity $\beta = v_e/c$.

For an allowed or super-allowed transition, the hadronic matrix element is independent of the kinetic energy of the electron. Generally, this matrix

---

[3] In (2.15), we have not considered the spin of the leptons yet. The two possible spin orientations of each lepton would give another factor of 2×2 if the spins were independent, which is not the case. The parity-violating V–A structure of the weak interaction correlates the spin of the electron with that of the neutrino in a maximum way. And we have just neglected the nucleus concerning energy and momentum, which we cannot do for the angular momenta. The conservation of angular momentum connects the lepton spins to the change of the spin of the decaying nucleus. Therefore, the accounting for the possible spin states is usually done in the calculation of the nuclear part M_{had} of the matrix element M.

element can be divided into a vector current or Fermi part ($\Delta I_{\text{had}} = 0$) and an axial current or Gamow–Teller part ($\Delta I_{\text{had}} = \pm 1$). The hadronic matrix element for tritium is [4]

$$|M_{\text{had}}^2(\text{tritium})| = 5.55\hbar^6 . \tag{2.20}$$

Using (2.12), (2.15), (2.17), and (2.18) and the energy difference $\varepsilon = E_0 - E$, the β spectrum of an allowed or super-allowed decay is given by

$$\begin{aligned}
\frac{d^2 N}{dt\,dE} &= \frac{G_{\text{F}}^2 \cos^2 \Theta_{\text{C}}}{2\pi^3 \hbar^7} |M_{\text{had}}^2| F(E, Z+1) p(E+m)\varepsilon \\
&\quad \times \sqrt{\varepsilon^2 - m^2(\nu_{\text{e}})}\Theta(\varepsilon - m(\nu_{\text{e}})) \\
&= AF(E, Z+1)p(E+m)\varepsilon \\
&\quad \times \sqrt{\varepsilon^2 - m^2(\nu_{\text{e}})}\Theta(\varepsilon - m(\nu_{\text{e}})) .
\end{aligned} \tag{2.21}$$

Equation (2.21) holds only for the decay of a bare, infinitely heavy nucleus. For the more realistic case of an atom or a molecule, the possible excitation of the electron shell due to the sudden change of the nuclear charge by one unit has to be taken into account. The atom or molecule will end up in a specific state of excitation energy V_j with a probability W_j. The corresponding excitation probabilities can be calculated in the sudden approximation from the overlap of the primary electron wave function Ψ_0 with the wave functions of the daughter ion $\Psi_{\text{f},j}$:

$$W_j = |\langle \Psi_0 | \Psi_{\text{f},j} \rangle|^2 \tag{2.22}$$

Equation (2.21) is thus modified into a sum of β spectra of amplitude W_j with different endpoint energies $E_{0,j} = E_0 - V_j$:

$$\begin{aligned}
\frac{d^2 N}{dt\,dE} &= A\,F(E, Z+1)\,p\,(E+m) \\
&\quad \times \sum_j W_j \varepsilon_j \sqrt{\varepsilon_j^2 - m^2(\nu_{\text{e}})}\Theta(\varepsilon_j - m(\nu_{\text{e}})) .
\end{aligned} \tag{2.23}$$

The energy differences ε_j are then defined by $\varepsilon_j = E_0 - V_j - E$.

In case of neutrino mixing, the spectrum is a sum of the components of decays into mass eigenstates:

$$\begin{aligned}
\frac{d^2 N}{dt\,dE} &= A\,F(E, Z+1)\,p\,(E+m) \\
&\quad \times \sum_j W_j \varepsilon_j \left(\sum_i |U_{ei}|^2 \sqrt{\varepsilon_j^2 - m^2(\nu_i)}\Theta(\varepsilon_j - m(\nu_i)) \right) .
\end{aligned} \tag{2.24}$$

When this spectrum is convoluted with an experimental resolution function which is much wider than the mass differences $|m(\nu_i) - m(\nu_j)|$ (which has

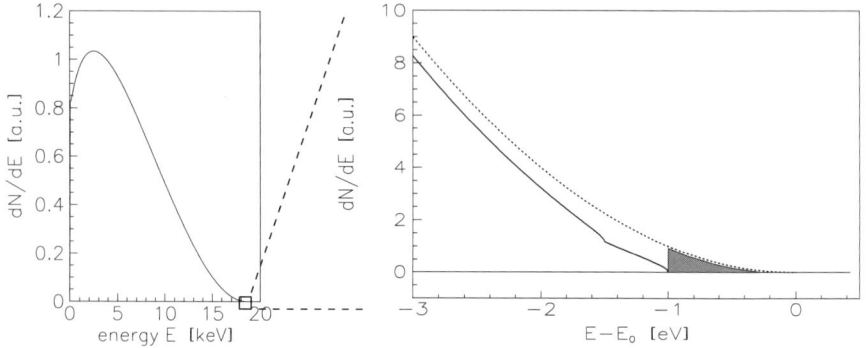

Fig. 2.1. Tritium β spectrum according to (2.24), not taking into account the electronic final-state distribution. *Left*, full β spectrum; *right*, expanded region around tritium endpoint E_0 for $m(\nu_e) = 0$ (*dashed line*) and for two neutrino mass states with arbitrarily chosen values $m(\nu_1) = 1.0\,\text{eV}$, $|U_{e1}|^2 = 0.7$, and $m(\nu_2) = 1.5\,\text{eV}$, $|U_{e2}|^2 = 0.3$ (*solid line*). The *gray shaded area* corresponds to a fraction 2×10^{-13} of all tritium β decays

always been the case so far), the resulting spectrum can be analyzed in terms of a single mean squared electron neutrino mass

$$m^2(\nu_e) = \sum_i |U_{ei}|^2 m^2(\nu_i),\qquad(2.25)$$

and (2.23) applies again.

The square-root term of (2.21) shows that the neutrino mass influences the β spectrum only at the upper end below E_0, and its relative influence decreases in proportion to $m^2(\nu_e)/\varepsilon^2$ (see Fig. 2.1), leading far below the endpoint to a small, constant offset proportional to $-m^2(\nu_e)$.

Figure 2.1 defines the requirements of a direct neutrino mass experiment which investigates a β spectrum: the task is to resolve the tiny change in the spectral shape due to the neutrino mass in the region just below the endpoint E_0, where the count rate is about to vanish. Therefore, high energy resolution is required, combined with large source strength and acceptance and a low background rate.

Tritium β Decay Experiments. The majority of the direct laboratory results on $m(\nu_e)$ published originate from the investigation of tritium β decay, while one single result from ^{187}Re has been reported at conferences.[4] In the long history of tritium β decay experiments, about a dozen experimental results have been reported, starting with the experiment of Curran in the late 1940s, which yielded $m^2(\nu_e) < 1\,\text{keV}$ [21]. A pioneering breakthrough

---

[4] There are also results from investigations of electron capture [19] and bound-state β decay [20], which are about two orders of magnitude less stringent on the neutrino mass.

was obtained by Bergkvist in the early 1970s [22] by constructing a high-luminosity, high-resolution $\sqrt{2}\pi$ magnetic spectrometer, resulting in an upper limit on the neutrino mass of 55 eV. This was also the first work which took into account the influence of the excitations of the electron shell.

After a claim of a discovery of a nonzero neutrino mass around 30 eV at the beginning of the 1980s by a group from the Institute of Theoretical and Experimental Physics (ITEP) in Moscow [23] a series of new experiments were started to check this result. The ITEP group used a thin film of tritiated valine as the β source, combined with a new type of magnetic spectrometer. This Tretyakov spectrometer had a superior luminosity and energy resolution compared with previous spectrometers. The first results testing the ITEP claim came from experiments at the University of Zürich [24] and the Los Alamos National Laboratory (LANL) [25]. Both of those groups used similar Tretyakov-type spectrometers, but more advanced tritium sources than that of the ITEP group. The Zürich group used a solid source of tritium implanted into carbon and, later, a self-assembling film of tritiated hydrocarbon chains. The LANL group developed a gaseous molecular-tritium source, thereby avoiding solid-state corrections. Both experiments disproved the ITEP result. The reason for the ITEP "mass signal" was twofold: the energy loss correction was probably overestimated, and a measurement of the ^{3}He–T mass difference [26] confirming the endpoint energy of the ITEP result turned out later to be significantly wrong [27].

In the 1990s, tritium β decay experiments again yielded controversial results. Figure 2.2 shows the final results of the experiments at LANL and Zürich together with the results from other, more recent measurements with magnetic spectrometers at the University of Tokyo, Lawrence Livermore National Laboratory, and Bejing. The sensitivity to the neutrino mass has improved a lot, but the observed values of $m^2(\nu_e)$ populate the unphysical region of negative $m^2(\nu_e)$. In the case of two experiments, significantly negative results for the mass were obtained. In 1991 and 1994, two new experiments started taking data at Mainz and at Troitsk; these use a new type of electrostatic spectrometer, called a MAC-E filter, which is superior in energy resolution and luminosity with respect to the magnetic spectrometers used previously. However, the early data of those experiments confirmed the large negative $m^2(\nu_e)$ values of the LANL and Livermore experiments when the data were analyzed over the last 500 eV of the β spectrum below the endpoint E_0. However, a new feature was observed. The large negative values of $m^2(\nu_e)$ disappeared when only small intervals below the endpoint E_0 were analyzed (see Fig. 2.6). This effect, which could only be investigated with the high-resolution MAC-E filters, pointed towards an energy loss process that had been underestimated or not taken into account, seemingly present in all experiments. The only common feature of the various experiments seemed to be the calculations of the excitation energies V_i of the daughter ions and of the probabilities W_i. Various theory groups checked these calculations in de-

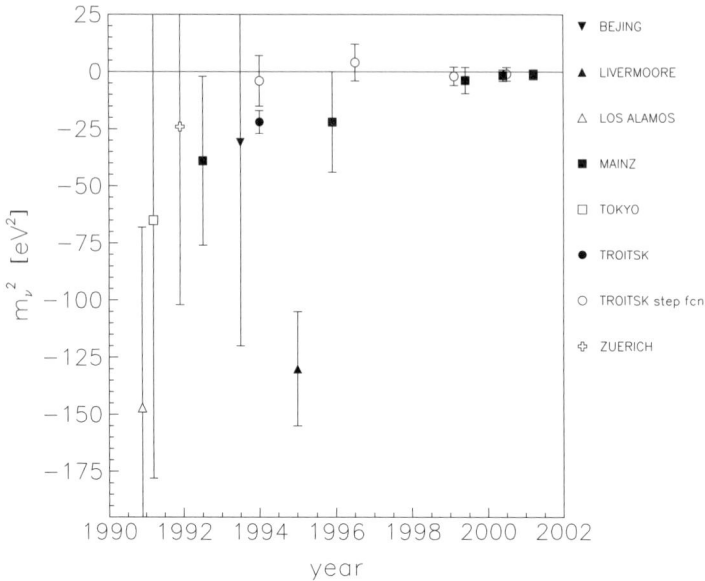

Fig. 2.2. Results of tritium β decay experiments in terms of the observed value of $m^2(\nu_e)$ over the last decade. The experiments at Los Alamos, Zürich, Tokyo, Beijing, and Livermore [7–11], which have been completed, used magnetic spectrometers; the ongoing experiments at Mainz and Troitsk [12–18] are using electrostatic spectrometers of the MAC-E Filter type (see text)

tail. The expansion was calculated to one further order and new, interesting insight into this problem was obtained, but no significant changes were found (see [28] and references therein).

Then the Mainz group found the origin of the energy loss process in its experiment that was missing from the analysis. The Mainz experiment uses as the tritium source a film of molecular tritium quench-condensed onto an aluminum or graphite substrate. Although the film was prepared as a homogeneous thin film with a flat surface, detailed studies showed that the film underwent a temperature-activated roughening transition to an inhomogeneous film by formation of microcrystals, leading to an unexpectedly large inelastic-scattering probability.

The Troitsk experiment, on the other hand, uses a windowless gaseous molecular tritium source, similar to that in the LANL apparatus. Here, the influence of large-angle scattering of electrons magnetically trapped in the tritium source was not considered in the first analysis. The Troitsk group stated that if this effect was taken into account it gave a correction large enough to make the negative values of $m^2(\nu_e)$ disappear.

It is very likely that for the experiments at LANL and Livermore also, experimental effects caused the negative values of $m^2(\nu_e)$. A missing or underestimated experimental correction leads to a negative value of $m^2(\nu_e)$. This can be understood by the following consideration. For $\varepsilon \gg m(\nu_e)$, (2.21) can be expanded into

$$\frac{dN}{dE} \propto \varepsilon^2 - m^2(\nu_e)/2 \ . \tag{2.26}$$

On the other hand, the convolution of a β spectrum (2.21) with a Gaussian of width σ leads to

$$\frac{dN}{dE} \propto \varepsilon^2 + \sigma^2 \ . \tag{2.27}$$

Therefore, in the presence of a neglected experimental broadening with a Gaussian width σ, one expects a shift of the result for $m^2(\nu_e)$ of

$$\Delta m^2(\nu_e) \approx -2\sigma^2 \ , \tag{2.28}$$

which gives rise to a negative value of $m^2(\nu_e)$.

MAC-E Filter. The significant improvement in the sensitivity to the ν mass obtained by the Troitsk and Mainz experiments is due to the use of MAC-E filters. This new type of spectrometer is based on early work by Kruit and Read [29] and was later redeveloped for application to the study of tritium β decay independently at Mainz and Troitsk [30,31]. The MAC-E filter combines high luminosity with a low background and a high energy resolution. Both features are essential to measuring the neutrino mass from the endpoint region of a β decay spectrum. The name "MAC-E filter" stands for Magnetic Adiabatic Collimation followed by an Electrostatic Filter.

The main features of the MAC-E filter are illustrated in Fig. 2.3. Two superconducting solenoids produce a guiding magnetic field. The β electrons, which start from the tritium source in the left solenoid and are emitted into the forward hemisphere, are guided magnetically, in a cyclotron motion along the magnetic field lines, into the spectrometer, thus resulting in an accepted solid angle of nearly 2π. On their way into the center of the spectrometer, the magnetic field B drops by several orders of magnitude. Therefore, the magnetic gradient force transforms most of the cyclotron energy $E_\perp$ into longitudinal motion. This is illustrated in Fig. 2.3 at the bottom by the momentum vector. Owing to the slowly varying magnetic field, the momentum transforms adiabatically, keeping the magnetic moment μ constant: in the the nonrelativistic approximation,

$$\mu = \frac{E_\perp}{B} = \text{const.} \tag{2.29}$$

This transformation can be summarized as follows: the β electrons, emitted isotropically at the source, are transformed into a broad beam of electrons flying almost parallel to the magnetic field lines.

This parallel beam of electrons is energetically analyzed by applying an electrostatic potential generated by a system of cylindrical electrodes. Those

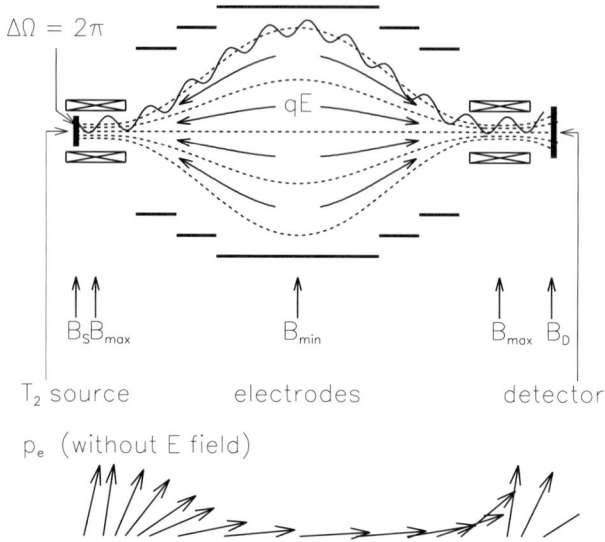

Fig. 2.3. Principle of the MAC-E filter. *Top*, experimental setup; *bottom*, momentum transformation due to adiabatic invariance of the orbital magnetic moment μ in the inhomogeneous magnetic field

electrons which have enough energy to pass the electrostatic barrier are reaccelerated and collimated onto a detector; all other electrons are reflected. Therefore, the spectrometer acts as an integrating high-energy-pass filter. The relative sharpness of this filter is given simply by the ratio of the minimum magnetic field $B_{\min}$ in the analyzing plane in the middle to the maximum magnetic field between the β electron source and the spectrometer, $B_{\max}$:

$$\frac{\Delta E}{E} = \frac{B_{\min}}{B_{\max}} \ . \tag{2.30}$$

By scanning the electrostatic retarding potential, the β spectrum can be measured.

The experiments at Mainz and Troitsk use similar MAC-E filters, which differ slightly in size. The diameter and length of the Mainz spectrometer are 1 m and 4 m, respectively, and those of the Troitsk spectrometer are 1.5 m and 7 m, respectively. The major differences between the two setups are in the tritium sources.

The Troitsk Neutrino Mass Experiment. The windowless gaseous tritium source of the Troitsk experiment [17] is essentially a tube 5 cm in diameter filled with T_2, with a column density of $\rho d \approx 10^{17}$ molecules/cm^2. The source is connected to the ultrahigh vacuum of the spectrometer by a series of differential-pumping stations.

From their first measurement in 1994 on, the Troitsk group has reported the observation of a small but significant anomaly in its experimental spectra starting a few eV below the β endpoint E_0. This anomaly appears as a sharp step in the count rate [16]. Since a MAC-E filter is integrating, this step should correspond to a narrow line in the primary spectrum, with a relative intensity of about 10^{-10} of the total decay rate. In 1998 the Troitsk group reported that the position of this line oscillates with a periode of 0.5 years, between 5 eV and 15 eV below E_0 [17]. In 2000, the anomaly did not follow the 0.5 year periodicity anymore, but still existed [39]. In total, the Troitsk experiment has taken 200 days of tritium data, and in almost all of the runs this anomaly has been observed.

The reason for such an anomaly with such features is not clear. Detailed investigations are continuing at Troitsk. In addition, synchronous measurements with the Mainz experiment have been performed. In 2001, the Troitsk group improved the differential pumping between the gaseous tritium source and the spectrometer, lowered the electric field strength in a critical region, and improved the vacuum. The first two runs of 2001 either gave no indication of an anomaly or showed only a small effect with an amplitude of 2.5×10^{-3} /s, to be compared with the previously observed effects with amplitudes between 2.5×10^{-3} /s and 13×10^{-3} /s. These findings also support the assumption that the Troitsk anomaly is due to an still unidentified experimental artefact [32].

By fitting a standard β spectrum to the data, the Troitsk group obtained significantly negative values of $m^2(\nu_e)$ of -10 to -20 eV^2 (see filled circle in Fig. 2.2). Describing the anomaly phenomenologically by adding a monoenergetic line, free in amplitude and position, to a standard β spectrum results in values of $m^2(\nu_e)$ compatible with zero [17] (see open circles in Fig. 2.2). After this correction, the average over all runs until 2001 amounts to [32]

$$m^2(\nu_e) = -2.3 \pm 2.5 \pm 2.0\,\mathrm{eV}^2\,,$$

which corresponds – under the assumption that the run-by-run correction with an additional line is correct – to an upper limit [32] of

$$m(\nu_e) \leq 2.2\,\mathrm{eV} \qquad (95\%\ \mathrm{C.L.})\,.$$

The Mainz Neutrino Mass Experiment. The Mainz experiment uses a film of molecular tritium quench-condensed onto a graphite (HOPG) substrate. The film has a diameter of 17 mm and a typical thickness of 40 nm, which is measured by laser ellipsometry. The problem of the roughening transition mentioned above has been investigated by the Mainz group in cooperation with Leiderer's condensed-matter group at Konstanz, Germany, using conversion electron spectroscopy and scattered-light techniques on the various hydrogen isotopes [33, 34]. The following results were obtained. The roughening transition follows an Arrhenius-type law. Thus, it cannot be avoided, but it can be drastically slowed down by using lower temperatures. A T$_2$ film at

2 K has a time constant of order 10 yr [34], i.e. much longer than the typical duration of a measurement.

In the years 1995–1997, the Mainz setup was upgraded with a new cryostat, providing a temperature of the tritium film below 2 K, to avoid the roughening transition. Also, a new, tilted pair of superconducting solenoids was installed (see Fig. 2.4). Consequently, β particles from the source are still guided magnetically into the spectrometer, whereas tritium molecules evaporating from the source are trapped in the bend of a liquid-helium-cooled tube covered with graphite. This measure eliminated source-correlated background and allowed us to increase the source strength significantly. The upgrade of the Mainz setup was completed by the application of high-frequency pulses to one of the electrodes in between measurements every 20 s, and a full automation of the apparatus and remote control. The former improvement lowers and stabilizes the background; the latter improvement allows long-term measurements.

Figure 2.5 shows the endpoint region of the Mainz 1998 and 1999 data [15], in comparison with the earlier Mainz 1994 data [13]. An improvement in the signal-to-background ratio by a factor of 10 and a significant enhancement in the statistical quality of the data are clearly visible. A fit with $m^2(\nu_e)$ fixed to zero fits perfectly the latter data set over the last 15 eV of the β spectrum. This limits any persistent spectral anomaly in this range to an amplitude below 10^{-3} /s (as against a total flux of 10^8 /s entering the spectrometer). A spectral anomaly like the fluctuating anomaly reported by the Troitsk group [16, 17], on the other hand, would reaches an amplitude of up to 10^{-2} /s.

The main systematic uncertainties of the Mainz experiment originate from the physics and properties of the quench-condensed tritium film: the inelastic scattering of β electrons within the tritium film, the excitation of neighboring molecules due to the β decay, and the self-charging of the tritium film by radioactivity.

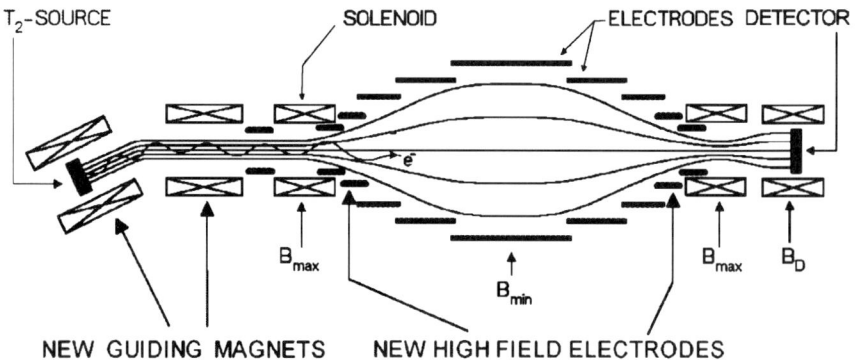

Fig. 2.4. The upgraded Mainz setup, shown schematically. The outer diameter is 1 m; the distance from source to detector is 6 m

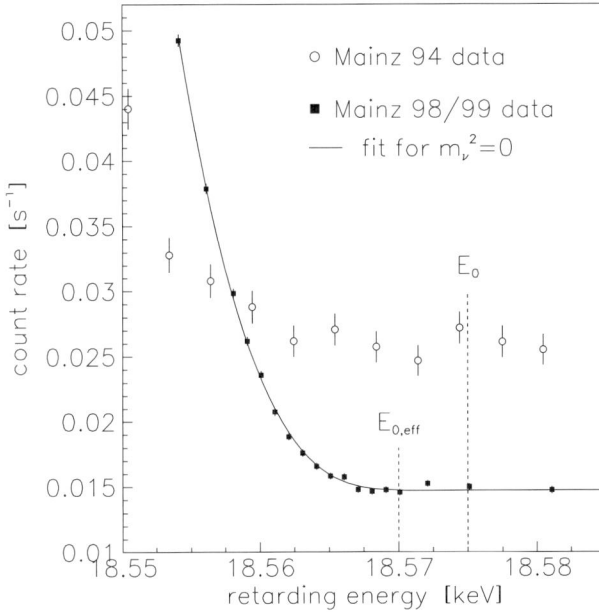

Fig. 2.5. Averaged count rate of the Mainz 1998 and 1999 data [15] (*points*), with fit (*line*), in comparison with earlier Mainz data from 1994 [13], as a function of the retarding energy near the endpoint E_0, and the effective endpoint $E_{0,\text{eff}}$. The position of the latter takes into account the width of the response function of the setup and the mean rotation–vibration excitation energy of the electronic ground state of the $^3\text{HeT}^+$ daughter molecule

These systematic uncertainties were studied in detail by various investigations [35–37], and the knowledge of the corresponding corrections was significantly improved.

Figure 2.6 shows the fit results of the combined Mainz 1998 and 1999 data for $m^2(\nu_e)$ as a function of the lower limit of the fit interval. The monotonic trend towards negative values of $m^2(\nu_e)$ for larger fit intervals, as observed for the Mainz 1991 and 1994 data [12, 13], has vanished. This shows that the dewetting of the T_2 film from the graphite substrate [33, 34] was indeed the reason for this behavior. Now this effect is safely suppressed by the much lower temperature of the T_2 film. The data do not show any indication of other residual distortions. The energy interval below the endpoint that yields the smallest combined statistical and systematic uncertainty in the neutrino mass corresponds to the last 70 eV below the endpoint E_0, and gives [15]

$$m^2(\nu_e) = -1.6 \pm 2.5 \pm 2.1 \,\text{eV}^2 \,,$$

which is compatible with a zero neutrino mass. Considering its uncertainties, this value corresponds to an upper limit on the electron neutrino mass of [15]

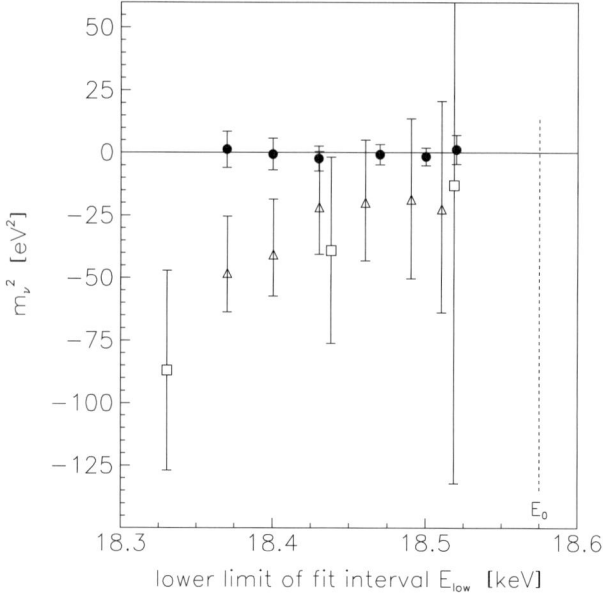

Fig. 2.6. Mainz fit results for $m^2(\nu_e)$ as a function of the lower limit of the fit interval (upper limit always 18.66 keV, well above E_0), for data from 1991 [12] (*open squares*), from 1994 [13] (*open triangles*), and from the last four runs of 1998 and 1999 [15] (*filled circles*)

$$m(\nu_e) \leq 2.2\,\text{eV} \qquad (95\ \%\ \text{C.L.})\,.$$

Further data have been taken at Mainz in 2000 and 2001, reducing slightly the uncertainties and bringing $m^2(\nu_e)$ closer to zero, but without improving the limit on $m(\nu_e)$ [32,38].

Indications of a "Troitsk-like" anomaly at Mainz were observed only once, in summer 1998. This single coincidence did not appear in previous or later runs [14]. Of special interest are the Mainz 2000 runs, although they were done under less favorable conditions. In particular, part of the data was taken in parallel with the Troitsk experiment. Whereas the Troitsk experiment again found indications of the anomaly [39], at Mainz no indication of a Troitsk-like anomaly was found [40]. In summary, the Mainz data clearly indicate that the anomaly observed at Troitsk is caused by some experimental artifact [32].

β Decay Experiments with Cryogenic Bolometers. Owing to the complicated electronic structure of ^{187}Re and its β decay (see Sect. 2.2.3), the advantage of the seven times lower endpoint energy E_0 of ^{187}Re compared with tritium can only be exploited if the β spectrometer measures the entire released energy, except that of the neutrino. This situation can be realized by using a cryogenic bolometer, which at the same time contains the ^{187}Re β emitter, as the β spectrometer (see Fig. 2.7).

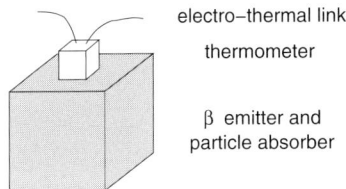

electro–thermal link

thermometer

β emitter and
particle absorber

Fig. 2.7. Principle of a cryogenic bolometer for direct neutrino mass measurements, consisting of a β-emitting crystal, which serves at the same time as the particle and energy absorber. The energy release ΔW gives rise to a temperature increase $\Delta T = \Delta W/C$, where C is the heat capacity; the temperature rise is measured by a thermometer. The electric readout wires of the thermometer link the whole bolometer to a thermal bath

One disadvantage of this method is the fact that one always measures the entire β spectrum. Even in the case of the very low endpoint energy of ^{187}Re, the relative fraction of events in the last eV below E_0 is only of order 10^{-10} (compare Fig. 2.1). Considering the long time constant of the signal of a cryogenic bolometer (typically several hundred microseconds), only large arrays of cryogenic bolometers can deliver the signal rate needed.

At present, two groups are working on ^{187}Re β decay experiments, in Milan [41] and Genoa [42]. Although cryogenic bolometers with an energy resolution of 5 eV have been produced with other absorbers, this has yet not been achieved for rhenium. The two groups are using different ways to produce the crystals used in the experiments. The MANU2 experiment in Genoa has succeeded in preparing crystals from metallic rhenium. This group has reported a limit on $m(\nu_e)$ of 26 eV [43]. The Genoa group understands its measured spectra well and has seen, for the first time, the oscillation pattern of the β environmental fine structure [44], which describes the interference between the outgoing electron wave function with its own after being coherently scattered on the crystal. This effect is similar to the interference on crystals observed near the photon absorption edges for photoelectrons under the name X-ray environmental fine structure (XEFS). The Genoa group expects a sensitivity to $m(\nu_e)$ of 10 eV in the near future. A significant further improvement could be obtained by improving the energy resolution of the crystals, by using new, superconducting transition thermometers. The MiBeta experiment in Milan uses $AgReO_4$ crystals with a typical energy resolution of 35 eV. There is no result for $m(\nu_e)$ yet; the expected sensitivity is similar to that for the Genoa experiment.

2.3 Search for Neutrinoless Double β Decay

The second way to look for neutrino masses is to search for neutrinoless double β decay, $0\nu\beta\beta$ [45–49]. In contrast to the direct neutrino mass determination based on the investigation of the kinematics of weak decays, $0\nu\beta\beta$ decay requires that neutrinos are identical to their antiparticles. This method is sensitive to the phases of the neutrino mixing matrix, as outlined in the following. For these reasons, the search for neutrinoless double β decay is complementary to the direct neutrino mass investigations.

2.3.1 Double β Decay

In 1937 Majorana discussed the possibility that the neutral neutrinos could be their own antiparticles ("Majorana particles") [50]. Earlier, Goeppert-Mayer had calculated that double-β lifetimes were of order 10^{20} yr [51].

After indirect proof by geochemical methods [52], a double β decay of ^{82}Se was directly observed for the first time in 1987 [53], yielding in both cases halflives in accordance with theoretical expectations. This twofold conversion of a neutron into a proton (or vice versa) in one nucleus at the same time is a weak process of second order which emits two electrons and two electron antineutrinos (or two positrons and two electron neutrinos):

$$\beta^-\beta^- : \quad (Z, A) \to (Z + 2, A)^{2+} + 2e^- + 2\overline{\nu}_e ,$$
$$\beta^+\beta^+ : \quad (Z, A) \to (Z - 2, A)^{2-} + 2e^+ + 2\nu_e . \tag{2.31}$$

This second-order decay can be observed if single β decay is not allowed energetically or is highly forbidden. Double-β candidates are provided by pairs of even–even nuclei; this can be explained by the two separate mass parabolas for even–even and odd–odd nuclei. About a dozen double-β emitters have been detected experimentally (see Table 2.1).

Let us consider now the case in which one virtually emitted antineutrino (or neutrino for $\beta^+\beta^+$ decay) at one decay vertex is absorbed at the other vertex as a neutrino (see Fig. 2.8). This process is called neutrinoless double β decay:

$$0\nu\beta^-\beta^- : \quad (Z, A) \to (Z + 2, A)^{2+} + 2e^- ,$$
$$0\nu\beta^+\beta^+ : \quad (Z, A) \to (Z - 2, A)^{2-} + 2e^+ . \tag{2.32}$$

The existence of this process requires neutrinos to be identical to their antiparticles, i.e. Majorana particles (see Chap. 1). This process violates conservation of lepton number by two units. Therefore, this process cannot be described in the framework of the Standard Model, but it is predicted in most theories going beyond it.

In addition, the left-handed structure of the weak interaction requires that the antineutrino is emitted at the first vertex as a right-handed particle, but it has to be absorbed at the second vertex as a left-handed particle. Therefore, a helicity flip of the neutrino is necessary. To see the dependence

Table 2.1. Experimentally measured $2\nu\beta\beta$ half-lives (lower limit for ^{136}Xe), from [49] (where details and further references may be found).

Isotope	$T_{1/2}^{2\nu}$ (yr)
^{48}Ca	$(4.2 \pm 1.2) \times 10^{19}$
^{76}Ge	$(1.3 \pm 0.1) \times 10^{21}$
^{82}Se	$(9.2 \pm 1.0) \times 10^{19}$
^{96}Zr†	$(1.4^{+3.5}_{-0.5}) \times 10^{19}$
^{100}Mo	$(8.0 \pm 0.6) \times 10^{18}$
^{116}Cd	$(3.2 \pm 0.3) \times 10^{19}$
^{128}Te$^{(1)}$	$(7.2 \pm 0.3) \times 10^{24}$
^{130}Te$^{(2)}$	$(2.7 \pm 0.1) \times 10^{21}$
^{136}Xe	$> 8.1 \times 10^{20}$ (90% C.L.)
^{150}Nd†	$7.0^{+11.8}_{-0.3} \times 10^{18}$
^{238}U$^{(3)}$	$(2.0 \pm 0.6) \times 10^{21}$

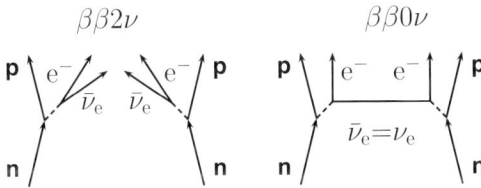

Fig. 2.8. Double β decays; *left*, $2\nu\beta\beta$; *right*, $0\nu\beta\beta$

on the neutrino mass,[5] we write down the part of the two lepton currents $J_\alpha J_\beta$ of the transition matrix element M for the ordinary double β decay and use the neutrino mass states (2.1) directly:

$$M_{\alpha\beta} \propto J_\alpha J_\beta \tag{2.33}$$

$$= \langle \bar{e} | \gamma_\alpha (1 - \gamma_5) | \nu_e \rangle \langle \bar{e} | \gamma_\beta (1 - \gamma_5) | \nu_e \rangle \tag{2.34}$$

$$= \left(\sum_i U_{ei} \langle \bar{e} | \gamma_\alpha (1 - \gamma_5) | \nu_i \rangle \right) \left(\sum_j U_{ej} \langle \bar{e} | \gamma_\beta (1 - \gamma_5) | \nu_j \rangle \right) . \tag{2.35}$$

We can turn the bracketed quantity denoting the second current into its transpose:

---

[5] The existence of weak right-handed currents would also allow the coupling of a right-handed neutrino at the second vertex, without requiring a nonzero neutrino mass. However, a rather general theorem predicts nonzero neutrino masses for all cases in which $0\nu\beta\beta$ decay takes place [54].

$$J_\alpha J_\beta = \left(\sum_i U_{ei} \langle \bar{e} | \gamma_\alpha (1 - \gamma_5) | \nu_i \rangle \right) \left(\sum_j U_{ej} \langle \overline{\nu_j^c} | (1 - \gamma_5) \gamma_\beta | e^c \rangle \right) .$$

(2.36)

To describe a neutrinoless double β decay, we have to contract two neutrinos of (2.36) into an inner spin-1/2 propagator of the neutrinos (which needs to be summed over all neutrino spins). For this purpose, we have to transform the antineutrino states $\nu_i{}^c$ into neutrino states ν_i, which gives rise to Majorana phases $e^{i\alpha_i'}$. Considering in addition the orthogonality of the different neutrino mass eigenstates ν_i, we obtain

$$\sum_i U_{ei} | \nu_i \rangle \sum_j U_{ej} \langle \overline{\nu_j^c} | = \sum_i U_{ei} | \nu_i \rangle \sum_j U_{ej} e^{i\alpha_i'} \langle \overline{\nu_j} | \qquad (2.37)$$

$$= \sum_i U_{ei}^2 e^{i\alpha_i'} | \nu_i \rangle \langle \overline{\nu_i} | \qquad (2.38)$$

$$\rightarrow \sum_i U_{ei}^2 e^{i\alpha_i'} \frac{\gamma^\mu p_\mu + m(\nu_i)}{p^2 - m^2(\nu_i)} . \qquad (2.39)$$

The numerator of (2.39) has two terms. Since γ_5 and γ^μ anticommute, the first term, proportional to $\gamma^\mu p_\mu$, vanishes between the two left-handed projectors $(1 - \gamma_5)$, whereas the second term, proportional to $m(\nu)$, survives. Therefore the leptonic part of the 0$\nu\beta\beta$ decay matrix element M is proportional to an effective neutrino mass:

$$M \propto \left| \sum_i U_{ei}^2 e^{i\alpha_i'} m(\nu_i) \right| := m_{ee} . \qquad (2.40)$$

When we include the nuclear part of the matrix element M_{nucl} and the phase space factor $G^{0\nu}$, we obtain the following for the half-life:

$$\left(T_{1/2}^{0\nu\beta\beta} \right)^{-1} = G^{0\nu} |M_{\text{nucl}}|^2 m_{ee}{}^2 . \qquad (2.41)$$

Although the measured nuclear matrix elements of normal 2$\nu\beta\beta$ decay can be calculated quite successfully, the calculation of the nuclear part of the matrix element for 0$\nu\beta\beta$ decay is a difficult task, and gives rise to a systematic uncertainty with regard to m_{ee} of about a factor of 2. The reason for this is the virtual neutrino exchange in the case of 0$\nu\beta\beta$ decay, which transfers momenta up to $\approx 100\,\text{MeV}$ ($\approx \hbar/R_{\text{nucleus}}$), possibly exciting intermediate states with $L \leq 5$.

In addition to the Majorana phases α_i', the values U_{ei}^2 are in general complex, owing to a possible CP-violating phase δ in the lepton sector. The latter is in principle observable in oscillation experiments; the former Majorana phases α_i' are not. We combine both phases into new phases α_i:

$$m_{ee} = \left| \sum_i e^{i\alpha_i'} U_{ei}^2 m(\nu_i) \right| = \left| \sum_i e^{i\alpha_i} |U_{ei}^2| m(\nu_i) \right| . \qquad (2.42)$$

In the case of CP conservation, each phase gives only a sign:

$$e^{i\alpha_i} = \pm 1 \; . \tag{2.43}$$

Equation (2.42) has to be compared with the average neutrino mass measured in β decay,

$$m^2(\nu_e) = \sum_i |U_{ei}^2| m^2(\nu_i) \; . \tag{2.44}$$

Equation (2.44) describes a real average with coefficients $0 \leq |U_{ei}^2| \leq 1$ originating from an incoherent sum, whereas in (2.42), cancellation can take place, since it describes a coherent sum over all mass eigenstates contributing to the inner neutrino propagator. In principle, neutrinos could be massive Majorana particles while m_{ee} vanishes owing to cancellation effects.

Since the value of U_{e3} is very small (see Chap. 4), especially in the case of quasi-degenerate neutrino masses ($m(\nu_1) \approx m(\nu_2) \approx m(\nu_3)$) and large mixing for solar neutrinos ($\sin 2\Theta_\odot \rightarrow 1$), the 0$\nu\beta\beta$ decay signal is strongly reduced in the case where $e^{i\alpha_1} e^{i\alpha_2} = -1$. The following relation [55] connects the "effective neutrino mass" m_{ee} measured in 0$\nu\beta\beta$ decay (2.42) with the "electron neutrino mass" $m(\nu_e)$ determined in direct mass measurements by investigating β decays (2.44) and the two-flavor mixing angle $\Theta_\odot$ of solar neutrino oscillation (see Chap. 4):

$$m_{ee} \leq m(\nu_e) \leq \frac{m_{ee}}{|\,|\cos 2\Theta_\odot|\,(1 - |U_{e3}^2|) - |U_{e3}^2|\,|} \; . \tag{2.45}$$

It should be noted that in theories beyond the Standard Model, such as Supersymmetry, additional, new particles can appear in the inner propagator, which would require us to enlarge (2.40).

2.3.2 Double-β-Decay Experiments

The signature of a 0$\nu\beta\beta$ decay in the case of a $\beta^-\beta^-$ decay (or $\beta^+\beta^+$ decay) is the emission of two decay electrons (or positrons) which carry the whole available energy Q, since no neutrinos are emitted. The sum energy spectrum of the two electrons should exhibit a sharp line at the Q-value of the double β decay (see Fig. 2.9). The phase space factor for 0$\nu\beta\beta$ decay is larger than that for 2$\nu\beta\beta$ decay and scales with Q^5. For $m_{ee} = 1\,\mathrm{eV}$, the corresponding half-lives for 0$\nu\beta\beta$ decay range from 10^{22} yr to 10^{25} yr, depending on the phase space factor and the nuclear matrix elements (see Table 2.2). The long half-life requires experiments to use large masses and to have a sufficiently small background. Therefore, all such experiments have been installed in underground laboratories with several thousands of meter water-equivalent of overburden to shield muons. In the case of experiments with moderate energy resolution or ultrahigh sensitivity, not only is environmental background a problem, but also the background from the normal double β decay plays

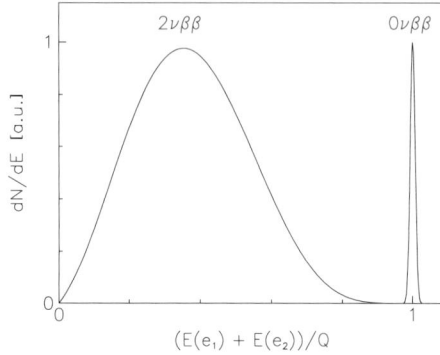

Fig. 2.9. Schematic illustration of sum energy spectra for 2νββ and 0νββ decays. The 0νββ decay signal was convoluted with an arbitrarily chosen experimental Gaussian resolution function

Table 2.2. Limits on 0νββ decay at 90% confidence level (except where noted), from [49] (where details may be found). The limits and ranges of $m(\nu)$ are those deduced by the authors of the papers cited, using their choice of matrix elements

Isotope	$T_{1/2}^{0\nu}$ (yr)	m_{ee} (eV)	Reference
^{48}Ca	$> 9.5 \times 10^{21} (76\%)$	< 8.3	[61]
^{76}Ge	$> 1.9 \times 10^{25}$	< 0.35	[62]
	$> 1.6 \times 10^{25}$	$< 0.33\text{–}1.35$	[63]
^{82}Se	$> 2.7 \times 10^{22} (68\%)$	< 5	[56]
^{100}Mo	$> 5.5 \times 10^{22}$	< 2.1	[57]
^{116}Cd	$> 7 \times 10^{22}$	< 2.6	[64]
^{128,130}Te	$T_{1/2}(130)/T_{1/2}(128)$	$< 1.1\text{–}1.5$	[65]
	$= (3.52 \pm 0.11) \times 10^{-4}$		
	(geochemical)		
^{128}Te	$> 7.7 \times 10^{24}$	$< 1.1\text{–}1.5$	[65]
^{130}Te	$> 1.4 \times 10^{23}$	$< 1.1\text{–}2.6$	[60]
^{136}Xe	$> 4.4 \times 10^{23}$	$< 1.8\text{–}5.2$	[59]
^{150}Nd	$> 1.2 \times 10^{21}$	< 3	[66]

a role. A way to enhance the signal-to-background ratio is to use isotopically enriched material.

There are several experimental approaches to search for 0νββ decay. One way to obtain a sufficient background suppression is to identify both of the electrons emitted from a thin foil and measure their energy. The pioneering University of California at Irvine (UCI) group used a time projection chamber (TPC) to do this with ^{82}Se, ^{100}Mo, and ^{150}Nd [56]. The Electron Gamma-Ray Neutrino Telescope (ELEGANT) experiments use several methods to perform spectroscopy on the two double-β-decay electrons of ^{48}Ca, ^{100}Mo, and ^{116}Cd [57]. The Neutrino Ettore Majorana Observatory (NEMO) detectors use foils of ^{82}Se, ^{94,96}Zr, ^{100}Mo, and ^{116}Cd and tracking chambers [58].

An alternative way is to use the double-β-emitter source as an active detector: one realization of this principle uses xenon as a counting gas in a TPC. The Gotthard xenon experiment is searching for the double β decay of ^{136}Xe [59]. The cryogenic bolometer technique allows one, in principle, to use various double-β emitters as active sources. The MiBeta experiment at Gran Sasso [60] currently uses 20 TeO$_2$ crystals with a total weight of 6.8 kg. The energy resolution in the signal region is 8 keV. Although four of the 20 crystals are isotopically enriched, the high natural abundance of ^{130}Te does not in fact require isotopic enrichment. The third way to perform an active-source experiment is to use the semiconductor material germanium, which contains the double-β-decay isotope ^{76}Ge with a 7.8% natural abundance. Table 2.2 lists the experimental results obtained in the search for 0νββ decay.

At present, the most sensitive experiment is the Heidelberg–Moscow experiment, which uses about 11 kg of enriched germanium with an isotopic abundance of ^{76}Ge of 86% [62]. Five detectors have been running for ten years in the Gran Sasso underground laboratory. The use of good shielding, low-level materials, and pulse shape discrimination has led to a very low background in the signal region of $b = 0.06\,(\mathrm{yr\,kg\,keV})^{-1}$ (see Fig. 2.10). Description of the various background lines by a detailed background model allowed the researchers to subtract the line-induced background. The remaining background below the signal region is mainly due to 2νββ decay (see Fig. 2.11).

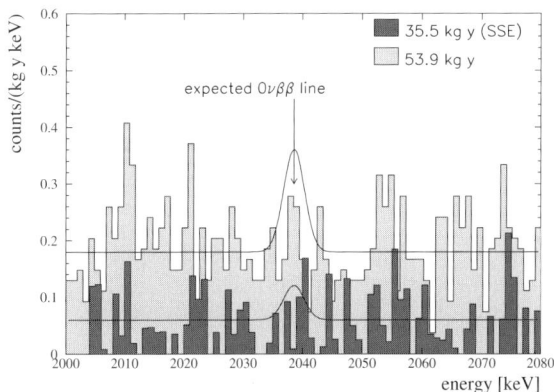

Fig. 2.10. Summed spectra of all five Germanium detectors in the Heidelberg–Moscow experiment in the region of interest for 0νββ decay, after 53.9 kg yr of measurement (*light gray*), and "single-side events" after pulse shape discrimination to reduce photon-induced background after 35.5 kg yr of measurement (*dark*). The *curves* correspond to the excluded signals for $T_{1/2}^{0\nu} \geq 1.3 \times 10^{25}$ yr (90% C.L.) and $T_{1/2}^{0\nu} \geq 1.9 \times 10^{25}$ yr (90% C.L.). From [62].

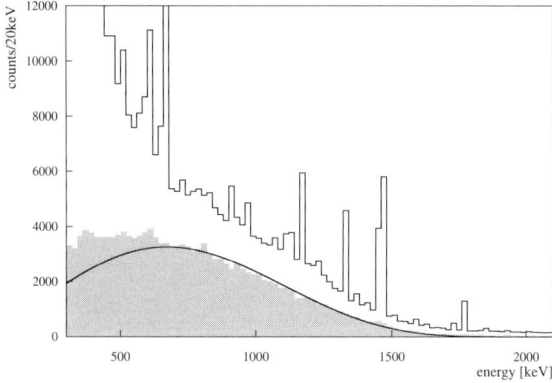

Fig. 2.11. Summed spectra of all five germanium detectors in the Heidelberg–Moscow experiment after 47.7 kg yr of measurement (*histogram*), residual spectrum after subtracting all identified background components (*gray-shaded histogram*), and fitted 2νββ decay signal (*solid line*), from [62]

Until recently, the Heidelberg–Moscow collaboration reported a lower limit on the half-life of

$$T_{1/2}^{0\nu\beta\beta} \geq 1.9 \times 10^{25} \text{ yr} \quad (90\% \text{ C.L.}), \tag{2.46}$$

and an upper limit on the effective neutrino mass m_{ee} of

$$m_{ee} \leq 0.35 \text{ eV} \quad (90\% \text{ C.L.}) \tag{2.47}$$

(not considering the systematic uncertainty in the nuclear matrix element of about a factor of 2) [62]. Very similar values are being obtained from the IGEX collaboration, which also uses detectors containing enriched ^{76}Ge [63]. Considering (2.45), the germanium double-β-decay experiments and the direct tritium β decay experiments have reached about the same sensitivity for the electron neutrino mass $m(\nu_e)$.

In December 2001, a subgroup of the Heidelberg–Moscow collaboration published evidence for a 0νββ decay signal [67]. This statement was based on nearly the same data as that of the previous analysis, but this time the existence of several peaks around the signal region was assumed. This assumption decreases the average background rate, and a signal peak at the expected position shows up at a significance of 2–3σ giving a best value of $m_{ee} = 0.39$ eV. This result provoked several critical comments (see [68–71]), which stated that a much better understanding of the background was needed before this claim could be accepted. Therefore, further investigation and an independent check are needed.

The current 0νββ decay experiments are not sensitive enough to check the Heidelberg–Moscow claim, but two next-generation experiments have just started to take data. NEMO 3 [72], a large follow-up of the two previous

NEMO experiments, started to run in spring 2002. NEMO 3 uses double-β-decay active foils with a total mass of about 10 kg, in between drift chambers operating in Geiger mode in a magnetic field. CUORICINO [73], a follow-up of the MiBeta TeO$_2$ cryogenic bolometers, but with a better energy resolution, lower background, and higher total mass, has started to take data in 2002. Both experiments aim for a sensitivity to m_{ee} well below 1 eV.

2.4 Summary and Outlook

As shown in Chap. 4, there is strong evidence for neutrino masses from experiments with atmospheric, solar and reactor neutrinos, clearly indicating that neutrinos have nonzero masses. Oscillation experiments are sensitive to the neutrino mixing matrix $U_{\alpha i}$ and differences between squared neutrino masses Δm_{ij}^2, but not to neutrino masses directly. At least one neutrino mass has to be determined to fix the neutrino mass scale. As shown in Chaps. 1, 3, and 6, the scale of the neutrino masses is of crucial importance for cosmology and astrophysics, as well as for particle physics. The sensitivity which should be aimed for is at least a few tenths of an eV, which would clarify whether neutrino masses are relevant for cosmology and would check all models with quasi-degenerate neutrino masses. Of course, a sensitivity improvement by at least another order of magnitude is needed to test all possible scenarios and to be sure of finding the neutrino masses.

One important way to determine the neutrino mass scale is to search for neutrinoless double β decay. This method is important on its own, since it is the only way to probe the question of whether Neutrinos are Majorana particles ($\bar{\nu} = \nu$) or of Dirac type ($\bar{\nu} \neq \nu$). A few double-β-decay experiments which will reach a sensitivity to m_{ee} of 0.1 eV or below [49] are under way or have been proposed. Common to all experiments is their large scale, with about 1 t total mass of material of high isotopic abundance, and their very stringent demands on background suppression. These experiments are exploring all possible ways to lower the background. The background reduction is limited by the 2νββ decay mode, requiring a good energy resolution, especially in cases with short 2νββ half-lives.

There are two projects using enriched germanium detectors. Special emphasis is put here on background suppression. The Majorana experiment (named after E. Majorana) aims to reach a very low background by standard low-level techniques and active vetoing of subdivided germanium detectors [74]. The GErmanium NItrogen Underground Setup (GENIUS) aims to put naked germanium detectors in a huge tank of liquid nitrogen to shield the detector with this ultrapure liquid [75]. Owing to the low energy threshold, it will serve as a dark-matter search experiment at the same time. The GENIUS test facility has been funded; its operation has started in 2003. The Enriched Xenon Observatory (EXO) is an ambitious project for a xenon TPC, currently in the research and development phase. The idea of EXO is to obtain a very low background by not only measuring the signal of the two β electrons but also identifying the daughter barium ions by laser spectroscopy in coincidence [76]. The Cryogenic Underground Observatory for Rare Events (CUORE) is a large array of cryogenic bolometers made from TeO_2 crystals, with a total mass of 0.76 t, based on the Mi-Beta experiment. The prototype experiment CUORICINO has started taking data in 2002 [77]. The Mo Observatory Of Neutrinos (MOON) plans to use ^{100}Mo as the double-β-decay source [78]. This detector could also serve as a solar-neutrino detector, since an inverse β decay into ^{100}Tc caused by a solar ν_e will be followed shortly afterwards by a β decay into ^{100}Ru with a lifetime of 16 s, which could serve as a coincidence for background suppression.

In Sect. 2.3.1, we described how neutrinoless double β decay and direct ν mass measurements give complementary information with respect to the neutrino mass and mixing matrix. Considering in addition the possible cancellation effects summarized in (2.45), a direct neutrino mass experiment is mandatory. Since a nearby supernova cannot give a sub-eV sensitivity to the neutrino mass, a laboratory β decay experiment with a sensitivity one order of magnitude higher is needed. The cryogenic bolometers have bright prospects in many fields, but in the near future they will probably not reach a sub-eV sensitivity to the neutrino mass.

Therefore, the KArlsruhe TRItium Neutrino (KATRIN) experiment, a next-generation tritium β decay experiment with sub-eV sensitivity to the neutrino mass $m(\nu_e)$, has been proposed [79]. The international collaboration includes most of the previous and present tritium β decay groups. The experiment is based on the superior MAC-E filter principle explored by the Mainz and Troitsk experiments. A huge spectrometer with a diameter of 10 m and an energy resolution of $\Delta E = 1$ eV, in combination with a windowless gaseous tritium source, will give small enough systematics as well as the resolution and the luminosity to reach a sensitivity on $m(\nu_e)$ of 0.2 eV [80]. This value of 0.2 eV corresponds to an upper limit with 90% C.L. if a non-zero neutrino mass would not be found. A neutrino mass of 0.30 eV (0.35 eV) would be detected with 3σ (5σ) significance. For such small neutrino masses it is enough to investigate the last few ten eV below the β endpoint E_0. Owing the energy thresholds, the electronic excitations of the T^3He^+ ion and the inelastic scattering processes in T_2 play only a minor role. The start of data taking by KATRIN is planned for 2007.

Acknowledgments

We would like to thank Herbert Dreiner, Guido Drexlin, and Ernst Otten for helpful discussions.

References

1. D.E. Groom et al. (Particle Data Group), Eur. Phys. J. C **15**, 1 (2000).
2. K. Assamagan et al., Phys. Rev. D **53**, 6065 (1996).
3. R. Barate et al., Eur. Phys. J. C **2**, 3 (1998).
4. R.G.H. Robertson and D.A. Knapp, Annu. Rev. Nucl. Sci. **38**, 185 (1988).
5. E. Holzschuh, Rep. Prog. Phys. **55**, 1035 (1992).
6. J.F. Wilkerson and R.G.H. Robertson, in *Current Aspects of Neutrino Physics*, ed. D.O. Caldwell (Springer, Berlin, Heidelberg, 2001), p. 39.
7. R.G.H. Robertson et al., Phys. Rev. Lett. **67**, 957 (1991).
8. E. Holzschuh et al., Phys. Lett. B **287**, 381 (1992).
9. H. Kawakami et al., Phys. Lett. B **256**, 105 (1991).
10. C.R. Ching et al., Int. J. Mod. Phys. **A10**, 2841 (1995).
11. W. Stoeffl and D.J. Decman, Phys. Rev. Lett. **75**, 3237 (1995).
12. C. Weinheimer et al., Phys. Lett. B **300**, 210 (1993).
13. H. Backe et al., in *Proceedings of Neutrino 96*, ed. K. Enquist, K. Huitu, J. Maalampi (World Scientific, Singapore, 1997), p. 259.
14. C. Weinheimer et al., Phys. Lett. B **460**, 219 (1999).
15. J. Bonn et al., Nucl. Phys. B (Proc. Suppl.) **91**, 273 (2001).
16. A.I. Belesev et al., Phys. Lett. B **350**, 263 (1995).
17. V.M. Lobashev et al., Phys. Lett. B **460**, 227 (1999).
18. V.M. Lobashev et al., Nucl. Phys. B (Proc. Suppl.) **91**, 280 (2000).
19. S. Yasumi et al., Phys. Lett. B **334**, 229 (1994).
20. M. Jung et al., Phys. Rev. Lett. **69**, 2164 (1992).
21. S.C. Curran et al., Phil. Mag. **40**, 53 (1949).
22. K.E. Bergkvist, Nucl. Phys. B **39**, 317 (1972).
23. V.A. Lubimov et al., Phys. Lett. B **94**, 66 (1980); S. Boris et al., Phys. Rev. Lett. **58**, 2019 (1987).
24. M. Fritschi et al., Phys. Lett. B **173**, 485 (1986).
25. J.F. Wilkerson et al., Phys. Rev. Lett. **58**, 2023 (1987).
26. E.T. Lippmaa et al., Sov. Phys. Dokl. **30**, 393 (1985).
27. R.S. Van Dyck et al., Phys. Rev. Lett. **70**, 2888 (1993).
28. A. Saenz et al., Phys. Rev. Lett. **84**, 242 (2000).
29. P. Kruit and F.H. Read, J. Phys. E **16**, 313 (1983).
30. A. Picard et al., Nucl. Instrum. Meth. B **63**, 345 (1992).
31. V.M. Lobashev, Nucl. Instrum. Meth. A **240**, 305 (1985).
32. C. Weinheimer, Nucl. Phys. B (Proc. Suppl.) **118**, 279 (2003).
33. L. Fleischmann et al., J. Low Temp. Phys. **119**, 615 (2000).
34. L. Fleischmann et al., Eur. Phys. J. B **16**, 521 (2000).
35. V.N. Aseev et al., Eur. Phys. J. D **10**, 39 (2000).
36. H. Barth et al., Prog. Part. Nucl. Phys. **40**, 353 (1998).
37. B. Bornschein et al., J. Low Temp. Phys. **131**, 69 (2003)
38. C. Kraus et al., Nucl. Phys. B (Proc. Suppl.) **118**, 482 (2003)
39. V.M. Lobashev, Prog. Part. Nucl. Phys. **48**, 123 (2002).
40. J. Bonn et al., Prog. Part. Nucl. Phys. **48**, 113 (2002).
41. A. Nucciotti et al., Nucl. Instrum. Meth. A **444**, 77 (2000).
42. M. Galeazzi et al., Phys. Rev. C **63**, 014302 (2001).
43. F. Gatti, Physics B (Proc. Suppl.) **91**, 293 (2001).
44. F. Gatti et al., Nature **397**, 137 (1999).

45. H. Ejiri, Phys. Rep. **338**, 265 (2000).
46. A. Faessler and F. Simkovic, J. Phys. G **24**, 2139 (1998).
47. K. Zuber, Phys. Rep. **305**, 295 (1998); arXiv:hep-ph/9811267
48. P. Vogel, in *Current Aspects of Neutrino Physics*, ed. D.O. Caldwell, (Springer, Berlin, Heidelberg, 2001), p. 177.
49. S.R. Elliott and P. Vogel, Annu. Rev. Nucl. Part. Sci. **52**, (2002); arXiv:hep-ph/0202264.
50. E. Majorana, Nuovo Cim. **14**, 171 (1937).
51. M. Goeppert-Mayer, Phys. Rev. **48**, 512 (1935).
52. T. Kirsten et al., Phys. Rev. Lett. **50**, 474 (1983).
53. S.R. Elliott et al., Phys. Rev. Lett. **59**, 2020 (1987).
54. J. Schechter and J.W.F. Valle, Phys. Rev. D **25**, 2951 (1982).
55. Y. Farzan, O.L.G. Peres, and A.Yu. Smirnov, Nucl. Phys. B **612**, 59 (2001).
56. S.R. Elliott et al., Phys. Rev. C **46**, 1535 (1992).
57. H. Ejiri et al., Phys. Rev. C **63**, 065501 (2001).
58. R. Arnold et al., Nucl. Phys. A **658**, 29 (1999).
59. R. Luescher et al., Phys. Lett. B **434**, 407 (1998).
60. A. Alessandrello et al., Phys. Lett. B **486**, 13 (2000).
61. Ke You et al., Phys. Lett. B **265**, 53 (1991).
62. H.V. Klapdor-Kleingrothaus et al., Eur. Phys. J. A **12**, 147 (2001).
63. C.E. Aalseth et al., Phys. Rev. C **59**, 2108 (1999); arXiv:hep-ex/0202026
64. F.A. Danevich et al., Phys. Rev. C **62**, 045501 (2000).
65. T. Bernatowicz et al., Phys. Rev. C **47**, 806 (1993).
66. A. De Silva et al., Phys. Rev. C **56**, 2451 (1997).
67. H.V. Klapdor-Kleingrothaus, A. Dietz, H.L. Harney, and I.V. Krivosheina (Heidelberg–Moscow Collaboration), Mod. Phys. Lett. A **16**, 2409 (2001); arXiv:hep-ph/0201231.
68. C.E. Aalseth et al., arXiv:hep-ex/0202018.
69. F. Feruglio et al., arXiv:hep-ph/0201291.
70. H.V. Klapdor-Kleingrothaus, arXiv:hep-ph/0205228.
71. H.L. Harney, arxiv:hep-ph/0205293.
72. R. Arnold et al., Nucl. Instrum. Meth. A **474**, 93 (2001).
73. A. Alessandrello et al., Nucl. Phys. (Proc. Suppl.) **87**, 78 (2000).
74. C.E. Aalseth et al., arXiv:hep-ex/0201021.
75. H.V. Klapdor-Kleingrothaus et al., J. Phys. G **24**, 483 (1998).
76. M. Danilov et al., Phys. Lett. B **480**, 12 (2000).
77. S. Pirro et al., Nucl. Instrum. Meth. A **444**, 71 (2000).
78. H. Ejiri et al., Phys. Rev. Lett. **85**, 2917 (2000).
79. A. Osipovicz et al., arXiv:hep-ph/0109032.
80. C. Weinheimer, in *Proceedings of the 10th Int. Workshop on Neutrino Telescopes*, ed. M. Baldo Ceolin, Venice, Italy, 2003, arXiv:hep-ex/0306057.

3 Astrophysical and Cosmological Constraints on Neutrino Masses

Kimmo Kainulainen and Keith A. Olive

We review some astrophysical and cosmological properties and implications of neutrino masses and mixing angles. These include constraints based on the relic density of neutrinos, limits on their masses and lifetimes, limits from big bang nucleosynthesis on mass parameters, and the relation of neutrinos to supernovae and high-energy cosmic rays.

3.1 Introduction

The role of neutrinos in cosmology and astrophysics cannot be understated [1]. They play a critical role in the physics of the early universe, at temperatures scales of order $1\,\mathrm{MeV}$, and strongly determine the abundances of the light elements produced in big bang nucleosynthesis. They almost certainly play a key role in supernova explosions and, if they have mass, could easily contribute to the overall mass density of the universe. At the present time, the only indicators of neutrino masses are from astrophysical sources, i.e. the inferred oscillations of neutrinos produced in the sun, and of those produced in cosmic-ray collisions in the atmosphere. Indeed, their elusive character has meant that a great deal of information about neutrino properties can be gained by studying their behavior in astrophysical and cosmological environments. Here, we shall try to elucidate some of the constraints on neutrino masses.

In our discussion below, we shall assume that the early universe is well described by a standard Friedmann–Lemaitre–Robertson–Walker metric

$$ds^2 = dt^2 - R^2(t) \left(\frac{dr^2}{1 - kr^2} + r^2 \left(d\theta^2 + \sin^2 \theta \, d\varphi^2 \right) \right) . \qquad (3.1)$$

We assume further that thermal equilibrium was established at some early epoch and that we can describe the radiation by a black-body equation of state, $p = \rho/3$, at a temperature T. Solutions to Einstein's equations allow one to determine the expansion rate of the universe, defined by the Hubble parameter, in terms of the energy density of radiation, the curvature, and the cosmological constant. In the early universe, the latter two quantities can be neglected, and we write

$$H^2 \equiv \left(\frac{\dot{R}}{R}\right)^2 = \frac{8\pi G_N \rho}{3} , \tag{3.2}$$

where the energy density is

$$\rho = \left(\sum_B g_B + \frac{7}{8}\sum_F g_F\right)\frac{\pi^2}{30}T^4 \equiv \frac{\pi^2}{30}N(T)\,T^4 . \tag{3.3}$$

The present neutrino contribution to the total energy density, relative to the critical density (for a spatially flat universe), is

$$\Omega_\mathrm{v} = \frac{\rho_\mathrm{v}}{\rho_\mathrm{c}} , \tag{3.4}$$

where $\rho_\mathrm{c} = 1.06 \times 10^{-5}\,h^2\,\mathrm{GeV/cm^3}$ and $h = (H/[100\,(\mathrm{km/s})/\mathrm{Mpc}])$ is the scaled Hubble parameter. For a recent review of standard big bang cosmology, see [2].

3.2 The Cosmological Relic Density of Stable Neutrinos

The simplicity of the standard big bang model allows one to compute in a straightforward manner the relic density of any stable particle if that particle was once in thermal equilibrium with the thermal radiation bath. At early times, neutrinos were kept in thermal equilibrium by their weak interactions with electrons and positrons. Equilibrium is achieved whenever some rate Γ is larger than the expansion rate of the universe, or $\Gamma_i > H$. Recalling that the age of the universe is determined by H^{-1}, this condition is equivalent to requiring that, on average, at least one interaction has occurred over the lifetime of the universe. On dimensional grounds, one can estimate the thermally averaged low-energy weak-interaction scattering cross section as

$$\langle \sigma v \rangle \sim g^4 T^2 / m_\mathrm{W}^4 \tag{3.5}$$

for $T \ll m_\mathrm{W}$. Recalling that the number density scales as $n \propto T^3$, we can compare the weak-interaction rate $\Gamma \sim n\langle \sigma v \rangle$ with the expansion rate given by (3.2) and (3.3). Neutrinos will be in equilibrium when $\Gamma_\mathrm{wk} > H$, or

$$T^3 > \sqrt{8\pi^3 N/90}\,m_\mathrm{W}^4/M_\mathrm{P} , \tag{3.6}$$

where $M_\mathrm{P} = G_N^{-1/2} = 1.22 \times 10^{19}$ GeV is the Planck mass. For $N = 43/4$ (accounting for photons, electrons, positrons, and three neutrino flavors), we see that equilibrium is maintained at temperatures greater than $\mathcal{O}(1)$ MeV (for a more accurate calculation, see [3]).

The decoupling scale of $\mathcal{O}(1)$ MeV has an important consequence for the final relic density of massive neutrinos. Neutrinos more massive than 1 MeV will begin to annihilate prior to decoupling and, while in equilibrium, their number density will become exponentially suppressed. Lighter neutrinos decouple like radiation, on the other hand, and hence do not experience the suppression due to annihilation. Therefore, the calculations of the number densities of light ($m_\nu \lesssim 1\,\mathrm{MeV}$) and heavy ($m_\nu \gtrsim 1\,\mathrm{MeV}$) neutrinos differ substantially.

The number of density of light neutrinos with $m_\nu \lesssim 1\,\mathrm{MeV}$ can be expressed at late times as

$$\rho_\nu = m_\nu Y_\nu n_\gamma \,, \tag{3.7}$$

where $Y_\nu = n_\nu/n_\gamma$ is the density of neutrinos relative to the density of photons, which today is 411 photons per cm^3. It is easy to show that in an adiabatically expanding universe, $Y_\nu = 3/11$. This suppression is a result of the e^+e^- annihilation which occurs after neutrino decoupling and heats the photon bath relative to the neutrinos. In order to obtain an age of the universe $t > 12\,\mathrm{Gyr}$, one requires that the matter component is constrained by

$$\Omega h^2 \leq 0.3 \,. \tag{3.8}$$

From this, one finds the strong constraint (upper bound) on Majorana neutrino masses [4]

$$m_{\mathrm{tot}} = \sum_\nu m_\nu \lesssim 28\,\mathrm{eV} \,, \tag{3.9}$$

where the sum runs over neutrino mass eigenstates. The limit for Dirac neutrinos depends on the interactions of the right-handed states (see discussion below). As one can see, even very small neutrino masses, of order $1\,\mathrm{eV}$, may contribute substantially to the overall relic density. The limit (3.9) and the corresponding initial rise in $\Omega_\nu h^2$ as a function of m_ν are displayed in Fig. 3.1 (at the low-mass end, where $m_\nu \lesssim 1\,\mathrm{MeV}$).

The calculation of the relic density for neutrinos more massive than $\sim 1\,\mathrm{MeV}$ is substantially more involved. The relic density is now determined by the freeze-out of neutrino annihilations, which occurs at $T \lesssim m_\nu$, after annihilations have begun to seriously reduce the number density of neutrinos [5]. The annihilation rate is given by

$$\Gamma_{\mathrm{ann}} = \langle \sigma v \rangle_{\mathrm{ann}} n_\nu \sim \frac{m_\nu^2}{m_Z^4}(m_\nu T)^{3/2}\, e^{-m_\nu/T} \,, \tag{3.10}$$

where we have assumed, for example, that the annihilation cross section is dominated by $\nu\bar\nu \to f\bar f$ via Z-boson exchange[1] and that $\langle \sigma v \rangle_{\mathrm{ann}} \sim m_\nu^2/m_Z^4$. When the annihilation rate becomes slower than the expansion rate of the universe, the annihilations freeze out and the relative abundance of neutrinos

---

[1] While this is approximately true for Dirac neutrinos, the annihilation cross section of Majorana neutrinos is p-wave suppressed and is proportional to the final-state fermion masses rather than m_ν.

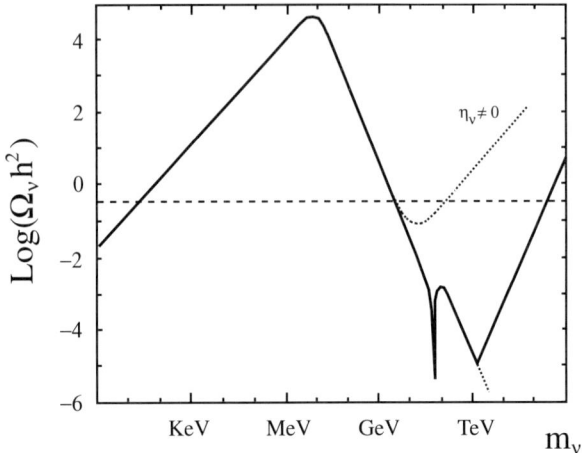

Fig. 3.1. Summary plot of the relic density of Dirac neutrinos (*solid line*), including the effect of a possible neutrino asymmetry of $\eta_N = 5 \times 10^{-11}$ (*dotted line*)

becomes fixed. Roughly, $Y_\nu \sim (m \langle \sigma v \rangle_{\mathrm{ann}})^{-1}$ and hence $\Omega_\nu h^2 \sim \langle \sigma v \rangle_{\mathrm{ann}}^{-1}$, so that parametrically $\Omega_\nu h^2 \sim 1/m_\nu^2$. As a result, the constraint (3.8) now leads to a *lower* bound [5–7] on the neutrino mass, of about $m_\nu \gtrsim 3$–$7\,\mathrm{GeV}$, depending on whether it is a Dirac or a Majorana neutrino. This bound and the corresponding downward trend $\Omega_\nu h^2 \sim 1/m_\nu^2$ can be seen in Fig. 3.1. The result of a more detailed calculation is shown in Fig. 3.2 [7] for the case of a Dirac neutrino. The two curves show a slight sensitivity to the temperature

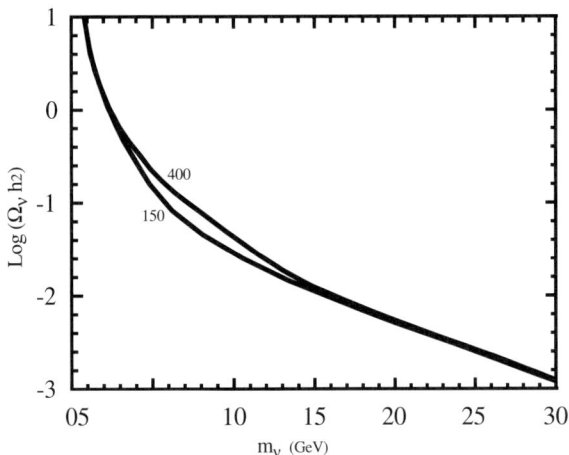

Fig. 3.2. The relic density of heavy Dirac neutrinos due to annihilation [7]. The curves are labeled by the assumed quark–hadron phase transition temperature in MeV

scale associated with the quark–hadron transition. The result for a Majorana-mass neutrino is qualitatively similar. Indeed, any particle with roughly weak-scale cross sections will tend to give an interesting value of $\Omega h^2 \sim 1$.

The deep drop in $\Omega_\nu h^2$ visible in Fig. 3.1 at around $m_\nu = M_Z/2$, is due to a very strong annihilation cross section at the Z-boson pole. For yet higher neutrino masses the Z-annihilation channel cross section drops as $\sim 1/m_\nu^2$, leading to a brief interval of an increasing trend in $\Omega_\nu h^2$. However, for $m_\nu \gtrsim m_W$, the cross section regains its parametric form $\langle \sigma v \rangle_{\mathrm{ann}} \sim m_\nu^2$ owing to the opening up of a new annihilation channel to W-boson pairs [8], and the density drops again as $\Omega_\nu h^2 \sim 1/m_\nu^2$. The tree-level W-channel cross section breaks the unitarity at around a few TeV [9], however, and the full cross section must be bound by the unitarity limit [10]. This behaves again as $1/m_\nu^2$, whereby $\Omega_\nu h^2$ has to start increasing again, until it becomes too large again at 200–400 TeV [9, 10] (or perhaps somewhat earlier as the weak interactions become strong at the unitarity-breaking scale).

3.3 Neutrinos as Dark Matter

On the basis of the leptonic and invisible widths of the Z boson, experiments at LEP have determined that the number of neutrinos is $N_\nu = 2.9841 \pm 0.0083$ [11]. Conversely, any new physics must fit within these brackets, and thus LEP excludes additional neutrinos (with standard weak interactions) with masses $m_\nu \lesssim 45$ GeV. Combined with the limits displayed in Figs. 3.1 and 3.2, we see that the mass density of ordinary heavy neutrinos is bound to be very small; $\Omega_\nu h^2 < 0.001$ for masses $m_\nu > 45$ GeV up to $m_\nu \sim \mathcal{O}(100)$ TeV.

A bound on neutrino masses even stronger than (3.9) can be obtained from the recent observations of active–active mixing in both solar- and atmospheric-neutrino experiments. The inferred evidence for ν_μ–ν_τ and $\nu_e - \nu_{\mu,\tau}$ mixings is on the scales $m_\nu^2 \sim 1$–10×10^{-5} and $m_\nu^2 \sim 2$–5×10^{-3}. When this is combined with the upper bound on the mass of an electron-like neutrino $m_\nu < 2.8$ eV [12] and the LEP limit on the number of neutrino species, one finds the following constraint on the sum of neutrino masses:

$$0.05 \, \mathrm{eV} \lesssim m_{\mathrm{tot}} \lesssim 8.4 \, \mathrm{eV} \ . \tag{3.11}$$

Conversely, the experimental and observational data imply that the cosmological energy density of all light, weakly interacting neutrinos can be restricted to the range

$$0.0005 \lesssim \Omega_\nu h^2 \lesssim 0.09 \ . \tag{3.12}$$

Interestingly, there is now also a lower bound due to the fact that at least one of the neutrino masses has to be larger than the scale $m^2 \sim 10^{-3}$ eV^2 set by the atmospheric neutrino data. Combined with the results on the relic mass density of neutrinos and the LEP limits, the bound (3.12) implies that the ordinary weakly interacting neutrinos, once the standard dark-matter candidate [13], can be ruled out completely as a dominant component of dark matter.

However, this conclusion can be avoided if neutrinos are Dirac particles and have a nonzero asymmetry, since then the relic density could be governed by the asymmetry rather than by the annihilation cross section. Indeed, it is easy to see that the neutrino mass density corresponding to an asymmetry $\eta_\nu \equiv (n_\nu - n_{\bar\nu})/n_\gamma$ is given by [14]

$$\rho = m_\nu \eta_\nu n_\gamma \,, \tag{3.13}$$

which implies

$$\Omega_\nu h^2 \simeq 0.004 \, \eta_{\nu 10} \, (m_\nu/\mathrm{GeV}) \,, \tag{3.14}$$

where $\eta_{\nu 10} \equiv 10^{10} \eta_\nu$. We have shown the behavior of the energy density of neutrinos with an asymmetry by the dotted line in Fig. 3.1. At low m_ν, the mass density is dominated by the symmetric, relic abundance of both neutrinos and antineutrinos which have already frozen out. At higher values of m_ν, annihilations suppress the symmetric part of the relic density, until $\Omega_\nu h^2$ eventually becomes dominated by the linearly increasing asymmetric contribution. In the figure, we have assumed an asymmetry of $\eta_\nu \sim 5 \times 10^{-11}$ for neutrinos with a standard weak-interaction strength. In this case, $\Omega_\nu h^2$ begins to rise when $m_\nu \gtrsim 20\,\mathrm{GeV}$. Obviously, the bound (3.8) is saturated for $m_\nu = 75\,\mathrm{GeV}/\eta_{\nu 10}$.

There are also other cosmological considerations that give rise to interesting mass constraints on the eV scale. Indeed, light neutrinos were problematic in cosmology long before the improved mass limits leading to (3.12) were established, owing to their effect on structure formation. Light particles which are still relativistic at the time of matter domination erase primordial perturbations, owing to free streaming out to very large scales [15]. Given a neutrino with mass m_ν, the smallest surviving nonlinear structures are determined by the Jeans mass

$$M_{\mathrm J} = 3 \times 10^{18} \frac{M_\odot}{m_\nu^2(eV)} \,. \tag{3.15}$$

Thus, for eV-mass neutrinos, the large-scale structures, including filaments and voids [16, 17], must form first, and galaxies, whose typical mass scale is $\simeq 10^{12} M_\odot$, are expected to fragment out later. Particles with the above property are termed hot dark matter (HDM). It seemed that neutrinos were ruled out because they tend to produce too much large-scale structure [18], and galaxies formed too late [17, 19], at $z \lesssim 1$, whereas quasars and galaxies are seen out to redshifts $z \gtrsim 6$.

Subsequent to the demise of the HDM scenario, there was a brief revival of consideration of neutrino dark matter as part of a mixed dark-matter model, now using more conventional cold dark matter along with a small component of hot (neutrino) dark matter. The motivation for doing this was to recover some of the lost power on large scales that is absent in CDM models [20].

However, galaxies still form late in these models, and, more importantly, almost all evidence now points away from models with $\Omega_{\mathrm{m}} = 1$, where Ω_{m} is the total matter density, and strongly favors models with a cosmological constant (ΛCDM).

Combining the rapidly improving data on key cosmological parameters with the better statistics from large-redshift surveys has made it possible to go a step forward along this path. It is now possible to set stringent limits on the light-neutrino mass density $\Omega_{\mathrm{v}} h^2$ and hence on the neutrino mass on the basis of the power spectrum of the Ly α forest [21], $m_{\mathrm{tot}} < 5.5\,\mathrm{eV}$, and the limit is even stronger if Ω_m is less than 0.5. This limit has recently been improved by the 2dF galaxy redshift [22] survey by comparing the power spectrum of fluctuations derived from that work with structure formation models. If we focus on the the presently favored ΛCDM model, the neutrino mass bound becomes $m_{\mathrm{tot}} < 1.8\,\mathrm{eV}$ for $\Omega_m < 0.5$.

Finally, right-handed or sterile neutrinos may also contribute to dark matter. The mass limits for neutrinos with less than full weak-strength interactions are relaxed [23]. For Dirac neutrinos, the upper limit varies between 100 and 200 eV, depending on the strength of their interactions. For Majorana neutrinos, the limit is further relaxed to 200–2000 eV. This relaxation is primarily due to the dilution of the number density of super-weakly interacting neutrinos owing to entropy production by the decay and annihilation of massive states after their decoupling from equilibrium [24]. Such neutrinos make excellent warm-dark-matter candidates, even though the viable mass range for galaxy formation is quite restricted [25].

3.4 Neutrinos and Big Bang Nucleosynthesis

"Big bang nucleosynthesis" (BBN) is the cosmological theory of the origin of the light element isotopes D, ^{3}He, ^{4}He, and ^{7}Li [26]. The success of the theory when it is compared with the observational determinations of the light elements allows one to place strong constraints on the physics of the early universe at a time scale of 1–100 seconds after the big bang. ^{4}He is the most sensitive probe of deviations from the Standard Model, and its abundance is determined primarily by the neutron-to-proton ratio when nucleosynthesis begins at a temperature of $\sim 100\,\mathrm{keV}$ (to a good approximation, all neutrons are then bound to form ^{4}He). The ratio n/p is determined by the competition between the weak interaction rates which interconvert neutrons and protons,

$$\mathrm{p} + \mathrm{e}^- \leftrightarrow \mathrm{n} + \nu_{\mathrm{e}} , \qquad \mathrm{n} + \mathrm{e}^+ \leftrightarrow \mathrm{p} + \overline{\nu}_{\mathrm{e}} , \qquad \mathrm{n} \leftrightarrow \mathrm{p} + \mathrm{e}^- + \overline{\nu}_{\mathrm{e}} \qquad (3.16)$$

and the expansion rate, and is largely given by the Boltzmann factor

$$n/p \sim \mathrm{e}^{-(m_{\mathrm{n}} - m_{\mathrm{p}})/T_{\mathrm{f}}} , \qquad (3.17)$$

where $m_{\mathrm{n}} - m_{\mathrm{p}}$ is the difference between the masses of the neutron and the proton. As in the case of neutrinos discussed above, these weak interactions also freeze out at a temperature of roughly 1 MeV, when

$$G_{\mathrm{F}}^2 T_{\mathrm{f}}^5 \sim \Gamma_{\mathrm{wk}}(T_{\mathrm{f}}) = H(T_{\mathrm{f}}) \sim \sqrt{G_N N} T_{\mathrm{f}}^2 . \qquad (3.18)$$

The freeze-out condition implies the scaling $T_{\mathrm{f}}^3 \sim \sqrt{N}$. From (3.17) and (3.18), it is then clear that changes in N, caused for example by a change in the number of light neutrinos N_{v}, would directly influence n/p, and hence the ^{4}He abdundance. The dependence of the light-element abundances on N_{v} is shown in Fig. 3.3 [27], where the mass fraction of ^{4}He, Y, and the abundances by number of D, ^{3}He, and ^{7}Li are plotted as a function of the baryon-to-photon ratio η, for values of $N_{\mathrm{v}} = 2$–7. As one can see, an upper limit on Y combined with a lower limit on η will yield an upper limit on N_{v} [28].

Assuming no new physics at low energies, the value of η is the sole input parameter to BBN calculations. It is fixed by the comparison between the BBN predictions and the observational determinations of the isotopic abundances [29]. From the results for ^{4}He and ^{7}Li, one finds a relatively low value [29,30] of $\eta \sim 2.4 \times 10^{-10}$, corresponding to a low baryon density $\Omega_{\mathrm{B}} h^2 = 0.009$ with a 95% C.L. range of 0.006–0.017. The results for deuterium, on the other hand, imply a large value of η and hence a large baryon density: $\eta \sim 5.8 \times 10^{-10}$, and $\Omega_{\mathrm{B}} h^2 \sim 0.021$ with a 95% C.L. range

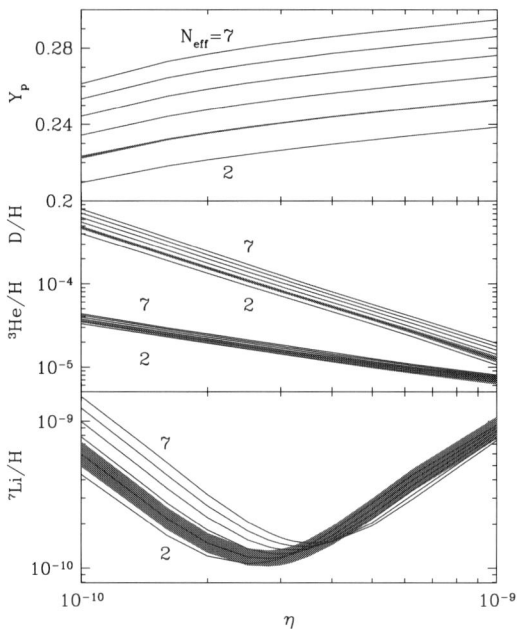

Fig. 3.3. The light-element abundances as a function of the baryon-to-photon ratio for different values of N_{v} [27]

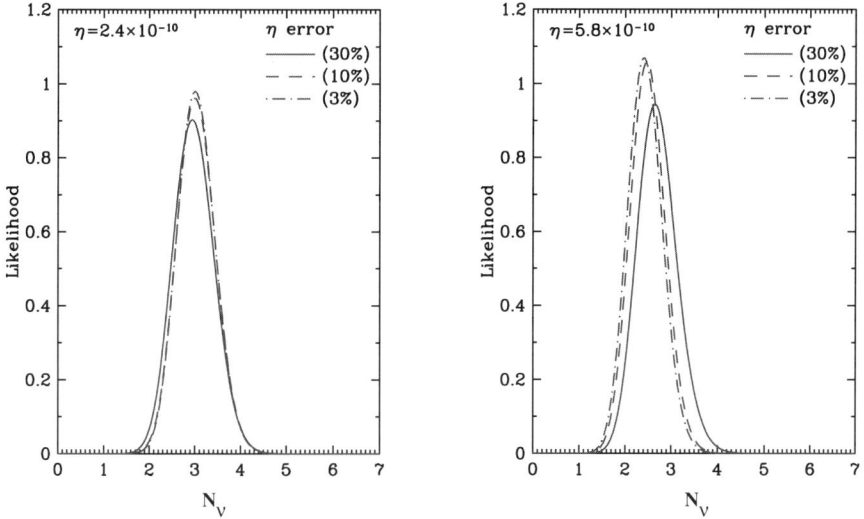

Fig. 3.4. *Left*: the distribution of N_ν, assuming a value of $\eta = 2.4 \times 10^{-10}$ obtained from the abundance of ^{4}He and ^{7}Li and the CBI measurement of the microwave background anisotropy [27]. The *curves* show the effect of the expected increased accuracy in the determination of η from the cosmic microwave background. *Right*: as on the left, but assuming a value of $\eta = 5.8 \times 10^{-10}$ obtained from the abundance of D and the DASI and BOOMERanG measurements of the microwave background anisotropy

of 0.018–0.027. The value of the baryon density has also been determined recently from measurements of microwave background anisotropies. The recent result from DASI [31] indicates that $\Omega_B h^2 = 0.022^{+0.004}_{-0.003}$, while a result from BOOMERanG-98 [32] indicates $\Omega_B h^2 = 0.021^{+0.004}_{-0.003}$ (using 1σ errors).

With the value of η fixed, one can use He abundance measurements to set limits on new physics. In particular, one can set upper limits on the number of neutrino flavors. Taking $Y_p = 0.238 \pm 0.002 \pm 0.005$ (see e.g. [33]), we show in Fig. 3.4 the likelihood functions for N_ν based on both the low and the high values of η [27]. The curves show the impact of an increasingly accurate determination of η, where the accuracy is assumed to vary from 30% to 3%. If one assumes a 20% uncertainty in η (the current uncertainty level), these calculations provide upper limits of

$$N_\nu < 3.9 \,, \qquad \eta = 2.4 \times 10^{-10} \,,$$
$$N_\nu < 3.6 \,, \qquad \eta = 5.8 \times 10^{-10} \tag{3.19}$$

at the 95% C.L. Although, as noted above, LEP has already placed a very stringent limit on N_ν, the limit (3.19) is useful, because it actually applies to the total number of degrees of freedom of new particles and is not tied specifically to neutrinos. In fact, more generally, the neutrino limit can be translated into a limit on the expansion rate of the universe at the time of BBN, which can be applied to a host of other constraints on particle properties.

3.4.1 BBN Limits on Neutrino Masses and Lifetimes

As discussed above, the prediction from nucleosynthesis theory for light-element abundances is sensitive to the changes in the expansion rate of the universe, which depends on the energy density of the universe during the BBN era (3.2). This extra energy density could be in the form of new mass-less degrees of freedom, in which case their number is directly constrained by (3.19). Equally well, the extra energy density could reside in the form of massive long-lived but unstable neutrinos, in which case nucleosynthesis provides interesting constraints on their masses and lifetimes.

We have already pointed out that the relic density of neutrinos depends strongly on whether they decouple while relativistic or nonrelativistic. Here, the calculations also depend on how the neutrino lifetimes relate to the BBN timescale of about 100 seconds. Also, in order to obtain reliable results for the light-element abundances, one must keep track of the induced perturbations (electron-neutrino heating) in the weak reaction rates (3.16) in addition to computing changes in the expansion rate. Nevertheless, even in this case it is customary to measure the change in helium abundance in units of equivalent effective neutrino degrees of freedom N_{eff}, such that the limit (3.19) can be applied to $N_{\mathrm{eff}}(\Delta Y(m_{\mathrm{v}}, \tau_{\mathrm{v}}))$.

When the neutrino lifetime is much larger than 100 seconds, neutrinos are effectively stable on a nucleosynthesis scale [34–37]. While accounting for changes in the rates in (3.16) is important for the detailed bounds, the

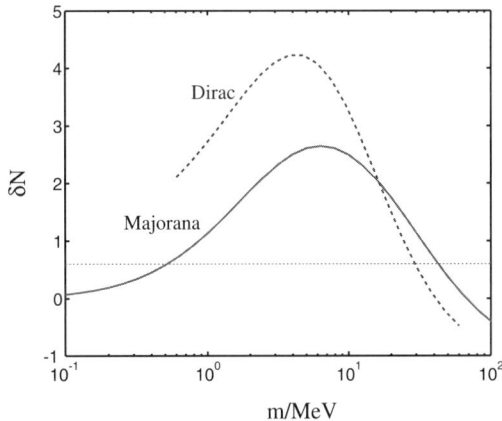

Fig. 3.5. Plot of the effective number of neutrino degrees of freedom during BBN for a Dirac neutrino (*dashed line*) and a Majorana neutrino (*solid line*) [35]

bulk behavior of $N_{\text{eff}}(m_{\text{v}})$ is dictated by the neutrino mass contribution to the energy density. Obviously, when $m_{\text{v}} \ll 0.1$ MeV, neutrinos are effectively massless during BBN, and $N_{\text{eff}}(m_{\text{v}}) \to 3$ when $m_{\text{v}} \to 0$. For masses in excess of 0.1 MeV, but below the neutrino decoupling temperature of a few MeV, their number density is unsuppressed and their mass density can be large during BBN, causing N_{eff} to increase. For even larger masses, however, the Boltzmann factor in (3.10) begins to suppress the mass density and eventually turns N_{eff} down again. This behavior is shown in Fig. 3.5 for massive Dirac- and Majorana-type tau neutrinos [35]. The bound (3.19) yields an excluded region of stable neutrino masses centered around a few MeV. For $N_{\text{v}} < 3.6$, the lower bound is $m_{\text{v}} > 42$ MeV (Majorana) and $m_{\text{v}} > 30$ MeV (Dirac) [35]. This is only relevant to ν_τ, and is complementary to the present laboratory limit on the τ-like neutrino, $m_{\text{v}} < 18$ MeV [38]. Owing to contributions from pion decays and inverse decays to neutrinos, the upper bound from BBN depends on the QCD phase transition temperature, T_{QCD}, and is also different for τ and μ neutrinos because of their different scattering rates off muons. Imposing again the constraint $N_{\text{v}} < 3.6$ and taking $T_{\text{QCD}} = 200$ MeV gives [39]

$$m_{\text{v}} \lesssim 230 \text{ keV} \qquad \mu\text{-like} ,$$
$$m_{\text{v}} \lesssim 290 \text{ keV} \qquad \tau\text{-like} . \qquad (3.20)$$

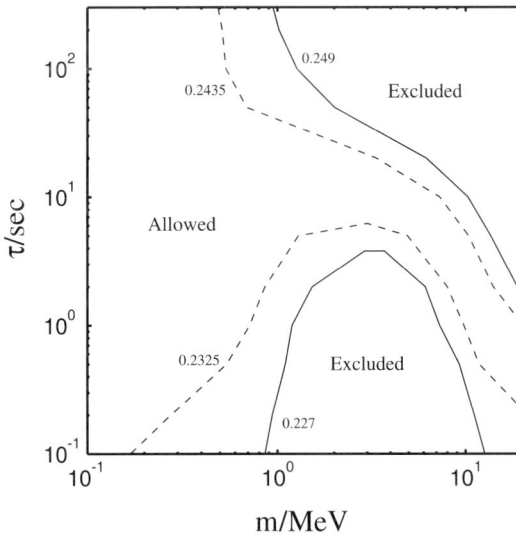

Fig. 3.6. Plot of BBN constraint on the mass and lifetime of an unstable tau neutrino. The contours are labeled with the corresponding values of Y_{p}, which deviate by 1σ (*dashed lines*) and 2σ (*solid lines*) from the observed value. The *upper right corner* is excluded owing to too much ^{4}He being produced and the *lower part* of the graph by too little being produced

The laboratory limit on the muon-like neutrino, for comparision, is $m_v \lesssim$ 170 keV [40]. To improve on this, the BBN limit would need to improved to $N_v \lesssim 3.4$ [39].

When the neutrino lifetime is smaller than or comparable to the nucleosynthesis timescale, one has to account for neutrino decay processes as well. This involves solving for the distributions of the final-state decay products, which might include new particles such as majorons, and their possible direct effect on BBN. (For example, energetic photons would cause the dissociation of the newly generated light nuclei.) Such calculations have been done by many groups [41, 42], and the results are given in exclusion plots in the mass–lifetime plane. Constraints are possible for masses of order 1 MeV and lifetimes of order 1 second. In Fig. 3.6, we show a constraint on the tau neutrino mass and lifetime as an example (data taken from [42]).

3.4.2 BBN Limits on Neutrino Mixing Parameters

Despite losing the competitive edge with respect to masses and lifetimes, BBN continues to put interesting limits on other neutrino mass parameters relevant to neutrino oscillations. The LEP limit of course applies only to neutrinos with weak interactions, while neutrinos without weak interactions, or *sterile* neutrinos, have been proposed in many different contexts over the years. At present, a prime motivation for introducing sterile neutrinos is to explain the LSND neutrino anomaly [43] in conjunction with the solar and atmospheric neutrino deficits.

BBN, on the other hand, is sensitive to any type of energy density that changes the expansion rate in the $\mathcal{O}(0.1\text{–}1)$ MeV range, irrespective of the interactions. It is therefore very interesting to observe that even if no sterile neutrinos were created at very early times, they could be excited by mixing effects in the early universe prior to nucleosynthesis. The basic mechanism is very simple. Suppose that an active state v_α ($\alpha = e, \mu, \tau$) *mixes* with a sterile state v_s. That is, the neutrino mass matrix, and hence the Hamiltonian, is not diagonal in the interaction basis. The mixing is further affected by the forward-scattering interactions with the background plasma, which are felt by the active state. As a result, even if a neutrino state was initially produced in a purely active projection, after some time t it becomes some coherent linear combination of both active and sterile states:

$$v(t) = c_e(t)v_e + c_s(t)v_s . \tag{3.21}$$

The coherent evolution of this state is interrupted by collisions, which effect a sequence of quantum mechanical measurements of the flavor content of the propagating state. Since the sterile state has no interactions, each measurement is complete, and collapses the wave function to the sterile state with a probability $P_{v_e \to v_s}(t) = |c_s(t)|^2$. As a result, the sterile states are populated at an average rate of roughly

$$\Gamma_{\nu_s} = \Gamma_\nu \langle |c_s(t)|^2 \rangle_{\mathrm{coll}} = \frac{1}{2} \sin^2 2\theta_m \Gamma_{\nu_\alpha} \ , \tag{3.22}$$

where Γ_{ν_α} is the weak-interaction rate of the active state ν_α, and we have assumed that the oscillation time is short in comparision with the collision timescale. The mixing angle in matter θ_m is given by [44]

$$\sin^2 2\theta_m = \frac{\sin^2 2\theta_0}{1 - 2\chi \cos 2\theta_0 + \chi^2} \ , \tag{3.23}$$

where $\sin 2\theta_0$ is the mixing angle in vacuum and $\chi \equiv 2p|V|/\delta m^2$, where δm^2 is the mass-squared difference between the vacuum mass eigenstates, $p \sim T$ is the momentum, and $|V|$ is the matter-induced effective potential in the Hamiltonian [44, 45]. The weak rate scales as $\Gamma_{\nu_\alpha} \sim T^5$. Moreover, $\chi \sim T^6$ at very high temperatures, which causes a strong suppression of mixing by matter, and hence $\Gamma_{\nu_s} \sim T^{-7}$. At very small temperatures, $\theta_m \to \theta_0$, on the other hand, and hence $\Gamma_{\nu_s} \sim T^5$. The rate is thus suppressed both at very large and at very small temperatures [46]. In the intermediate region of a few MeV, however, Γ_{ν_s} can exceed the expansion rate, bringing a significant amount of sterile neutrinos into equilibrium. An accurate treatment of the problem requires a numerical solution of the appropriate quantum kinetic equations, and the results depend on whether the mostly active state is the heavier ($\delta m^2 < 0$) or the lighter ($\delta m^2 > 0$) of the mixing states. We show the results of such a calculation in Fig. 3.7 [44]. The lines are labeled by constant effective number of degrees of freedom during BBN, $\delta N_\nu \equiv N_\nu - 3$. The most recent limits corresponding to (3.19) can be interpolated from the curves shown. The area above the curves is excluded by the BBN limit.

The BBN limit can be converted to an upper bound on the sterile-neutrino flux [47] in the atmospheric and solar neutrino observations. Using $N_\nu < 3.6$, one finds

$$\sin^2 \theta_{\mu s} \lesssim 0.03 \quad \text{(atmospheric)} \ ,$$
$$\sin^2 \theta_{es} \lesssim 0.06 \quad \text{(solar LMA)} \ , \tag{3.24}$$

whereas the bounds from the atmospheric and solar neutrino experiments are about an order of mangitude weaker: $\sin^2 \theta_{\mu s} \lesssim 0.48$ and $\sin^2 \theta_{es} \lesssim 0.72$.

The constraints shown in Fig. 3.7 and in (3.24) depend on the assumption that the primordial lepton asymmetry is not anomalously large [48]. In [49], it was suggested that a large effective asymmetry that violated this assumption could actually be generated by oscillations, given a particular neutrino mass and mixing hierarchy. For a while, those ideas generated a lot of interest, as they would have allowed one to reconcile all observed anomalies (including that of LSND) with the nucleosynthesis constraints. However, in such a scenario at least one of the active states would have to be much heavier than the two others, which is not allowed by the atmospheric and solar neutrino flux observations. As a result, the bounds (3.24) hold and, in particular, nucleosynthesis is very much

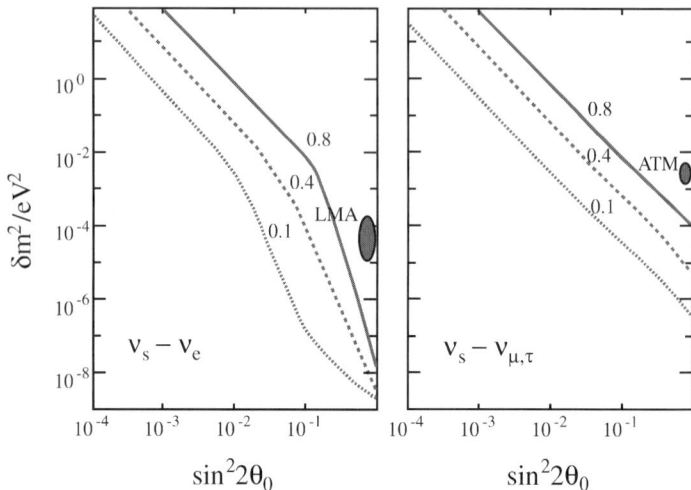

Fig. 3.7. The BBN constraints on the active–sterile neutrino mixing parameters [44] for $\nu_\alpha = \nu_e$ (*left*) and $\nu_\alpha = \nu_{\mu,\tau}$ (*right*). Regions to the right of the contours, labeled by the bound on the effective neutrino degrees of freedom $\delta N_{\rm eff} = N_{\rm eff} - 3$, are excluded. The currently accepted regions corresponding to the active–sterile mixing parameters for atmospheric neutrinos (ATM) and the large-mixing-angle (LMA) solar neutrino solutions are also shown

at odds with the possible existence of an LSND-type sterile state. To see this, observe that creating a large enough effective mixing between $\bar\nu_\mu$ and $\bar\nu_e$ to explain the anomaly [43] would require a sterile intermediate with $m_{\nu_s} \simeq 1\,{\rm eV}$ and $\sin^2 2\theta_{\mu e} \simeq (1/2)\sin^2 2\theta_{\mu s}\sin^2 2\theta_{se} \gtrsim 10^{-2}(\delta m^2/{\rm eV}^2)^{-2}$. In other words, at least one of the active–sterile mixings would have to satisfy $\sin^2 2\theta_{(\mu,e)s} \gtrsim 0.15(\delta m^2/{\rm eV}^2)^{-1}$, which is well within the BBN excluded regions shown in Fig. 3.7. (This requirement would be excluded even by $N_{\rm eff} \lesssim 3.9$, although we do not show the corresponding contour in Fig. 3.7).

It should be noted, finally, that active–active-type oscillations have hardly any effect on the expansion rate or the weak-interaction rates [50], and hence are not constrained in the above sense by BBN. However, large-mixing-angle active–active oscillations could equilibrate the lepton *asymmetries* prior to BBN. This consideration has been used to put strong bounds on muon and tau-lepton asymmetries [51], which exclude the possibility of degenerate nucleosynthesis.

3.5 Neutrinos and Supernovae

Neutrinos have long been known to play important role in the physics of supernovae. It is clear that by far the largest part, roughly 99%, of the gravitational binding energy of about 3×10^{53} erg involved in the explosion of a type II supernova is carried away by neutrinos, while just 1% powers the shock wave responsible for blowing out the mantle of the star, and only a tiny fraction, of about 0.1%, escapes in the form of the light responsible for the spectacular sights observed in the telescopes watching the sky.

The formation of the neutrino burst in the collapse of a type II supernova is rather well understood. The temporal structure of the burst and the energy spectrum of the emitted neutrinos can be computed fairly well [52]. It will be possible to use existing or planned large-scale neutrino detectors [53] to observe deviations from these signatures and to obtain interesting information about neutrino masses and mixing parameters [54], given a future observation of a galactic supernova. We show an example of a compilation of neutrino fluxes and spectra in Fig. 3.8.

A number of constraints on new physics and, in particular, on neutrino parameters have already been deduced from the well-known SN1987A event in the Small Magellanic Cloud. Of these, perhaps the most direct is the upper bound on the neutrino mass derivable from the maximum duration of the observed neutrino pulse of about 10 seconds. Given the initial energy spectrum of the neutrinos and the distance to the supernova, one can compute the expected spread in the arrival times of the neutrinos at the earth as a function of the neutrino mass. Comparing the predictions with the observations has been shown to yield the bound [55]

$$m_{\nu_e} \lesssim 6{-}20 \, \text{eV} \ . \tag{3.25}$$

The observed pulse length also leads to the classic *cooling argument*: any new physics that would enhance the neutrino diffusion such that the cooling

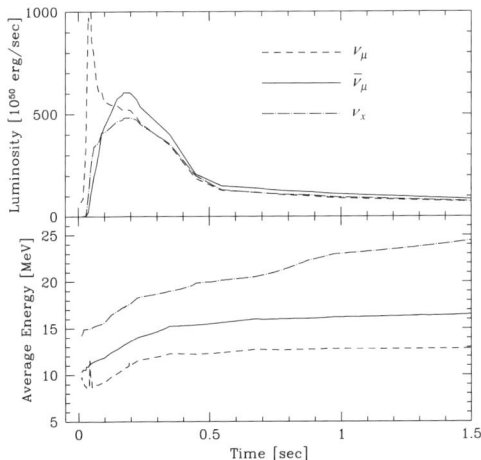

Fig. 3.8. Time evolution of neutrino luminosities and average energies: ν_x represents the spectrum of ν_μ, ν_τ, $\bar{\nu}_\mu$ and $\bar{\nu}_\tau$. Figure taken from [53]

time drops below the observed duration must be excluded. Cooling arguments have been used to set limits on various neutrino properties [56], such as active–sterile neutrino mixing [57] and neutrino magnetic moments. The magnetic-dipole-moment bound was recently revised by Ayala et al. [58] to

$$\mu_{\nu_e} \lesssim 1\text{--}4 \times 10^{-12} \mu_B , \qquad (3.26)$$

which is two orders of magnitude more stringent than the best laboratory bounds, and comparable to the bound derived from the cooling of red giants in globular clusters [59], which gives $\mu_{\nu_e} \lesssim 3 \times 10^{-12} \mu_B$.

At present, most of the activity concerning neutrinos in supernovae has focused on the effects of neutrino transport in supernova explosion dynamics rather than on finding constraints on neutrino mixing parameters or masses. Indeed, the details of the physics responsible for the actual visibly observed supernova explosion, including the blowing out of the stellar mantle, are not very well understood. In particular, the shock wave, which forms deep within the iron core as the infall of matter is reversed owing to the stiffening of the equation of state of nuclear matter, is typically found to be too weak to explode the star. This is believed to be due to energy loss from the shock and dissociating iron nuclei as the shock makes its way out from the core to the mantle. In the popular "delayed explosion scenario", the stalling shock wave is rejuvenated by energy transfer to the shock from the huge energy flux of neutrinos streaming freely away from the core. Recent numerical simulations including diffusive neutrino transport do not verify this expectation, however [60]; while neutrinos definitely help, they do not appear to solve the problem. These results are not conclusive, because the diffusive transport equations used so far [60] do not include all relevant neutrino interactions, most notably the nuclear bremsstrahlung processes [52, 61]. Furthermore, processes other than diffusive processes, such as convective flows in the core and behind the shock, appear to play an important role as well [62].

It is of course possible that a succesful supernova explosion requires help from some new physics to channel energy more efficiently to the shock, and neutrino oscillations have already been considered for this role [63]. The idea is that ν_μ and ν_τ interact more weakly, and hence escape more energetically from deeper in the core, than do electron neutrinos. Arranging the mixing parameters so that $\nu_{\mu,\tau}$ turn resonantly into ν_e in the mantle behind the shock could increase the energy deposited in the shock significantly. Unfortunately, the mass difference needed for the resonant transition would be very big:

$$\delta m^2 \sim 2EV_{\text{eff}} \simeq 1.5 \times 10^5 \rho_e E_\nu^{100} \text{ eV}^2 , \qquad (3.27)$$

where ρ_e is the electron density in units of $10^{10}\,\text{g/cm}^3$, which at the shock front is equal to about 10^{-3}, and E_ν^{100} is the neutrino energy in units of $100\,\text{MeV}$. So one obtains $m_\nu \gtrsim 10\,\text{eV}$, which is excluded by the present data.

A similar idea was behind the suggestion [64] that neutrino magnetic moments induce resonant transitions from ν_R (which escape energetically from deep in the core) to ν_L behind the mantle. This mechanism could actually be used to evade the bound (3.26), but as it demands a magnetic moment of order $\mu_\nu \sim 10^{-11}\mu_B$, it has problems in coping with the red-giant cooling bound [59].

Sterile neutrinos could also be relevant to supernovae by alleviating the problems with r-process nucleosynthesis, which is thought to be responsible for creation of the most heavy elements. In the standard supernova calculations, r-process nucleosynthesis is not effective, owing to excessively efficient de-neutronization by the process $\nu_e + n \rightarrow e^- + p$. If electron neutrinos mixed with a sterile state, however, this process could be made less effective, increasing the neutron density in the mantle, and hence improving the efficiency of the r-process [65].

Finally, there is also the old problem of the "kick" velocities of pulsars (neutron star remnants of supernova explosions). It has proven difficult to account for these velocities, which average around 450 km/s, on the basis of the normal fluid dynamics in asymmetric supernova explosions. The momentum carried away by neutrinos, $p_\nu \simeq E_\nu$, on the other hand, is about 100 times larger than the kinetic energy of a pulsar, so that a mere 1% asymmetry in the neutrino emission would be enough to power the pulsar velocities. Interesting attempts have been made to explain such asymmetric emission by an asymmetric distribution of inhomogeneities in the magnetic field of the supernova, combined with a large neutrino magnetic moment [64], or just the magnetic-field-induced deformation of the neutrino spheres [66]. However, the former possibility would again probably need too large a magnetic moment to work, and a detailed analysis of the latter possibility suggests that the a symmetric flux is very suppressed, requiring perhaps unrealistically large magnetic fields; according to [67], the field needs to be in excess of 10^{17} G, while the authors of [68] argue that a field of 10^{14-15} G would suffice. The true nature of the physics explaining the kick velocities may remain ambiguous for some time to come, but a neutrino solution looks definitely appealing from the point of view of pure energetics.

Before we conclude this section, we would like to mention several other astrophysical limits on neutrino properties. A sure limit on the mean-life/mass ratio is obtained from solar X- and γ-ray fluxes [69]. This limit is $\tau/m_{\nu_1} > 7 \times 10^9$ s/eV. This is far superior to the laboratory bound of 300 s/eV [70]. Other, much stronger limits ($> O(10^{15})$ s/eV) can be deduced from the lack of observation of γ-rays in coincidence with neutrinos from SN 1987A [71]. This latter limit applies to the heavier neutrino mass eigenstates, ν_2 and ν_3, as well.

3.6 Neutrinos and Cosmic Rays

One of the most interesting puzzles in astrophysics today concerns the observations of ultrahigh-energy cosmic rays (UHECRs), beyond the Greisen–Zatsepin–Kuzmin (GZK) cutoff

$$E_{\text{GZK}} \simeq 5 \times 10^{19} \, \text{eV} \,. \tag{3.28}$$

The problem is that cosmic rays at these energies are necessarily of extragalactic origin, since their gyromagnetic radius within the galactic magnetic field far exceeds galactic dimensions. However, the attenuation lengths of both protons and photons are rather small in comparision with intergalactic distances, and neither can have originated further than about 50 Mpc away from us, owing to their scattering off the intergalactic cosmic photon background. As a result, one would expect that the cosmic-ray spectrum would end abruptly around $E \sim E_{\text{GZK}}$ owing to scattering off the microwave background. This cutoff is represented by the dotted line in Fig. 3.9. In contrast, several groups, most notably the HiRes [72] and AGASA [73] collaborations, have reported events with energies well above the GZK cutoff: the latest compilation by AGASA, for example, contains ten events above the scale $E > 10^{20}$ eV observed since 1993.

The origin of UHECRs has been the subject of lively discussion over the last few years. Apart from the suggestion that they have an astrophysical origin, being accelerated in extreme environments at extragalactic distances

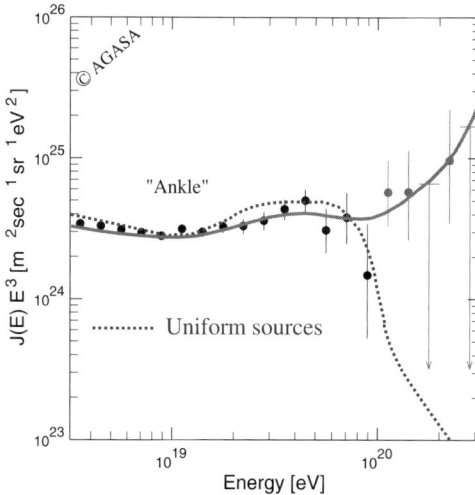

Fig. 3.9. Spectrum, scaled by E^3, of the highest-energy cosmic rays near the GZK cutoff. The *dotted line* corresponds to the expectation from uniformly distributed extragalactic sources, and the *solid line* shows the prediction of a Z-burst model [74]. (Figure modified from the original on the AGASA Web site [73])

(in Active Galactic Nuclei (AGNs), Gamma-Ray Bursts (GRBs), or blazars), they have been attributed to the decay products of very heavy particles or of topological defects. All these explanations have problems, however. Astrophysical explanations face the difficult task of accelerating particles to the extreme energies required, with little or no associated sub-TeV-scale photonic component (as no such component has ever been observed). This is in addition to the above-mentioned problem of the propagation of UHECRs over extragalactic distances. Decay explanations are somewhat disfavored by the growing evidence of doublets and triplets in the AGASA data, and by the correlation between the UHECR arrival directions and distant, compact blazars [75], which appears rather to point towards an astrophysical origin.

The attenuation problem for extragalactic UHECRs can be avoided, however, if they cross the universe in the form of a neutrino beam, since neutrinos travel practically freely over super-Hubble distances. Indeed, should this be the case, the initial ν_{UHE}'s could occasionally interact with the cosmological relic neutrino background close to us, giving rise to "Z-bursts" of hadrons and photons [76], which then would give rise to the observed UHECR events. Indeed, given a neutrino mass m_ν, such a collision has sufficient center-of-mass energy for resonant Z production if

$$E_\nu = \frac{M_Z^2}{2m_\nu} \simeq 4.2 \times 10^{21}\, \mathrm{eV} \left(\frac{\mathrm{eV}}{m_\nu} \right) . \tag{3.29}$$

The requirement for a super-GZK energy of the initial ν_{UHE} beam leads immediately to the interesting mass scale for neutrinos: $m_\nu \sim \mathcal{O}(1)\,\mathrm{eV}$.

Z-burst models have been extensively studied lately [77]. For example, in [74] it was shown that a Z-burst model with $m_\nu = 0.07\,\mathrm{eV}$, corresponding to a degenerate neutrino spectrum, could reproduce the AGASA data, including the spectral features such as the "ankle" and "bump" observed at $E \lesssim E_{\mathrm{GZK}}$. The best-fit model of [74] is shown by the solid line in Fig. 3.9. While the agreement with the AGASA data is good, it should be noted that this model predicts that cosmic-ray primaries are exclusively photons above $E \gtrsim 10^{20}\,\mathrm{eV}$, whereas the $E \simeq 3 \times 10^{20}\,\mathrm{eV}$ event observed by Fly's Eye was almost certainly not caused by a photon [78]. The model also predicts a large increase in the cosmic-ray flux above a few times $10^{20}\,\mathrm{eV}$ (as a direct result of the huge initial energy needed, $E_{\nu_{\mathrm{UHE}}} \simeq 6 \times 10^{22}\,\mathrm{eV}$), which exacerbates the outstanding problem of the origin of UHECRs. These problems could be ameliorated by assuming somewhat larger neutrino masses, and it has been argued by Fodor et al. [77] that the Z-burst scenario can be used to *constrain* the mass, plausibly to within $m_\nu \sim 0.08 - 1.3\,\mathrm{eV}$, in very good agreement with other mass determinations. The analysis of [77] was restricted to Z-resonance interactions, however, while for higher center-of-mass energies the cross section [8] for pair production of gauge bosons $\nu\nu \to ZZ, WW$ becomes important. These reactions have been shown to give accptable solutions with larger neutrino masses $m_\nu \gtrsim 3\,\mathrm{eV}$ [79].

In summary, although the origin of UHECRs remains a mystery today, it is almost certain that if it has to do with extragalactic sources, neutrino physics plays a very important role in the interpretation of these events.

Acknowledgments

K.K. would like to thank Steen Hannestad and Petteri Keranen for useful advice. The work of K.A.O. was supported partly by DOE grant DE-FG02-94ER-40823.

References

1. A.D. Dolgov, arXiv:hep-ph/0202122.
2. K. Olive and J.A. Peacock, to appear in Rev. Part. Prop. 2002, Phys. Rev. **D66**, 010001 (2002).
3. K. Enqvist, K. Kainulainen, and V. Semikoz, Nucl. Phys. B **374**, 392 (1992).
4. S.S. Gerstein and Ya.B. Zeldovich, JETP Lett. **4**, 647 (1966); R. Cowsik and J. McClelland, Phys. Rev. Lett. **29**, 669 (1972); A.S. Szalay and G. Marx, Astron. Astrophys. **49**, 437 (1976).
5. P. Hut, Phys. Lett. B **69**, 85 (1977); B.W. Lee and S. Weinberg, Phys. Rev. Lett. **39**, 165 (1977); M.I. Vysotsky, A.D. Dolgov, and Y.B. Zeldovich, Pisma Zh. Eksp. Teor. Fiz. **26**, 200 (1977).
6. E.W. Kolb and K.A. Olive, Phys. Rev. D **33**, 1202 (1986); Phys. Rev. D **34**, 2531 (1986) (erratum).
7. R. Watkins, M. Srednicki, and K.A. Olive, Nucl. Phys. B **310**, 693 (1988).
8. K. Enqvist, K. Kainulainen, and J. Maalampi, Nucl. Phys. B **317**, 647 (1989).
9. K. Enqvist and K. Kainulainen, Phys. Lett. B **264**, 367 (1991).
10. K. Griest and M. Kamionkowski, Phys. Rev. Lett. **64**, 615 (1990).
11. D. Abbaneo et al. (ALEPH Collaboration), arXiv:hep-ex/0112021.
12. C. Weinheimer et al., Phys. Lett. B **460**, 219 (1999).
13. D.N. Schramm and G. Steigman, Astrophys. J. **243**, 1 (1981).
14. P. Hut and K.A. Olive, Phys. Lett. B **87**, 144 (1979).
15. J.R. Bond, G. Efstathiou, and J. Silk, Phys. Lett. **45**, 1980 (1980); Ya.B. Zeldovich and R.A. Sunyaev, Sov. Astron. Lett. **6**, 457 (1980).
16. P.J.E. Peebles, Astrophys. J. **258**, 415 (1982); A. Melott, Mon. Not. R. Astron. Soc. **202**, 595 (1983); A.A. Klypin and S.F. Shandarin, Mon. Not. R. Astron. Soc. **204**, 891 (1983).
17. C.S. Frenk, S.D.M. White, and M. Davis, Astrophys. J. **271**, 417 (1983).
18. S.D.M. White, C.S. Frenk, and M. Davis, Astrophys. J. **274**, 61 (1983).
19. J.R. Bond, J. Centrella, A.S. Szalay, and J. Wilson, in *Formation and Evolution of Galaxies and Large Structures in the Universe*, eds. J. Andouze and J. Tran Thanh Van (Reidel, Dordrecht, 1983), p. 87.
20. M. Davis, F.J. Summers, and D. Schlegel, Nature **359**, 393 (1992); A. Klypin, J. Holtzman, J. Primack, and E. Regos, Astrophys. J. **416**, 1 (1993).
21. R.A. Croft, W. Hu, and R. Dave, Phys. Rev. Lett. **83**, 1092 (1999); arXiv: astro-ph/9903335.

22. Ø. Elgaroy et al., arXiv:astro-ph/0204152.
23. K.A. Olive and M.S. Turner, Phys. Rev. D **25**, 213 (1982).
24. G. Steigman, K.A. Olive, and D.N. Schramm, Phys. Rev. Lett. **43**, 239 (1979); K.A. Olive, D.N. Schramm, and G. Steigman, Nucl. Phys. B **180**, 497 (1981).
25. S.H. Hansen, J. Lesgourgues, S. Pastor, and J. Silk, arXiv:astro-ph/0106108.
26. K. Olive, G. Steigman, and T.P. Walker, Phys. Rep. **333**, 389 (2000); B.D. Fields and S. Sarkar, to appear in Rev. Part. Pro. 2002, Phys. Rev. **D66**, 010001 (2002).
27. R.H. Cyburt, B.D. Fields, and K.A. Olive, Astropart. Phys. **17**, 87 (2002); arXiv:astro-ph/0105397.
28. G. Steigman, D.N. Schramm, and J. Gunn, Phys. Lett. B **66**, 202 (1977).
29. R.H. Cyburt, B.D. Fields, and K.A. Olive, New Astron. **6**, 215 (2001).
30. B.D. Fields, K. Kainulainen, K.A. Olive, and D. Thomas, New Astron. **1**, 77 (1996).
31. C. Pryke et al., Astrophys. J. **568**, 46 (2002); arXiv:astro-ph/0104490.
32. C.B. Netterfield et al., arXiv:astro-ph/0104460.
33. B.D. Fields and K.A. Olive, Astrophys. J. **506**, 177 (1998); A. Peimbert, M. Peimbert, and V. Luridiana, Astrophys. J. **565**, 668 (2002); arXiv:astro-ph/0107189.
34. E.W. Kolb, M.S. Turner, A. Chakravorty, and D.N. Schramm, Phys. Rev. Lett. **67**, 533 (1991).
35. B.D. Fields, K. Kainulainen, and K.A. Olive, Astropart. Phys. **6**, 169 (1997); K. Kainulainen, in *Neutrino '96, Proceedings of the 17th International Conference on Neutrino Physics and Astrophysics*, Helsinki, 13 June 1996, eds. K. Enquist, K. Huitu, and J. Maalampi (World Scientific, Singapore, 1997), p. 461; arXiv:hep-ph/9608215.
36. S. Hannestad and J. Madsen, Phys. Rev. Lett. **76**, 2848 (1996); Phys. Rev. Lett. **77**, 5148 (1996) (erratum).
37. A.D. Dolgov, S.H. Hansen, and D.V. Semikoz, Nucl. Phys. B **524**, 621 (1998).
38. L. Passalacqua, Nucl. Phys. Proc. Suppl. **55C**, 435 (1997).
39. A.D. Dolgov, K. Kainulainen, and I.Z. Rothstein, Phys. Rev. D **51**, 4129 (1995).
40. K. Assamagan et al., Phys. Rev. D **53**, 6065 (1996).
41. M. Kawasaki, P. Kernan, H.S. Kang, R.J. Scherrer, G. Steigman, and T.P. Walker, Nucl. Phys. B **419**, 105 (1994); S. Dodelson, G. Gyuk, and M.S. Turner, Phys. Rev. D **49**, 5068 (1994); S. Hannestad, Phys. Rev. D **57**, 2213 (1998).
42. A.D. Dolgov, S.H. Hansen, S. Pastor, and D.V. Semikoz, Nucl. Phys. B **548**, 385 (1999).
43. A. Aguilar et al., Phys. Rev. D **64**, 112007 (2001).
44. K. Enqvist, K. Kainulainen and M.J. Thomson, Nucl. Phys. B **373**, 498 (1992).
45. D. Notzold and G. Raffelt, Nucl. Phys. B **307**, 924 (1988).
46. K. Kainulainen, Phys. Lett. B **244**, 191 (1990).
47. K. Kainulainen and A. Sorri, J. High Energy Phys. **0202**, 020 (2002).
48. K. Enqvist, K. Kainulainen, and J. Maalampi, Phys. Lett. B **244**, 186 (1990).
49. R. Foot and R.R. Volkas, Phys. Rev. Lett. **75**, 4350 (1995); R. Foot, M.J. Thomson, and R.R. Volkas, Phys. Rev. D **53**, 5349 (1996).
50. P. Langacker, S.T. Petcov, G. Steigman, and S. Toshev, Nucl. Phys. B **282**, 589 (1987).

51. A.D. Dolgov, S.H. Hansen, S. Pastor, S.T. Petcov, G.G. Raffelt, and D.V. Semikoz, Nucl. Phys. B **632**, 363 (2002); arXiv:hep-ph/0201287; K.N. Abazajian, J.F. Beacom, and N.F. Bell, arXiv:astro-ph/0203442; Y.Y. Wong, arXiv:hep-ph/0203180.

52. G.G. Raffelt, arXiv:astro-ph/0105250.

53. T. Totani, K. Sato, H.E. Dalhed, and J.R. Wilson, Astrophys. J. **496**, 216 (1998).

54. A.S. Dighe and A.Y. Smirnov, Phys. Rev. D **62**, 033007 (2000); J.F. Beacom, R.N. Boyd, and A. Mezzacappa, Phys. Rev. Lett. **85**, 3568 (2000).

55. T.J. Loredo and D.Q. Lamb, Ann. N.Y. Acad. Sci. **571**, 601 (1989); Phys. Rev. D **65**, 063002 (2002).

56. G. Raffelt, Annu. Rev. Nucl. Part. Sci. **49**, 163 (1999).

57. D. Notzold, Phys. Lett. B **196**, 315 (1987); K. Kainulainen, J. Maalampi, and J.T. Peltoniemi, Nucl. Phys. B **358**, 435 (1991).

58. A. Ayala, J. C. D'Olivo, and M. Torres, Phys. Rev. D **59**, 111901 (1999).

59. G.G. Raffelt, Phys. Rep. **320**, 319 (1999).

60. M. Rampp and H.T. Janka, Astrophys. J. **539**, L33 (2000); arXiv:astro-ph/0005438; M. Liebendorfer et al., Phys. Rev. D **63**, 103004 (2001).

61. S. Hannestad and G.G. Raffelt, Astrophys. J. **507**, 339 (1998).

62. G.G. Raffelt, arXiv:hep-ph/0201099.

63. G.M. Fuller et al., Astrophys. J. **389**, 517 (1992).

64. M.B. Voloshin, Phys. Lett. B **209**, 360 (1988).

65. C.J. Horowitz, arXiv:astro-ph/0108113; Y.-Z. Qian, arXiv:astro-ph/0203194.

66. A. Kusenko and G. Segre, Phys. Rev. Lett. **77**, 4872 (1996).

67. H.T. Janka and G.G. Raffelt, Phys. Rev. D **59**, 023005 (1999); arXiv:astro-ph/9808099.

68. A. Kusenko and G. Segre, Phys. Rev. D **59**, 061302 (1999); arXiv:astro-ph/9811144; M. Barkovich, J.C. D'Olivo, R. Montemayor, and J.F. Zanella, arXiv:astro-ph/0206471.

69. G.G. Raffelt, Phys. Rev. D **31**, 3002 (1985).

70. F. Reines, H.W. Sobel, and H.S. Gurr, Phys. Rev. Lett. **32**, 180 (1974).

71. F. Von Feilitzsch and L. Oberauer, Phys. Lett. B **200**, 580 (1988); E.L. Chupp, W.T. Vestrand, and C. Reppin, Phys. Rev. Lett. **62**, 505 (1989); E.W. Kolb and M.S. Turner, Phys. Rev. Lett. **62**, 509 (1989); S.A. Bludman, Phys. Rev. D **45**, 4720 (1992).

72. D.J. Bird et al. (HiRes Collaboration), Astrophys. J. **424**, 491 (1994).

73. M. Takeda et al., Phys. Rev. Lett. **81**, 1163 (1998); AGASA home page, www-akeno.icrr.u-tokyo.ac.jp/AGASA/.

74. G. Gelmini and G. Varieschi, arXiv:hep-ph/0201273.

75. P.G. Tinyakov and I.I. Tkachev, JETP Lett. **74**, 445 (2001).

76. T.J. Weiler, Astropart. Phys. **11**, 303 (1999).

77. Z. Fodor, S.D. Katz, and A. Ringwald, arXiv:hep-ph/0203198.

78. See e.g. F. Halzen, arXiv:astro-ph/0111059.

79. D. Fargion, P.G. De Sanctis Lucentini, M. Grossi, M. De Santis, and B. Mele, arXiv:hep-ph/0112014; D. Fargion, B. Mele, and A. Salis, Astrophys. J. **517**, 725 (1999).

4 Experimental Results on Neutrino Oscillations

Achim Geiser

A standard is the standard until replaced by the next ...

Neutrino physics, in particular the question of neutrino mass and oscillations, is currently giving rise to a minirevolution in particle physics. Here, we review the experimental evidence for and constraints on neutrino oscillations collected over the past decade. The experimental methods and measurements are described. The results are placed into their phenomenological context, including brief discussions of alternative interpretations and possible future improvements.

4.1 Introduction

... bin ich ... auf einen verzweifelten Ausweg verfallen, um den „Wechselsatz" der Statistik und den Energiesatz zu retten. Nämlich die Möglichkeit, es könnten elektrisch neutrale Teilchen ... existieren, welche den Spin 1/2 haben und das Ausschließungsprinzip befolgen und sich von Lichtquanten außerdem noch dadurch unterscheiden, daß sie nicht mit Lichtgeschwindigkeit laufen.

W. Pauli, December 1930

4.1.1 A Brief History of Neutrinos

When Wolfgang Pauli "invented" the neutrino in 1930 [1] to save energy–momentum conservation and Fermi statistics in nuclear beta decay, it seemed natural to him that neutrinos should be massive objects. Since then, the concept of neutrinos has undergone several, sometimes turbulent developments. Since neutrinos are rather elusive owing to their weak interaction with matter, many of their properties were determined indirectly long before direct measurements became accessible. For instance, it took 26 years before the first neutrino (from a reactor source) could be experimentally detected by Reines and Cowan [2] in 1956. Upper limits on the neutrino mass could be obtained much earlier from indirect measurements, eventually leading to the concept of massless neutrinos [3], which is being reexamined today. The fact that only left-handed neutrinos participate in weak interactions was first established indirectly from the measurement of the polarization of other particles involved in the interaction (e.g. [4]). Direct quantitative studies of this property were later used as a tool to obtain measurements of the fundamental coupling constants $\sin^2\theta_W$ and α_s [5]. The conjecture of the conservation of lepton number and lepton family number implies that a distinct neutrino should exist for each lepton family [6]. Currently, three charged leptons are known (e, μ, τ), and the corresponding expectation of three light-neutrino generations has been beautifully confirmed by the study of Z^0 decay at LEP and SLC [7]. The third of these neutrinos, the ν_τ, has only recently become accessible to direct detection [8]. At present, the most stringent upper limits [9] on neutrino masses, lifetimes, charges, and magnetic moments (except m_{ν_e}) arise from astrophysical observations. The confirmation and extension of these indirect measurements by direct means is an important goal of current neutrino research.

Until recently, there was no compelling evidence that neutrinos have mass. Consequently, the Standard Model treats neutrinos as massless and genuinely neutral stable particles. Reflecting parity violation, only left-handed neutrinos and right-handed antineutrinos exist in the Standard Model. The question of what would happen if neutrinos had mass after all and if one could rotate their spin was raised by Majorana in 1937 [10], and remains unsettled today! An important aspect of a nonzero neutrino mass is the question of neutrino oscillations. The possibility of such oscillations was first investigated by Pontecorvo [11] in 1958 in the form of neutrino–antineutrino oscillations, similar to the oscillations of neutral kaons. After the discovery of the muon neutrino, the concept was adapted to neutrino flavor oscillations (e.g. $\nu_\mu \leftrightarrow \nu_e$) [12]. Owing to the intimate relation between neutrino mass and neutrino oscillations, the observation of such oscillations is currently one of the most promising approaches to solving the neutrino mass puzzle.

The first indications of nonzero neutrino mass were obtained from the neutrino oscillation interpretation of a deficit in the rate at which neutrinos reach us from the sun, observed since 1970 [13]. These indications were followed by the observation of anomalies in the spectra of atmospheric neutrinos (created by cosmic-ray interactions in the Earth's atmosphere), beginning in 1988 [14]. The recent confirmation of these effects by the Super-Kamiokande [15, 16], SNO [17], and KamLAND [18] collaborations has triggered a turnaround in the general perception of the neutrino mass issue, such that a nonzero neutrino mass is now an almost universally accepted conjecture. However, this conjecture needs at least a minimal, if not a major, extension of the Standard Model (the first required by experimental evidence!). Since the data are scarce, many possibilities exist, and the question of *how* the Standard Model should be extended is one of the central issues of the current debate. A claim of direct evidence for $\nu_\mu \to \nu_e$ oscillations by the LSND collaboration [19], beginning in 1994, and indications of a potentially important role of massive neutrinos in cosmology (e.g. the formation of large-scale structures [20]) further enrich and complicate the issue.

This is a very exciting time for neutrino physics, with new data becoming available every few months. Here, we shall attempt a complete review of the experimental evidence for and constraints on neutrino oscillations collected within the last decade, and place them into their phenomenological context. Parts of this review are an abbreviated update of [21].

4.1.2 Neutrino Sources

Neutrino experiments can be performed either with neutrinos of astrophysical origin or with neutrinos from artificial, terrestrial sources. The former offer the advantage of a naturally available flux without the need for accelerators or reactors, while the latter allow high local fluxes with controlled experimental conditions.

Historically, the first neutrinos were detected at nuclear reactors [2]. Nuclear fission reactions produce neutron-rich nuclei, many of which disintegrate via β^- decay,

$$n \rightarrow pe^-\overline{\nu}_e . \tag{4.1}$$

These $\overline{\nu}_e$'s are emitted isotropically from the reactor core, and the total flux can reach up to 10^{21} $\overline{\nu}_e$ per second. The neutrino energy is of the order of a few MeV. A typical flux spectrum is shown in Fig. 4.1.

Somewhat higher neutrino energies can be achieved at low-energy proton accelerators. Here, a primary low-energy (e.g. 800 MeV) proton beam is dumped onto a target and produces pions as secondaries, which are stopped in the target. While most of the π^- are absorbed by nuclei, most of the π^+ decay according to the reaction

$$\pi^+ \rightarrow \mu^+\nu_\mu , \tag{4.2}$$

therefore yielding an isotropic flux of monoenergetic ν_μ's. The muons produced are stopped in turn, and their three-body decay

$$\mu^+ \rightarrow e^+\nu_e\overline{\nu}_\mu \tag{4.3}$$

yields a wide spectrum of ν_e and $\overline{\nu}_\mu$. Smaller components of the spectrum are obtained from pion and muon decays in flight. Again, a typical spectrum is shown in Fig. 4.1.

Finally, neutrino energies ranging from 1 GeV to 1 TeV can be obtained from dedicated neutrino beams at high-energy accelerators (see [22, 23] for historical articles). Here, protons are accelerated to energies between tens of GeV and 1 TeV in large proton synchrotron rings and sent to a relatively thin target, such that the secondaries produced in the hadronic showers (mainly pions and kaons) can escape the target. These secondaries are momentum and charge selected and focused into an evacuated tunnel, called the decay tunnel. There, part of the pions and kaons decay leptonically, and yield neutrinos as

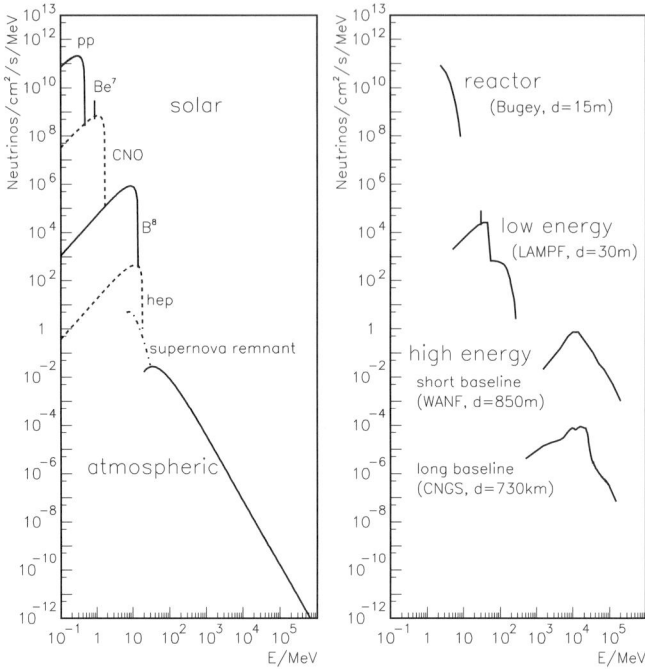

Fig. 4.1. Typical neutrino fluxes from astrophysical sources (*left*) and terrestrial sources (*right*) [21]. The terrestrial fluxes refer to dedicated experimental zones at the Bugey reactor, Los Alamos (LAMPF), CERN (WANF), and Gran Sasso (CNGS), whose distances from the source are also indicated

tertiary reaction products. Owing to the large Lorentz boost (typical γ factors are about 10–1000), the neutrinos are emitted approximately in the direction of their parent hadrons. Thus they form a reasonably focused "beam", with a typical opening angle of order 1 mrad. This focusing yields high neutrino fluxes in the direction of the beam line.

A sketch of a neutrino beam line is shown in Fig. 4.2, and typical spectra are shown in Figs. 4.1 and 4.3. The neutrino spectrum is dominated by muon neutrinos, since pions decay essentially through the reaction $\pi \to \mu\nu_\mu$.

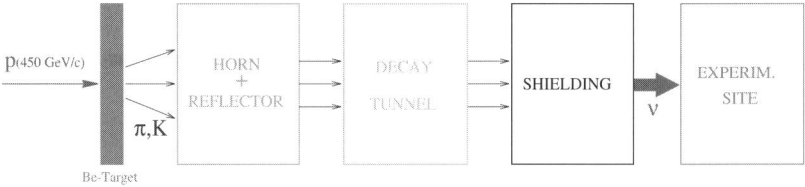

Fig. 4.2. Sketch of a typical high-energy neutrino beam line

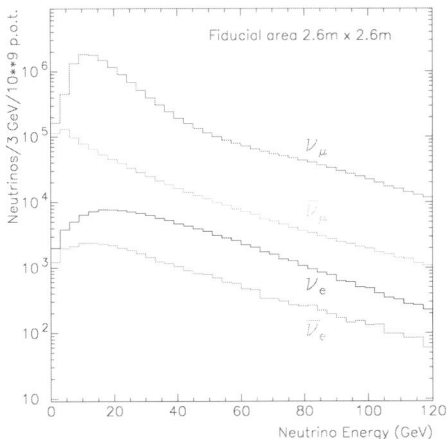

Fig. 4.3. Neutrino fluxes for the West Area Neutrino Facility (WANF) at CERN, integrated over an area of $2.6 \times 2.6\,\mathrm{m}^2$ at a distance of $850\,\mathrm{m}$ from the target, for 10^9 protons on target (p.o.t.) [24]

Depending on whether positive or negative particles are focused, the main beam component consists of either ν_μ or $\bar{\nu}_\mu$, while some "background" from the other polarity (of the order of a few percent) remains. A small (of order 1%) component of ν_e arises mainly from semileptonic K decays. Tau neutrinos can only be produced through decays of D_s and B mesons (because of the τ mass threshold) and are essentially absent.

Neutrino beams could also be produced by accelerating muons instead of protons, and letting them circulate in muon storage rings until they decay according to (4.3). Along the straight sections of such a ring, a highly pure beam of ν_e and $\bar{\nu}_\mu$ (or ν_μ and $\bar{\nu}_e$) would be produced, with energies ranging up to 50 GeV. Such "neutrino factories" [25] do not yet exist, but might be one of the next steps in current accelerator programs (Chap. 7).

As an alternative to these terrestrial sources, astrophysical neutrino sources play an important role in current neutrino-related research. Neutrinos left over from the Big Bang (cosmological neutrinos) are expected to have a "temperature" of $1.9\,\mathrm{K}$, and therefore a kinetic energy in the meV range, which is at present too small to be detected. They do, however, play a role in the evolution of the early universe [26].

The sun is an intense source of low-energy ($< 0.42\,\mathrm{MeV}$) electron neutrinos (solar neutrinos) based on the basic fusion reaction

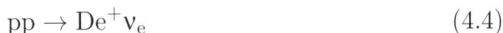

$$pp \rightarrow De^+\nu_e \qquad (4.4)$$

(these neutrinos are called "pp" neutrinos). In addition, electron neutrinos with energies up to 18 MeV are produced in side reactions, the most important of which involve the nuclei ^{7}Be and ^{8}B ("^{7}Be" and "^{8}B" neutrinos). The corresponding spectra are shown in Fig. 4.1. Solar neutrinos were first detected in 1970 [13], and play a major role in the current discussion about a nonzero neutrino mass (Sect. 4.3).

In contrast, interactions of primary cosmic-ray particles in the earth's atmosphere are an important source of high-energy neutrinos (atmospheric neutrinos). They were first detected in 1965 [27]. The processes yielding these neutrinos are essentially the same as the ones used to produce neutrino beams, except that the parent particles are not focused, and that pion and muon decays occur here at similar rates. Some specific features of these atmospheric neutrinos and their impact on our understanding of neutrino masses are discussed in Sect. 4.4. The corresponding neutrino spectrum is compared with the other spectra in Fig. 4.1.

Neutrinos from supernova collapses have a spectrum which lies between the solar and atmospheric neutrino spectra. Since supernovae are rare events the corresponding rate is discontinous in time. Only one supernova event has been seen in neutrinos so far (supernova 1987a [28]). However, the sum of all supernovae which have ever occurred yields a spectrum (Fig. 4.1) which might be detectable in the near future. Finally, ultrahigh-energy neutrinos (of the order of a TeV or higher) might originate from extragalactic sources. Such neutrinos might soon be observed with large-volume ice or underwater detectors [29, 30].

All neutrino sources have in common the feature that their fluxes decrease strongly with increasing average neutrino energy. However, this is compensated by the increasing interaction cross section and, in the case of neutrino beams, by the improved focusing at high energies. The lateral size of these "beams" is typically of order 1 m or more at the detector location, extending up to the kilometer range for "long-baseline" beams (Sect. 4.6.3).

4.2 Neutrino Oscillations and Their Detection

... there is a possibility of real transitions neutrino → antineutrino in vacuum, provided that the lepton (neutrino) charge is not conserved.

B. Pontecorvo, 1958

4.2.1 Neutrino Oscillations

The theory of neutrino oscillations is extensively discussed in Chaps. 1 and 5. Here, we shall briefly review the most important aspects, and define the terminology.

In the Standard Model, the neutrino is a massless, strictly neutral, stable particle. In general, if neutrinos have mass and form mixed states, neutrino oscillations can occur. Minimal extensions to the Standard Model that allow a neutrino mass without destroying gauge invariance can be obtained, for instance, by adding a standard Yukawa coupling term for neutrinos (a Dirac mass) or by introducing a "bare" right-handed neutrino mass term (a Majorana mass). The latter explicitly violates lepton number conservation, while both cases can lead to nonconservation of lepton family number.

In the context of Grand Unified Theories, many more (and much more complicated) possibilities exist, usually leading to both Dirac *and* Majorana mass terms. In such models it is possible to have oscillations between neutrinos and antineutrinos, as originally postulated by Pontecorvo [11]. Since angular momentum is conserved, left-handed neutrinos (right-handed antineutrinos) will oscillate into left-handed antineutrinos (right-handed neutrinos), which do not participate in standard weak interactions, and therefore appear "sterile". However, in many models (e.g. the seesaw model [31]) such oscillations are strongly suppressed by the introduction of very large Majorana masses and correspondingly heavy "right-handed" neutrinos, out of reach of current experiments.

In the following, we shall restrict the discussion to the more familiar case of the three known (active) neutrinos, yielding oscillations between different lepton families [32]. Potential additional, sterile neutrinos can be accounted for straightforwardly by extending this mechanism to four or more neutrinos.

In analogy to the quark case, the three flavor eigenstates can be expressed as a superposition of mass eigenstates:

$$\begin{pmatrix} \nu_e \\ \nu_\mu \\ \nu_\tau \end{pmatrix} = \begin{pmatrix} U_{e1} & U_{e2} & U_{e3} \\ U_{\mu1} & U_{\mu2} & U_{\mu3} \\ U_{\tau1} & U_{\tau2} & U_{\tau3} \end{pmatrix} \begin{pmatrix} \nu_1 \\ \nu_2 \\ \nu_3 \end{pmatrix} . \tag{4.5}$$

Here U is a unitary mixing matrix (the Maki–Nakagawa–Sakata matrix [33]) which, in analogy to the Cabibbo–Kobayashi–Maskawa matrix for quarks [34], has four independent parameters: three mixing angles and one CP-violating phase.

In an oscillation experiment, one starts with a neutrino flux of defined flavor eigenstates ν_α at $t_0 = x_0 = 0$ and tries to detect its composition in terms of the flavor states ν_β at a different point in space–time, for instance at a distance $x = L$. Since the space–time evolution between these two points is described by the non-flavor-diagonal mass eigenstates, the flavor composition changes, and the transition probability can be written as

$$P(\nu_\alpha \rightarrow \nu_\beta) = \left| \sum_i U_{\beta i} \exp\left[-i(E_i t - p_i x)\right] U_{\alpha i}^* \right|^2 . \tag{4.6}$$

For a fixed neutrino energy, one can express p_i and t in terms of $E_i = E_\nu$, m_i, and $x = L$. Neglecting CP violation, which does not yet play a role in current experiments (i.e. we assume that U is real), we obtain (in natural units), after some transformations,

$$P(\nu_\alpha \rightarrow \nu_\beta) = \delta_{\alpha\beta} - 4 \sum_{i=1}^{3} \sum_{j=i+1}^{3} U_{\alpha i} U_{\beta i} U_{\alpha j} U_{\beta j} \sin^2\left(\frac{\Delta m_{ij}^2 L}{2 E_\nu}\right) . \tag{4.7}$$

Thus, the matrix U defines the *amplitudes* of the resulting oscillations, while the oscillation *frequencies* are determined by the differences between the squared masses Δm_{ij}^2 for the mass eigenstates.

On one hand, this formula can be extended straightforwardly to more than three neutrino states. On the other hand, it is often useful for practical purposes to consider the simplified case where only two neutrino flavors a and b take part in the oscillations

$$\begin{pmatrix} \nu_a \\ \nu_b \end{pmatrix} = \begin{pmatrix} \cos\theta & \sin\theta \\ -\sin\theta & \cos\theta \end{pmatrix} \begin{pmatrix} \nu_1 \\ \nu_2 \end{pmatrix} . \tag{4.8}$$

In this two-flavor approximation, the oscillation probability in vacuum can be expressed by the well known formulae

$$P_{\nu_a \to \nu_b}(L/E) = \sin^2(2\theta) \, \sin^2(1.27 \, \Delta m^2 \, L/E) \, ,$$
$$P_{\nu_a \to \nu_a}(L/E) = 1 - \sin^2(2\theta) \, \sin^2(1.27 \, \Delta m^2 \, L/E) \, , \tag{4.9}$$

where L is the distance travelled in km, E is the neutrino energy in GeV, θ is the neutrino mixing angle, and Δm^2 is the difference between the mass-squared eigenvalues expressed in eV^2. The first formula applies to the appearance of ν_b, while the second describes the corresponding disappearance of ν_a. Note that there is no CP-violation parameter in the two-neutrino case. Three kinematical ranges can be distinguished:

- If $L/E \ll 1/\Delta m^2$ (E large or L small), the sine function can be approximated by a linear function, and the "oscillation" effect is small even if the mixing angle is large.
- If $L/E \sim 1/\Delta m^2$, the actual sinusoidal oscillation pattern is accessible to observation.
- If $L/E \gg 1/\Delta m^2$ (E small or L large), the sinusoidal oscillations become so fast that they will be unmeasurable. In this case, only an average oscillation effect, proportional to $\sin^2 2\theta$, can be observed.

For the more general case of three generations, the three mass eigenstates yield two independent Δm^2 scales in (4.7). The third scale is fixed by the other two, and is very close to the larger one. Any flavor transition can in principle occur on both scales. CP violation would manifest itself by differences between the differential transition rates of neutrinos and antineutrinos (CP-violation effects cancel in time-, space-, or energy-averaged rates). If one of the nondiagonal elements of the matrix U (e.g. U_{e3}) becomes very small or if the experimental L/E range is chosen such that only one of the Δm^2 values can contribute significantly, then the transition probability for each subset of two flavors can be approximated by a two-flavor formula similar to (4.9).

On the other hand, the situation can be complicated by the presence of additional degrees of freedom due to sterile neutrinos which participate in the oscillations. Furthermore, the presence of matter can significantly affect the oscillation parameters and distort the oscillation pattern.

4.2.2 Matter Effects

The properties of neutrinos, in particular their effective masses and mixing angles, can be affected by the presence of matter. A detailed discussion of these effects is outside the scope of this review, and the reader is referred to the literature [32, 35] for more information.

In a nutshell, the difference between the coherent forward-scattering amplitudes (on nucleons and electrons) for different kinds of neutrinos can be expressed phenomenologically in terms of a "refractive index" n, in analogy to the refractive index in standard optics. Sterile neutrinos do not interact at all and will therefore have $n = 1$. Forward scattering of ν_μ and ν_τ is possible via neutral-current (NC) reactions, and will induce a small nonzero value of $\varepsilon = n - 1$. The sign of ε will be opposite for neutrinos and antineutrinos. Electron (anti)neutrinos have an additional, somewhat larger forward-scattering amplitude through the charged-current (CC) reaction $\nu_e e^- \to e^- \nu_e$.

In matter, all formulae containing the neutrino momentum p must be modified by replacing p by np. This induces an additional phase factor in the neutrino wave function. If we work in the three-neutrino framework and arbitrarily take the muon neutrino as a reference (only *differences* in this phase factor are relevant), this yields an additional term in the neutrino mass matrix. In the flavor basis, the first diagonal mass matrix element m_{ee}^2 (the "electron neutrino" mass term if mixing is small) must be replaced by

$$m_{ee_m}^2 = m_{ee}^2 - 2p^2(n_e - n_\mu) = m_{ee}^2 + 2\sqrt{2}G_F N_e p \,, \qquad (4.10)$$

where G_F is the Fermi constant and N_e is the electron density. Diagonalization of the modified matrix will yield mass eigenstates which are different from the vacuum mass eigenstates.

Therefore, the effective neutrino mass changes when the neutrinos enter matter, and the size of this change depends on the neutrino energy (or momentum). The effective mixing angles are affected in a similar, energy-dependent way. Through this modification, two neutrino masses can become degenerate for specific values of the matter density. In this case, the oscillation parameters can go through a resonance which produces maximal oscillations even if the vacuum oscillation amplitude is small. If a neutrino traverses a large amount of material with a continously varying density (such as the interior of the sun) it is likely to go through this resonance point, effectively yielding almost complete flavor inversion (e.g. all ν_e become ν_μ and all ν_μ become ν_e). This is known as the MSW (Mikheyev–Smirnov–Wolfenstein) effect [36]. In the case of an approximately constant density (e.g. the earth), the masses and oscillation parameters will still be modified, although not necessarily resonantly. Three important points to be remembered are the following:

- ν_μ–ν_τ oscillations are unaffected by matter effects, since ν_μ and ν_τ have the same (NC) interactions with ordinary matter. All other kinds of oscillations *can* be affected, depending on the vacuum oscillation parameters, the matter distribution, and the neutrino energy.

- In general, matter effects are *different* for neutrinos and antineutrinos (because of the opposite signs of the scattering amplitudes), and could partially obscure or mimic CP-violation effects. If matter effects are important, it is useful to consider $\tan^2 \theta$ instead of $\sin^2 2\theta$ as the mixing-angle parameter, where $\tan^2 \theta < 1$ corresponds to the occurence of a matter resonance for neutrinos, and $\tan^2 \theta > 1$ corresponds to a resonance for antineutrinos.
- For short distances ($\ll$ one vacuum or resonance oscillation length) or large Δm^2 ($\gg 2\sqrt{2}G_F N_e p$), matter effects are negligible.

4.2.3 Detection of Neutrino Oscillations

On inspecting (4.7), one finds that neutrino oscillations can be detected in two fundamentally different ways:

- *appearance* of ν_β in a beam of ν_α, with $\beta \neq \alpha$;
- *disappearance* of ν_α into other neutrino flavors, which escape detection.

Since the neutrino flavor is *defined* by the charged lepton which is produced in its CC interactions, an appearance measurement implies that

- the neutrino energy is large enough to *produce* the final-state lepton, i.e. it is kinematically above the lepton mass threshold, and
- the detector is able to *distinguish* this final-state lepton from other final states that may occur.

On the other hand, a disappearance measurement requires a good control of the expected *neutrino flux*, since the signal will consist of deviations from the expected flux normalization and/or energy distribution. The same is true for appearance measurements if the relevant flavor is already present as an a priori background in the initial flux. The required knowledge can be obtained either from precise calculations of the expected flux or by using two detectors at different distances from the source (or several similar sources at different distances from the detector). In the latter case, the flux observed at a "near" distance (before oscillation effects become important) can be used to calibrate the expected flux at the "far" distance. So far, with one exception, positive evidence for neutrino oscillations has only been seen in disappearance measurements.

For either appearance or disappearance, oscillations can be observed in two ways:

- Indirectly, by integrating over a large range of L/E and detecting an *average* oscillation signal. This reduces the requirements on statistics and experimental resolution, but often does not allow a unique determination of the oscillation parameters.

— Directly, by observing the *differential* oscillation pattern (thereby observing the sinusoidal variation). This is needed to uniquely distinguish oscillations from other hypothetical flavor-changing processes (interactions and decays) that might yield similar average appearance or disappearance signals.

Almost all positive evidence for neutrino oscillations so far has been obtained from experiments yielding averaged measurements. New differential measurements are therefore needed to fully establish the oscillation phenomenon. Furthermore, the observation of CP violation in neutrino reactions will only be possible through the measurement of differential appearance rates.

Finally, sterile neutrinos could be "detected" in disappearance measurements using the fact that they will produce neither CC nor NC interactions, while active neutrinos below the charged-lepton mass threshold will "disappear" as far as their CC interactions are concerned, but still "appear" in NC interactions. Therefore, the observed NC/CC ratio will be enhanced in the latter case. Depending on the point of view, such an observation can be called a "CC disappearance" or an "NC appearance" measurement.

4.3 Solar Neutrino Experiments

The quest to understand energy production in the Sun frequently leads to fascinating discoveries about neutrinos.

J.N. Bahcall, 2001

The first indirect evidence for neutrino oscillations was obtained by Davis, from 1968 onwards, from the observation of a deficit in the neutrino rate from the sun detected in an experiment located in the Homestake mine in South Dakota, USA [13]. This observation, intimately related to our basic understanding of the fusion processes inside the sun alluded to by Bahcall [37], was subsequently confirmed by several other measurements. For detailed reviews, see [38, 39].

Solar neutrinos are produced by nuclear fusion processes in the core of the sun (Table 4.1), which yield exclusively electron neutrinos. The expected spectral composition of the solar neutrinos and its uncertainties are indicated in Fig. 4.4. The predicted solar-neutrino fluxes (which are also referred to as the standard solar model, SSM) are rather robust. The dominant pp flux is directly proportional to the total solar photon luminosity, which is very well known. The other contributions are constrained by the strong correlation between the occurrence of their corresponding nuclear reactions and the density and temperature profile of the sun. These profiles have been very accurately tested through the observation of the solar "vibrational" oscillation eigenmodes, called helioseismology [40].

Owing to the low neutrino energy, ν_μ and ν_τ CC reactions cannot occur. Hence only ν_e CC (+NC) interactions can be observed, while ν_μ and ν_τ (from

Table 4.1. Average and maximum neutrino energy for the various reactions contributing to the solar neutrino flux [41]

Source	Reaction	$\langle E_\nu \rangle$ (MeV)	Max. E_ν (MeV)
pp	$p + p \rightarrow d + e^+ + \nu_e$	0.267	0.423
pep	$p + e^- + p \rightarrow d + \nu_e$	1.445	1.445
hep	$^3He + p \rightarrow {}^4He + e^+ + \nu_e$	9.628	18.778
^7Be	$e^- + {}^7Be \rightarrow {}^7Li + \nu_e$	0.386	0.386
		0.863	0.863
^8B	$^8B \rightarrow {}^8Be^* + e^+ + \nu_e$	6.74	~ 15
^{13}N	$^{13}N \rightarrow {}^{13}C + e^+ + \nu_e$	0.706	1.198
^{15}O	$^{15}O \rightarrow {}^{15}N + e^+ + \nu_e$	0.996	1.732
^{17}F	$^{17}F \rightarrow {}^{17}O + e^+ + \nu_e$	0.998	1.736

potential oscillations) will produce NC interactions only. All CC solar neutrino observations are therefore disappearance measurements. In addition, different experimental techniques yield different thresholds for neutrino detection, which are also indicated in Fig. 4.4. In the following sections, we shall first present the current status of the experimental results, and then discuss the interpretation of these results in terms of neutrino oscillations or alternative solutions.

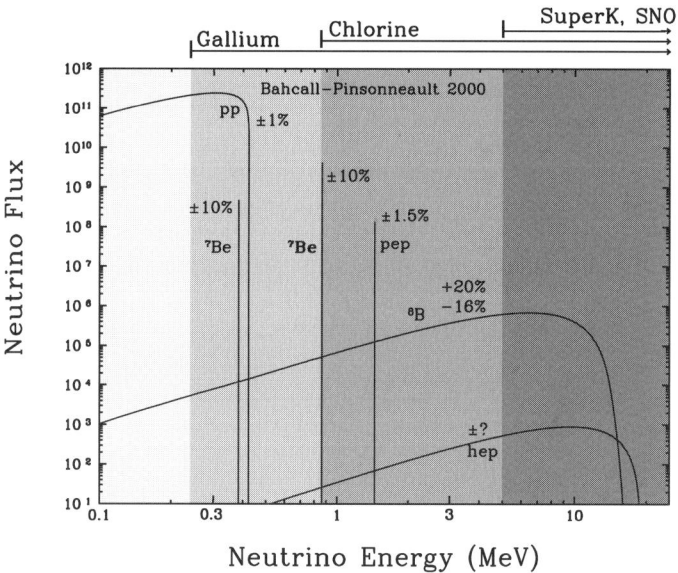

Fig. 4.4. BP2000 [42] predictions of the energy spectra of solar neutrinos produced in the pp-cycle reactions [43]. The continuum spectra are expressed in neutrinos $cm^{-2} s^{-1} MeV^{-1}$ at one astronomical unit, and the monochromatic lines (pep and ^7Be) are given in neutrinos $cm^{-2} s^{-1}$

4.3.1 Radiochemical Experiments

Two classes of experiments have yielded results so far. Radiochemical experiments use the reaction

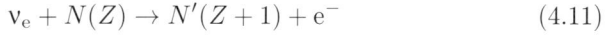

$$\nu_e + N(Z) \rightarrow N'(Z+1) + e^- \tag{4.11}$$

to produce semistable nuclei N' (typically a few per month in a target of nuclei N with a mass of the order of 100 tons), which are chemically extracted and detected through their radioactive decay. For historical reasons, neutrino rates are measured in solar neutrino units (SNU), corresponding to 10×10^{-36} events/atom/s. The observed rate is then compared with the expected flux, averaged over several weeks and over the whole energy spectrum. Obviously, only average oscillation signals can be obtained with this technique. Depending on the neutrino energy threshold of the process (4.11), different parts of the neutrino spectrum will be covered. The advantage of these measurements is that these thresholds are typically an order of magnitude lower than for the real-time experiments discussed in the next section.

The Homestake chlorine experiment [13, 44] is located 1480 m underground at the Homestake gold mine in South Dakota, USA. A tank of volume 6×10^5 litres volume is filled with tetrachloroethylene (C_2Cl_4). Neutrino detection is based on the reaction

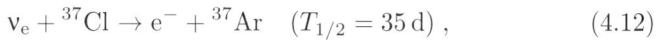

$$\nu_e + {}^{37}Cl \rightarrow e^- + {}^{37}Ar \quad (T_{1/2} = 35\,\mathrm{d}) , \tag{4.12}$$

which has a threshold of 0.81 MeV, above the endpoint of the pp energy spectrum and just below the ^{7}Be line. The few ^{37}Ar atoms produced in neutrino reactions are extracted every 1–3 months by purging the detector with ^{4}He. The measured rate, which arises mainly from ^{8}B and ^{7}Be neutrinos, plus a small contribution from pep neutrinos, is 2.56 ± 0.23 SNU [44], significantly below the predicted value of $7.6^{+1.3}_{-1.1}$ SNU. Owing to the lack of a suitable artificial neutrino source, the Homestake detector has not been fully calibrated experimentally. However, the efficiency of the extraction technique has been checked by doping the detector with a known amount of radioactive argon atoms.

A second group of experiments uses gallium as the target medium, exploiting the reaction

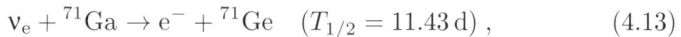

$$\nu_e + {}^{71}Ga \rightarrow e^- + {}^{71}Ge \quad (T_{1/2} = 11.43\,\mathrm{d}) , \tag{4.13}$$

for which the energy threshold (0.23 MeV) is well below the maximum energy of the pp neutrinos. This means that gallium experiments, by measuring low-energy solar neutrinos, are directly sensitive to the thermonuclear nature of the energy production mechanism in the sun.

The GALLEX experiment [45], now completed, was located at the Gran Sasso underground laboratory in Italy. It consisted of 30 tons of ^{71}Ga, in the form of a concentrated solution of gallium chloride ($GaCl_3-HCl$) in water. The ^{71}Ge atoms formed the volatile compound $GeCl_4$, which, at the end of each run (3–4 weeks), was swept out of the solution by means of a nitrogen stream. The GALLEX isotope extraction procedure was calibrated using two independent methods producing ^{71}Ge from radioactive sources [46]. The final results [45] give a solar neutrino flux of $77.5^{+7.6}_{-7.8}$ SNU, again well below the theoretical predictions of 128^{+9}_{-7} SNU.

In 1998, the experiment was upgraded and continues to run under the name of GNO (Gallium Neutrino Observatory), aiming at the measurement of long-term rate variations. The first flux measurement from GNO [47], $65.8^{+10.7}_{-10.2}$ SNU, is consistent with the GALLEX results.

In parallel with GALLEX/GNO, the SAGE experiment [48] is being carried out by a Russian–American collaboration at the underground Baksan Neutrino Observatory in the northern Caucasus mountains. The detector, which weighs 57 tons, uses gallium in its metallic form. The germanium produced in the reaction (4.13) is removed from the metallic gallium by liquid–liquid extraction into an $HCl-H_2O_2$ phase. The results reported by the SAGE collaboration are based on 10 years of exposure time (from January 1990 to October 1999), yielding a measured solar neutrino flux of $74^{+7.8}_{-7.4}$ SNU [48], in very good agreement with the GALLEX/GNO results. The SAGE detector has been calibrated by exposing 13 tons of metallic gallium to an intense ^{51}Cr source [49].

4.3.2 Real-Time Experiments

Complementary results are obtained from large water Cherenkov detectors. First, we shall discuss detectors filled with purified light water (H_2O). These detectors are Kamiokande [50] and its successor Super-Kamiokande [16] (Fig. 4.5), which are also used for detection of atmospheric neutrinos. These detectors measure Cherenkov light from neutrino–electron scattering,

$$\nu_x + e^- \rightarrow \nu_x + e^- \quad (x = e, \mu, \tau) , \tag{4.14}$$

with an energy threshold in the region of 5–10 MeV. They are therefore sensitive only to ^8B neutrinos, for which the flux is least well predicted. However, they are able to measure the resulting differential energy spectrum of the electrons in real time. This yields sensitivity to spectral distortions, and to short-term (e.g. day/night) variations. Also, they are sensitive to both NC and ν_e CC interactions, although the two are experimentally indistinguishable.

Both Kamiokande and Super-Kamiokande were/are located 1000 m underground in the Kamioka mine in Japan. Kamiokande started its operation in 1984 and was originally intended to study nucleon decay [52]. It was later upgraded to detect also low-energy events and succeeded in observing the first solar neutrinos in 1987. Kamiokande consisted of a cylindrical tank,

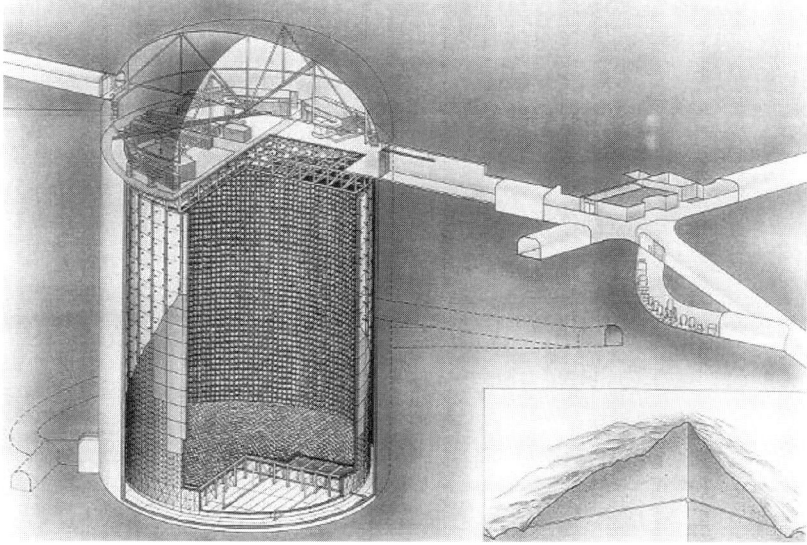

Fig. 4.5. The Super-Kamiokande detector [51]

filled with 4.5 kt of pure water. An inner volume containing 2.14 kt of water was defined by 980 inward-looking photomultiplier tubes (PMTs), used to detect the Cherenkov light produced by relativistic particles traversing the water. Kamiokande ended physics data-taking at the beginning of 1995.

Its successor Super-Kamiokande consists of a huge cylindrical tank filled with 50 kt of pure water. An inner volume (16.9 m in diameter, 36.2 m in height) is defined by an inner surface equipped with 11 146 PMTs of diameter 50 cm, yielding 40% surface coverage. This increased coverage improves the energy, position, and angular resolution. An outer volume, 2 m thick, equipped with 1185 PMTs (diameter 20 cm), surrounds the inner detector and serves as an active veto counter against gamma rays, neutrons, and through-going cosmic muons. The fiducial mass for the solar-neutrino analysis is 22.5 kt (the fiducial-volume boundaries are 2 m inside the inner surface). Data-taking has recently stopped, and could not be resumed as planned owing to an accident that has destroyed more than half of the photomultipliers [51].

The reaction (4.14) produces electrons whose direction is very closely related to the original neutrino direction. It is therefore possible to relate this direction to the position of the sun at the time of the interaction (Fig. 4.6). This proves the solar origin of the detected neutrinos, and allows a well-defined subtraction of the flat background. Moreover, the energy of the recoiling electron, whose spectrum can be measured (Fig. 4.7), yields a lower limit on the energy of the detected incoming neutrino. The energy scale for this measurement is calibrated at Super-Kamiokande using electrons from

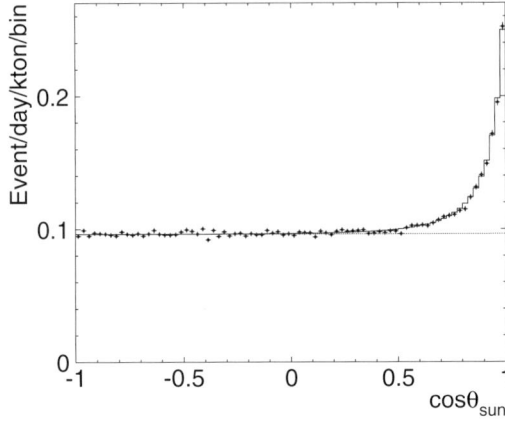

Fig. 4.6. Super-Kamiokande (1258 days) distribution of the cosine of the angle θ_{sun} between the recoil electron momentum and the vector from the sun to the earth [16]. The experimental data (*points*) are shown, together with the best fit to the solar neutrino signal plus background (*solid histogram*) and to the background only (*dotted line*)

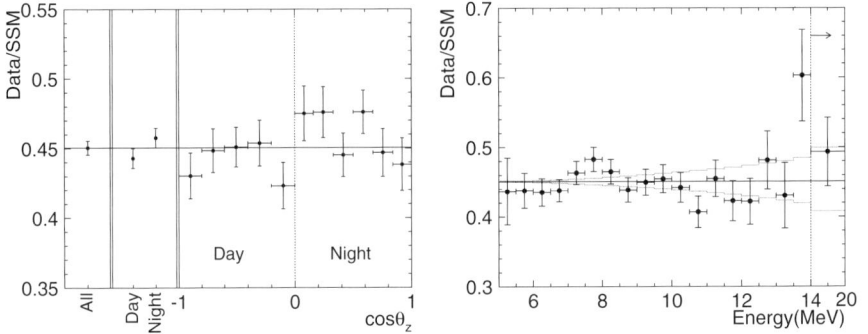

Fig. 4.7. Super-Kamiokande distributions [16] as a function of the solar zenith angle (*left*) and of the electron energy (*right*), normalized to the SSM prediction. The three curves on the right-hand graph show the expected shape uncertainty

a small linear accelerator placed near the detector tank [53], and electron and photon lines from ^{16}N decay [54]. The final energy threshold of 5 MeV (7.5 MeV for Kamiokande) limits the measurement to the ^{8}B neutrino flux (plus a small component from the hep reaction).

The overall value of the ^{8}B neutrino flux measured by Kamiokande, based on 2079 days of data-taking (from January 1987 to February 1995), is $2.80 \pm 0.19\,(\text{stat.}) \pm 0.33\,(\text{sys.}) \times 10^6\,\text{cm}^{-2}\,\text{s}^{-1}$ [50]. The current Super-Kamiokande result, based on 1258 days of data-taking, is $2.32 \pm 0.03\,(\text{stat.})^{+0.08}_{-0.07}\,(\text{sys.}) \times$

$10^6\,\mathrm{cm}^{-2}\,\mathrm{s}^{-1}$ [16]. For the hep neutrinos, the results set an upper limit of $4 \times 10^4\,\mathrm{cm}^{-2}\,\mathrm{s}^{-1}$, at 90% C.L.

Owing to the ability to perform real-time measurements, this flux can be measured separately for different times of the day, corresponding to different zenith angle positions of the sun with respect to the horizon (Fig. 4.7), or for different seasons of the year, corresponding to a small variation of the distance between the sun and the earth (Fig. 4.8). Despite the strong suppression of the rate, no significant shape variations with respect to the SSM expectations are observed.

A different experimental method, based on a heavy-water filling, is exploited by the SNO (Solar Neutrino Observatory) experiment [55], which started data-taking in 1999. SNO is a 1 kt heavy-water Cherenkov detector located 2 km underground in a working nickel mine in Sudbury, Canada. The detector is made of a spherical acrylic vessel 12 m in diameter containing ultrapure D_2O, surrounded by an ultrapure H_2O shield. The light produced in the water is detected by 9456 photomultipliers (20 cm in diameter), installed on a stainless steel structure surrounding the acrylic vessel.

In addition to the elastic-scattering (ES) reaction (4.14) exploited by (Super-)Kamiokande, SNO can detect ^8B solar neutrinos via pure CC and NC interactions resulting in the dissociation of the target deuterium, namely

$$\nu_e + d \rightarrow p + p + e^- \quad (\text{threshold} = 1.4\,\mathrm{MeV}) \tag{4.15}$$

and

$$\nu_x + d \rightarrow n + p + \nu_x \quad (\text{threshold} = 2.2\,\mathrm{MeV})\,. \tag{4.16}$$

As explained above, neutrino oscillations can significantly affect the measured NC/CC ratio. By comparing the rates for these processes with each

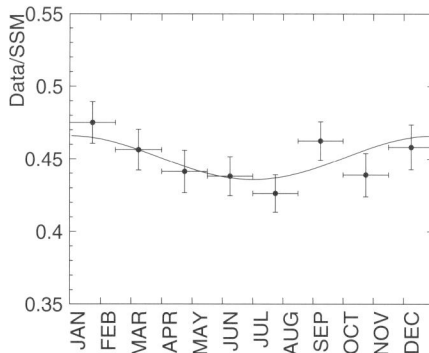

Fig. 4.8. Seasonal variation of the solar neutrino flux measured by Super-Kamiokande [16] (*points*), normalized to the average expected SSM flux. Also shown is the expected seasonal effect due to the eccentricity of the orbit of the earth without oscillations, but normalized to the data (*continuous line*)

other and with the rate for the ES reaction, which is, in the no-oscillation case, expected to be a superposition of about 83% CC + 17% NC, one can therefore get an additional handle for detection of neutrino oscillations. In addition, owing to the low threshold of the nuclear CC reaction (4.15), the corresponding electrons, again detected via their Cherenkov light, carry most of the energy of the detected neutrino. They therefore provide a better estimator of the initial neutrino energy. The resulting measured energy spectrum is shown in Fig. 4.9.

Detection of the NC reaction (4.16) requires the detection of the final-state neutron. To this end, SNO foresees three different methods, exploited in different phases of the experiment. The first phase, using heavy water only, has recently been completed. Neutrons were measured through the 6.25 MeV photons produced in the capture of neutrons by deuterium (25% efficiency). In the current phase, 2.5 tons of NaCl have been added to the water. This allows neutron detection through the 8.6 MeV photons produced in the capture of neutron by Cl (85% efficiency). Finally, in SNO's third phase, the salt will be removed and ^{3}He proportional counters will be installed, allowing direct detection of the neutrons (45% efficiency) with different systematics.

The sample collected during the first phase, when the sensitivity to NC interactions was lower, is essentially dominated by CC interactions, but is nevertheless sensitive to the NC contribution. The different contributions from CC, ES, and NC events were extracted by means of a maximum-likelihood function, which combined the information from three variables: the effective

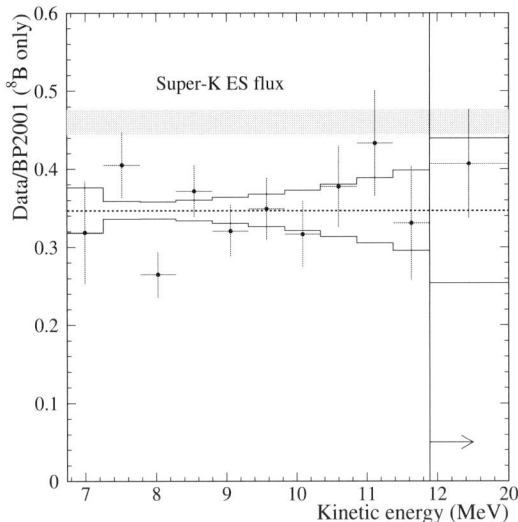

Fig. 4.9. SNO distribution [55] of the measured electron energy spectrum, normalized to the SSM prediction. The *bands* show the expected shape uncertainty (*unshaded band*) and the measured Super-Kamiokande ES flux (*shaded*)

kinetic energy of the event, the angle between the reconstructed direction of the event and the instantaneous sun-to-earth direction and a volume-weighted radial variable related to the position of the event inside the detector.

In a first analysis, using events with kinetic energy above 6.75 MeV only (i.e. above the NC threshold), the effective ^{8}B neutrino fluxes measured from CC and ES interactions in SNO were determined and compared with the Super-Kamiokande ES result [55]. Evidence for flavor change (i.e. a nonzero $\nu_\mu + \nu_\tau$ component) was obtained at the 3σ level from this analysis.

In a second analysis, the energy threshold was lowered to 5 MeV, so that the measurements were directly sensitive to the NC contribution [17]. The cross sections measured with 306.4 days of data-taking are

$$\Phi_{\mathrm{SNO}}^{\mathrm{CC}}(\nu_e) = (1.76^{+0.06}_{-0.05\mathrm{stat}}\ {}^{+0.09}_{-0.09\mathrm{sys}}) \times 10^6\,\mathrm{cm}^{-2}\,\mathrm{s}^{-1}\,, \qquad (4.17)$$

$$\Phi_{\mathrm{SNO}}^{\mathrm{ES}}(\nu_e + \sim 0.2(\nu_\mu + \nu_\tau)) = (2.39^{+0.24}_{-0.23\mathrm{stat}}\ {}^{+0.12}_{-0.12\mathrm{sys}}) \times 10^6\,\mathrm{cm}^{-2}\,\mathrm{s}^{-1}\,, \quad (4.18)$$

$$\Phi_{\mathrm{SNO}}^{\mathrm{NC}}(\nu_e + \nu_\mu + \nu_\tau) = (5.09^{+0.44}_{-0.43\mathrm{stat}}\ {}^{+0.46}_{-0.43\mathrm{sys}}) \times 10^6\,\mathrm{cm}^{-2}\,\mathrm{s}^{-1} \qquad (4.19)$$

if the standard ^{8}B shape of the neutrino spectrum is used. These cross sections, when combined yield a non-ν_e component of

$$\Phi_{\mathrm{SNO}}^{\mu\tau} = (3.41^{+0.45}_{-0.45\mathrm{stat}}\ {}^{+0.48}_{-0.45\mathrm{sys}}) \times 10^6\,\mathrm{cm}^{-2}\,\mathrm{s}^{-1}\,, \qquad (4.20)$$

5.3σ from 0, which provides strong evidence for flavor transformation in solar ν_e's (Fig. 4.10). The total flux measured with the NC reaction is consistent

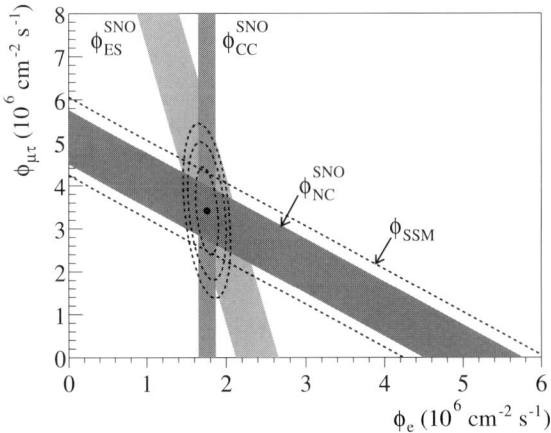

Fig. 4.10. Flux of ^{8}B solar neutrinos which are of μ or τ flavor vs. flux of those of e flavor, as measured by SNO [17]. The diagonal bands show the total ^{8}B flux as predicted by the SSM [42] (*dashed lines*) and that measured with the NC reaction by SNO (*shaded bands*). The intercepts of the bands with the axes represent the $\pm 1\sigma$ errors. The combined fit result (error ellipses) is consistent with neutrino flavor transformation without distortion of the ^{8}B spectrum

with solar models. An even more significant result is expected from the on-going and future direct measurements of the NC rate from phases 2 and 3 of the SNO experiment.

4.3.3 Discussion of Results

All solar neutrino experiments see a neutrino rate which is significantly suppressed with respect to expectations based on the solar model (Fig 4.11). Even without the SNO measurement, attempts to explain this suppression by deficiencies of the solar model used in the calculations can be easily refuted [56] in a model-independent way. On one hand, the (Super-)Kamiokande experiments actually *measure* the effective ^{8}B neutrino flux to be about half the expected value. This fully saturates the measured chlorine rate. On the other hand, the pp contribution to the flux measured by the gallium experiments is so well constrained by the total solar luminosity and by helioseismology measurements that its error is almost negligible compared with the error from the other contributions. This, together with half the ^{8}B flux (as measured), saturates the observed gallium rate. Therefore, it is easily deduced from Fig. 4.11 that the only remaining way to explain the measured gallium and chlorine rates is to assume that the ^{7}Be neutrino rate is essentially completely suppressed. This, however, is impossible, since the ^{8}B neutrinos originate from the same process cycle as the ^{7}Be ones, and would have to be completely suppressed too, contradicting the (Super-)Kamiokande measurement. The dis-

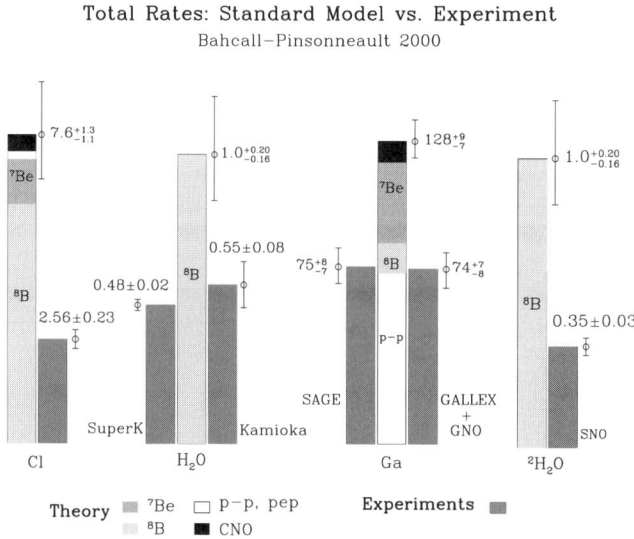

Fig. 4.11. Measured vs. predicted solar neutrino rates [43] for all of the solar neutrino experiments. The predicted rates for the various components of the neutrino spectrum are taken from the BP2000 model [42]

crepancy between the SNO CC and Super-Kamiokande ES rates discussed above further strengthens the contradiction. Similar arguments hold even if one or two of the experimental measurements are completely discarded. Therefore, it is considered established that the observed rate suppressions must be due to a new physics effect beyond standard electroweak interactions, leading to the disappearance of ν_e.

The fact that the suppression factors are *different* for the different kinds of experiments indicates an energy and detection dependence of the neutrino disappearance mechanism. Such a dependence could be generated by neutrino oscillations into other active neutrino flavors (ν_μ, ν_τ, or a mixture thereof, as suggested by SNO), either through the matter effects discussed in Sect. 4.2.2, occurring inside the sun and/or the earth, or through vacuum oscillations between the sun and the earth. The energy suppression patterns for various solutions are shown in Fig. 4.12. All these solutions can explain the observed rate suppression pattern. A detailed discussion of all their features can be found in [38] and Chap. 5. Here, we present a simplified discussion of some of their properties, by means of which they can be distinguished in principle. The further constraints imposed by the recent reactor neutrino results from KamLAND [18] are not included here, and will be discussed in Sect. 4.5.

The small-angle MSW solution (SMA, $\Delta m^2 \sim 10^{-5}\,\text{eV}^2$, $\sin^2 2\theta \sim 10^{-2}$) actually predicts total suppression of the ^{7}Be neutrinos, and was many people's favorite for a long time. However, it also predicts a significant distortion of the effective ^{8}B flux spectrum, and possibly a significant day/night

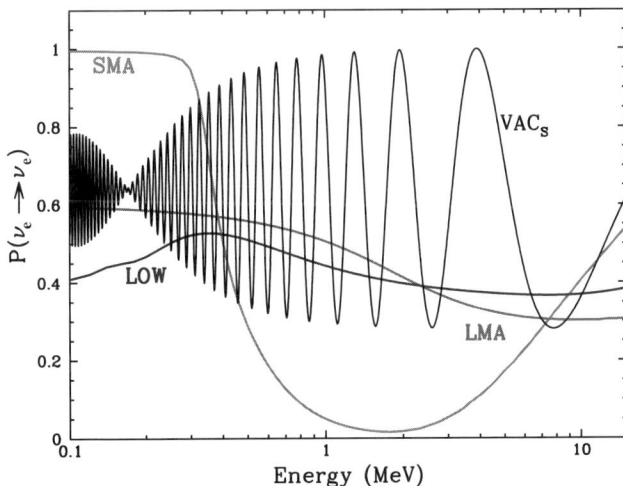

Fig. 4.12. Electron neutrino flux suppression as a function of energy for various neutrino oscillation solutions [57]. The labels SMA, LMA, LOW, and VAC are explained in the text

difference due to matter effects in the earth, neither of which is observed (Fig. 4.7). This solution is therefore disfavored.

The several variants of the vacuum solution (VAC, $\Delta m^2 \sim 10^{-10}\,\mathrm{eV}^2$, $\sin^2 2\theta \sim 0.5-1$) typically predict again a distortion of the $^8\mathrm{B}$ spectrum, and/or a deviation from the standard seasonal dependence due to variations of the earth/sun distance (Fig. 4.8). Initial indications of such deviations [58] have almost vanished, and do not favor this solution anymore.

The large-angle MSW solution (LMA, Δm^2 of the order of a few times $10^{-5}\,\mathrm{eV}^2$, $\sin^2 2\theta \sim 0.8$) predicts a flat, strong rate suppression at high energies, and a somewhat weaker suppression at lower energies. Some day/night effect is expected, but it may be small enough to be undetectable. This is currently the most favored solution.

Finally, the low-Δm^2 matter solution (LOW, $\Delta m^2 \sim 10^{-7}\,\mathrm{eV}^2$, $\sin^2 2\theta \sim$ 0.9) predicts an almost flat suppression inside the sun for all energies. An energy dependence is introduced through strong matter effects in the earth, leading to a large day/night dependence. However, this day/night dependence is concentrated at low energies, and can therefore be detected neither by the real-time experiments (because of the energy threshold) nor by the radiochemical experiments (because of the time integration). This solution was disfavored by the rate measurements for many years, but has recently become one of the prime candidates again.

Variants of the vacuum and SMA solutions are also possible for oscillations into sterile neutrinos. However, these would not alter the NC/CC ratio, since both the CC and NC rates would be suppressed. These are therefore strongly disfavored [59] but maybe not completely excluded [60] by the inclusion of the recent SNO results [61].

Figure 4.13 shows the surviving parameter space obtained from a recent analysis [62] combining all the information discussed above except the most recent SNO NC measurement. At the 3σ level, only the LMA and LOW solutions and parts of the vacuum solution for oscillations into active neutrinos survive if the overall *absolute* minimum (LMA solution) is taken as a reference. However, the *local* minima of the SMA and sterile-neutrino solutions (not shown) still feature a goodness of fit at the 5–10% level, and should not be completely discarded yet. The inclusion of the NC measurement increases the preference for the LMA solution [17, 61] and further disfavors the alternatives.

So far, we have discussed solutions based on neutrino oscillations, which are the solutions which require the smallest deviation from classical Standard Model physics. Alternative solutions based on flavor transitions initiated by neutrino magnetic moments (resonant spin–flavor precession), new flavor-changing interactions, or violation of the equivalence principle are beyond the scope of this summary, but can offer equally viable solutions [63]. The neutrino oscillation interpretation, although very plausible, can therefore not be considered to be fully experimentally established yet.

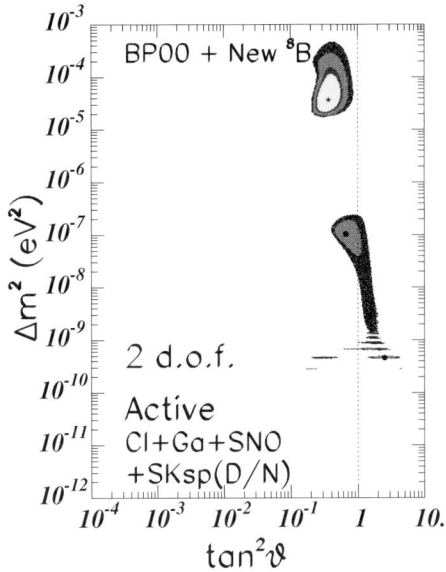

Fig. 4.13. Allowed regions in the Δm^2 vs. $\tan^2 \theta$ plane from a recent global solar-neutrino analysis [62] performed before the result from KamLAND [18]. The *shaded areas* denote the 68%, 90%, 95%, and 99% C.L. regions. The *points* denote local minima corresponding to the LMA, LOW, and VAC solutions discussed in the text

New experiments are constraining the possible solutions further. The Borexino experiment [64] is aimed at an explicit measurement of the flux of the ^{7}Be line in real time. It can check the various solar neutrino solutions via the absolute ^{7}Be rate, and through day/night and seasonal variations. Phases 2 and 3 of the SNO experiment [55] should settle even more firmly the question of whether the ν_e's disappear into active or sterile neutrinos. Most recently, the first results from the KamLAND experiment [18] have beautifully confirmed the LMA MSW solution using reactor neutrinos (see Sect. 4.5). If the tentative observation of an actual oscillation pattern is further established, it will finally rule out not only the other solar neutrino oscillation solutions, but also most nonoscillation scenarios.

4.4 Atmospheric Neutrino Experiments

We conclude that the atmospheric neutrino data, especially from Super-Kamiokande, give evidence for neutrino oscillations.

T. Kajita, for the Super-Kamiokande collaboration, June 1998

The sentence above, expressed at the Neutrino 1998 conference in Takayama [65], marked the turning point for the perception of neutrino oscillations as almost an established mainstream physics effect. Extensive reviews of atmospheric neutrino physics can be found in [66]. Recent updates are given in [67, 68].

Atmospheric neutrinos are generated through interactions of primary cosmic rays (mainly protons and iron nuclei) in the upper atmosphere. The processes that yield neutrinos (π and K decay) are the same as the ones used to produce high-energy neutrino beams. Atmospheric neutrinos are therefore a natural variant of these beams. The combination of an isotropic primary cosmic-ray flux with the spherical symmetry of the target (the atmosphere) ensures that the total neutrino flux must be up/down symmetric with respect to the horizon [69]. Small deviations from this symmetry due to the earth's magnetic field are relevant only at low energies, and corresponding measurements agree with calculations [70]. A detector placed near the earth's surface is therefore simultaneously a medium-baseline (~ 20 km, from above) and very long-baseline (up to 13 000 km, from below) experiment (Fig. 4.14).

Experiments to detect atmospheric neutrinos are placed underground to shield them from the cosmic-ray muon background. The neutrino flux is roughly proportional to E^{-2} (Fig. 4.1). At sub-GeV energies, the flavor composition in the absence of oscillations can be approximated by $\nu_\mu : \bar\nu_\mu : \nu_e : \bar\nu_e = 2 : 2 : 1 : 1$. At higher energies, the $\nu_\mu : \nu_e$ ratio increases gradually with E. Neutrinos from below traverse several different regions of roughly uniform matter densities. These include the earth's mantle (density $\simeq 4.5$ g/cm^3) and the earth's core (density $\simeq 11.5$ g/cm^3). Neutrinos from

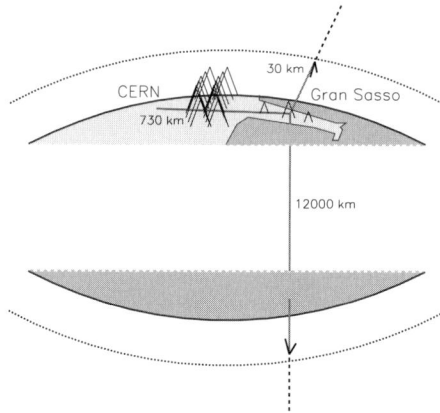

Fig. 4.14. Sketch of the earth showing the principle of an atmospheric-neutrino measurement. An atmospheric-neutrino detector near the earth's surface (at the Gran Sasso laboratory in Italy in this figure) simultaneously measures neutrinos from a large range of distances. Examples of a downgoing (~ 30 km) and an upgoing ($\sim 12\,000$ km) neutrino, emerging from atmospheric shower cascades, are shown. A typical long-baseline neutrino beam (the CNGS beam from CERN) pointing to the same location is also shown for comparison

above are unaffected by potential matter effects. By using both CC and NC events which occur inside the detector, and muons from CC events occurring below the detector, a neutrino energy range from about 200 MeV to 1 TeV can be explored. Events occuring above the detector cannot be used, owing to the large background from cosmic-ray muons.

4.4.1 Main Experimental Results

The first indication of an anomaly in the observation of atmospheric neutrinos came from the detection of the ν_μ/ν_e flavor ratio [14, 15, 71, 76]. Measurements of the ratio of ratios $R' = (\mu/e)_{\text{data}}/(\mu/e)_{\text{MonteCarlo}}$, in which most systematic errors cancel, are shown in Fig. 4.15.

Recently, there were three underground experiments which were accumulating atmospheric neutrino data (and in some cases still are). The largest one, Super-Kamiokande [15], has also been used for the detection of solar neutrinos (see Sect. 4.3 and Fig. 4.5). It is a 50 kt water Cherenkov detector deployed in a configuration with two concentric cylindrical volumes. The inner volume is the 22.5 kt fiducial region, while the surrounding outer volume is used to veto entering tracks and to tag exiting tracks. The flavor tagging of events is based upon the relative sharpness or diffuseness of Cherenkov rings, with muon tracks yielding sharp rings and electrons yielding diffuse ones [65].

MACRO [72] and Soudan 2 [73] are tracking calorimeter detectors. MACRO is a large-area, planar tracker located in the Gran Sasso laboratory in Italy. It is optimized for tracking in the vertical direction and is sufficiently massive (about 5.3 kt) to be effective as a neutrino detector. Charged-particle tracking is carried out using horizontal layers of streamer tubes with wire and stereo-strip readout. Three horizontal planes and also vertical walls of liquid scintillator counters provide timing information with a resolution of about 0.5 ns.

Soudan 2 is a fine-grained iron tracking calorimeter of total mass 963 tons in the Soudan mine in the USA. Its tracking elements are plastic drift tubes,

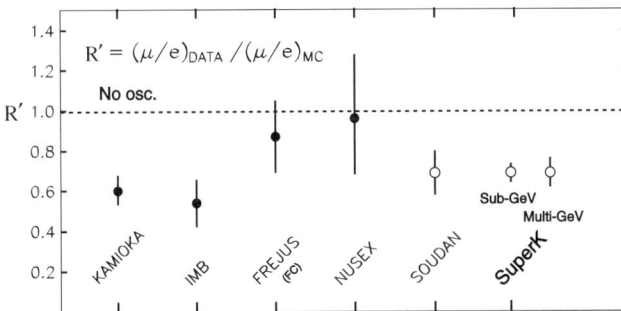

Fig. 4.15. Measurements of the atmospheric-neutrino ratio of flavor ratios [67]

sandwiched with steel sheets. The calorimeter is surrounded on all sides by an active shield of two or three layers of proportional tubes.

The various measurements of R' shown in Fig. 4.15, obtained by different experiments with different techniques and systematics, give a consistent picture, and are statistically dominated by the Super-Kamiokande result. They are consistent with an interpretation in terms of neutrino flavor oscillations, which either deplete the ν_μ flux or enhance the ν_e flux, or do both.

To decide between these possibilities and to support the neutrino oscillation hypothesis, more detailed measurements are needed. One such measurement is that of the dependence of the neutrino flux on the zenith angle θ_z, as illustrated in Fig. 4.16a. The curves depict the ν_μ survival probability for a "representative" set of oscillation parameters of $\sin^2 2\theta = 1.0$ and $\Delta m^2 = 5 \times 10^{-3}$ eV^2. These curves reflect (4.9), where L has been converted to the equivalent zenith angle $\cos\theta_z$ via the relation

$$L = \sqrt{(R_e \cos\theta_z)^2 + (2R_e + h_a)h_a} - R_e \cos\theta_z \ . \tag{4.21}$$

Here, $R_e = 6370$ km is the radius of the earth, and $h_a \sim 20$ km is the effective height at which the neutrino is produced. The case $\cos\theta_z = 1.0$ corresponds to vertically downgoing neutrinos, and $\cos\theta_z = -1.0$ to vertically upgoing neutrinos. For a value of E_ν of 250 MeV, the first oscillation swing severely depletes the downward-going flux, and rapid oscillations deplete the flux incident from below the horizon; the net result is a substantial average depletion at all incident angles. At energies above 1 GeV, however, the depletion

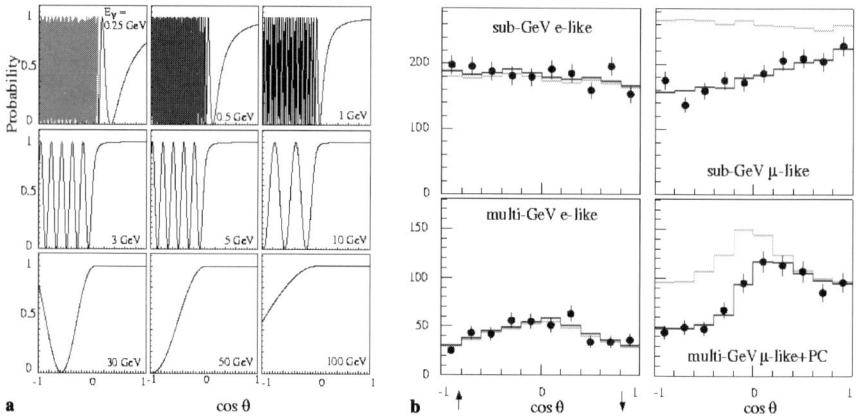

Fig. 4.16. (**a**) Survival probability curves for monoenergetic, isotropic fluxes of muon neutrinos with $\nu_\mu \to \nu_\tau$ oscillations and $\sin^2 2\theta = 1.0$, $\Delta m^2 = 5 \times 10^{-3}$ eV^2, for nine values of E_ν [74]. (**b**) Distributions of single-ring e-like and μ-like events as a function of $\cos\theta_z$ obtained by Super-Kamiokande. the expectations for no oscillations and for $\nu_\mu \to \nu_\tau$ oscillations are shown by *gray-line* and *solid-line histograms*, respectively [67]

moves almost entirely to the ν_μ flux incident from below the horizon. For $E_\nu \sim 100\,\text{GeV}$ or more, the pattern shifts so that the oscillation is beyond the observable range, and ν_μ depletion ceases because the diameter of the earth is not large enough to accommodate the first oscillation swing. In practice, such patterns will be smeared out and averaged by the finite detector resolution.

Figure 4.16b shows the same distribution as measured by the Super-Kamiokande collaboration, separately for ν_e and ν_μ events, and integrated over two different energy ranges: sub-GeV ($E_\nu < 1.4\,\text{GeV}$) and multi-GeV ($E_\nu > 1.4\,\text{GeV}$). Only events with a primary vertex inside the detector fiducial volume and a single Cherenkov ring have been used for this measurement. The neutrino direction is inferred from the direction of the detected charged lepton. Owing to the angular smearing in the neutrino scattering process, the angular resolution is good for high-energy neutrinos, but rather poor at low energies. The observed ν_e flux agrees well with the predicted flux at all angles and energies. However, the ν_μ flux in the multi-GeV region appears severely depleted for upgoing events ($\cos\theta < 0$), while it is consistent with expectations for the downgoing events. The pattern for sub-GeV events is less pronounced, in accordance with the reduced angular resolution.

Since the up and down neutrino fluxes are expected to be essentially symmetric on the basis of simple geometrical arguments, the suppression of upward-going muon-like events must be attributed to a new physics effect. This pattern is very suggestive of oscillations of ν_μ into some other neutrino flavor which is not predominantly ν_e [75], since otherwise the ν_e spectra would also be distorted. Indeed, the hypothesis of $\nu_\mu \rightarrow \nu_\tau$ oscillations, also indicated in Fig. 4.16b, yields a very good fit to the data, while the no-oscillation hypothesis is disfavored by more than 10 standard deviations ($\Delta\chi^2 > 100$ [67]). It was this measurement (although with somewhat poorer statistics) which led the Super-Kamiokande collaboration to claim "evidence for ν_μ oscillations" in 1998 [15]. A hint of this effect had been detected earlier by the Kamiokande collaboration [14]. The effect has been qualitatively confirmed by the Soudan 2 collaboration [67, 76] with a much smaller detector mass, but different systematics.

Note the subtlety that the final state that the ν_μ should oscillate into is not mentioned in the original Super-Kamiokande claim: oscillations into ν_τ or into a sterile neutrino ν_s cannot be distinguished at this level [77], since most ν_τ are below the τ production threshold and will therefore behave like sterile neutrinos concerning their CC interactions. Some other clues as to how to distinguish the two possibilities will be discussed below.

Similar measurements, although with less statistical significance and larger systematic errors, can also be made at super-Kamiokande with upward-going muons from events which occurred below the detector [78]. Such measurements have also been performed by Kamiokande, IMB, and Baksan [79], and by the MACRO collaboration [80].

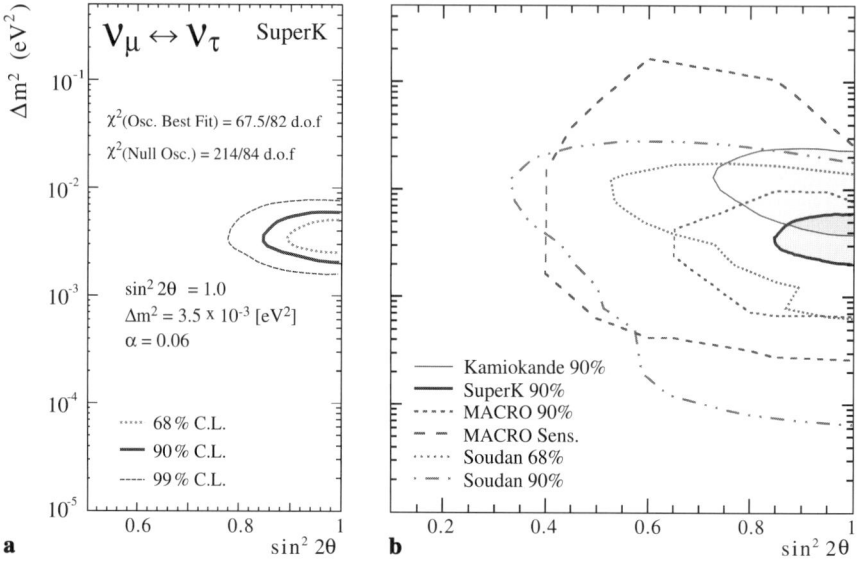

Fig. 4.17. (a) Allowed regions obtained by Super-Kamiokande on the basis of χ^2 fitting to fully contained and partially contained single-ring events, plus upward stopping muons and upward through-going muons. (b) Allowed regions of the oscillation parameters obtained from Kamiokande, Super-Kamiokande, MACRO, and Soudan 2 [67]

Figure 4.17 shows the resulting confidence-level contours for the oscillation parameters, assuming ν_μ–ν_τ oscillations, when all measurements are combined for each experiment. Clearly, all experiments are compatible with close to maximal oscillations ($\sin^2 2\theta = 1$) and Δm^2 of the order of a few times 10^{-3} eV^2.

The most recent Super-Kamiokande results yield a best-fit central value of $\Delta m^2 = 2.4 \times 10^{-3}$ eV^2 for maximal mixing [68].

4.4.2 Alternative Explanations

Unfortunately, none of the experiments which have yielded indications of neutrino oscillations have so far succeeded in measuring an actual sinusoidal oscillation pattern. Figure 4.18a shows the L/E distribution from Super-Kamiokande [15] compared with the expectation for neutrino oscillations and with a functional form suggested by a recent neutrino decay model [81]. Once the detector resolution is taken into account, the two hypotheses are essentially indistinguishable [81]. Other exotic processes whose effect could mimic an average oscillation signal (e.g. large extra dimensions, flavor changing neutral currents, and decoherence) are reviewed in [82]. Therefore, a more precise measurement of the oscillation pattern is necessary to actually prove the oscillation hypothesis for atmospheric neutrinos.

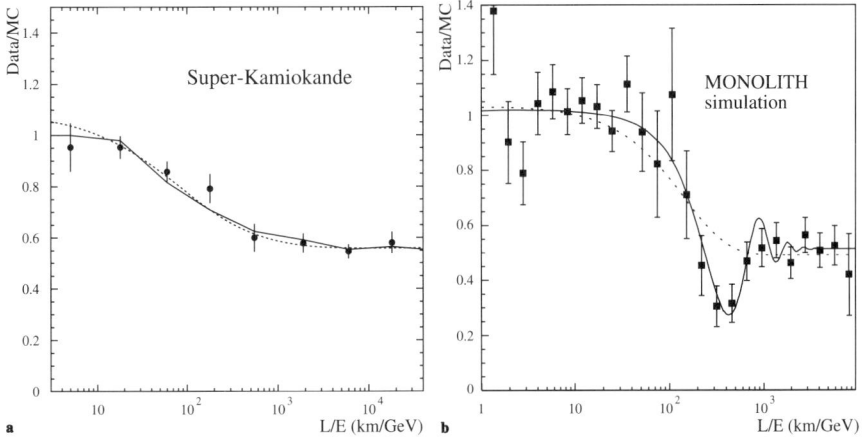

Fig. 4.18. (a) L/E distribution from Super-Kamiokande [15] compared with the best-fit oscillation hypothesis (*continuous line*) and with a parameterization corresponding to the neutrino decay model of [81] (*dashed line*). The distributions are smoothed out by the detector resolution. (b) L/E distribution to be expected from MONOLITH [83] for $\Delta m^2 = 3 \times 10^{-3}$ eV^2, compared with the best-fit oscillation hypothesis (*oscillating line*) and with the corresponding best fit of the neutrino decay model of [81] (*dotted line*, showing a smooth threshold effect)

The proposed MONOLITH experiment [83] was explicitly designed to address this question. Aiming for a similar mass to that of Super-Kamiokande, significantly larger acceptance at high neutrino energies, and better L/E resolution, the experiment is optimized to observe the whole of the first oscillation minimum, i.e. half an oscillation period, in ν_μ disappearance. Therefore, the oscillation hypothesis can be clearly distinguished from other hypotheses which yield a pure threshold behavior (Fig. 4.18b). Furthermore, the sensitivity range of 2×10^{-4} eV$^2 < \Delta m^2 < 0.1$ eV^2 [83] comfortably covers the full range of the oscillation parameters in Fig. 4.17. This is in contrast to long-baseline accelerator experiments (Sect. 4.6.3), which can do similar measurements at the highest allowed Δm^2 if a low-energy beam is used [84], but can observe only a quarter of an oscillation period or less in the lower Δm^2 range. Currently, it is not clear whether a MONOLITH-like experiment will be built in the near future.

Assuming that the oscillation hypothesis will be confirmed, we now come back to the question of sterile-neutrino final states. ν_μ–ν_s oscillations can be distinguished from ν_μ–ν_τ oscillations in atmospheric-neutrino experiments through

– an asymmetry in the up/down rate for NC [85] or inclusive [86] events;
– a study of the NC/CC ratio [87];
– a distortion of the energy dependence of the oscillations due to matter effects, which are present for ν_s but not for ν_τ [88];

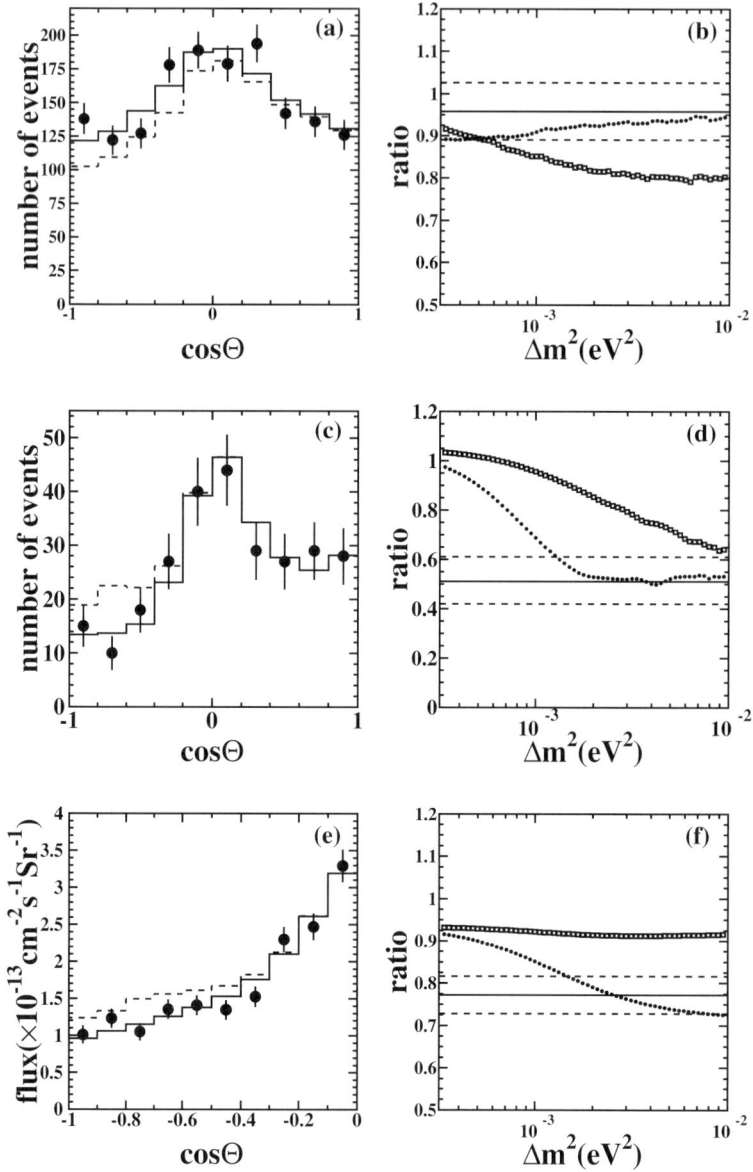

Fig. 4.19. Super-Kamiokande zenith angle distributions [89] for an NC-enriched sample (**a**), a high-energy partially contained sample (**c**), and the upward-going muon sample (**e**). The data (*points*) and the expectations for ν_μ–ν_τ oscillations (*continuous line*) and for ν_μ–ν_s oscillations (*dashed line*, including matter effects) are shown. (**b**), (**d**), and (**f**) depict the corresponding expected up/down ratios (*solid symbols*, ν_μ–ν_τ oscillations; *open symbols*, ν_μ–ν_s oscillations) as a function of Δm^2, compared with the measured value (error band indicated by *dashed lines*)

– a difference between the oscillation parameters for ν_μ and $\bar{\nu}_\mu$, again only possible for ν_s, owing to matter effects [88];
– the lack of an excess of NC-like events from below at high energies, owing to ν_τ CC events followed by hadronic τ decay [83].

The first three items have already been studied by Super-Kamiokande [89] (Fig. 4.19). None of them yields a significant result by itself over the full allowed Δm^2 range. When the three methods are combined, the ν_μ–ν_s hypothesis is disfavored at the 99% confidence level. This result is confirmed by the MACRO analysis of upward-going muons [80]. Therefore, the ν_μ–ν_τ hypothesis seems to be reasonably well established. However, this statement is only valid for the pure two-flavor scenario, and does not apply to possible more complicated mixtures. More experimental evidence is therefore necessary to settle the question finally. This is one of the primary goals of the new long-baseline accelerator beam experiments currently in preparation (Sect. 4.6.3 and Chap. 7). The difference between ν_μ and $\bar{\nu}_\mu$ oscillations caused by matter effects can only be addressed by new atmospheric neutrino experiments or by very long-baseline beams, with MONOLITH-type magnetized detectors [90].

4.5 Reactor Experiments

An experiment has been performed to detect the free neutrino.
F. Reines and C.L. Cowan, 1953

Nuclear fission reactions at reactors produce neutron-rich nuclei which decay through the classical β decay process

$$n \rightarrow p e^- \bar{\nu}_e \tag{4.22}$$

and yield an intense isotropic flux of electron antineutrinos (Fig. 4.1). The inverse CC reaction, proposed by Pontecorvo in 1946 [91] and pioneered experimentally by Reines and Cowan [92],

$$\bar{\nu}_e + p \rightarrow e^+ + n + 1.8\,\mathrm{MeV} \ , \tag{4.23}$$

led to the first direct detection of a neutrino [2] and continues to be the primary mode for the detection of reactor neutrinos.

As with solar neutrinos, oscillation searches are only possible in the disappearance mode in this energy range. No signal has been observed so far. In classical reactor experiments [93–95], the detector is positioned typically at 10–90 m from the reactor core to profit from the highest possible neutrino flux. Both the absolute neutrino flux and its spectral shape have been found to be in good agreement with expectations, yielding limits on ν_e disappearance in the range $\Delta m^2 > 10^{-2}\,\mathrm{eV}^2$.

In order to cover the parameter region of atmospheric neutrino oscillations, "long-baseline" reactor experiments have recently been performed using detectors positioned at about 1 km from the reactor source. Both the CHOOZ [96] and the Palo Verde [97] experiments are based on the reaction (4.23), followed by a correlated (e$^+$, n) signature in a Gd-loaded liquid scintillator. Gd loading reduces the neutron capture time, and gives rise to a high-energy gamma cascade with an energy of up to 8 MeV. At a distance of about 1 km from the reactor, the detector response is about 5 events per day per ton of scintillator. The CHOOZ experiment, shielded from the cosmic-ray background by its location in an existing deep tunnel near the CHOOZ reactor in France, consists of a homogeneous central target of 5 tons of Gd-loaded scintillator surrounded by detection and veto volumes. The Palo Verde experiment, being in a shallow underground laboratory near the Palo Verde reactor in the USA, is made from finely segmented detector cells, as required for more powerful background rejection.

Both experiments measure electron spectra whose shape and normalization agree with nonoscillation expectations. Figure 4.20 shows the limits obtained by the CHOOZ [96] and Palo Verde [97] collaborations from the absence of a signal. The absence of a strong ν_e–ν_μ oscillation effect further constrains the parameter space allowed for atmospheric ν_μ–ν_e oscillations. Figure 4.21 illustrates this for the case of three-flavor oscillations with one relevant Δm^2 scale. The allowed residual ν_e admixture in the atmospheric neutrino oscillations is less than 10% unless Δm^2 is below 1×10^3 eV^2.

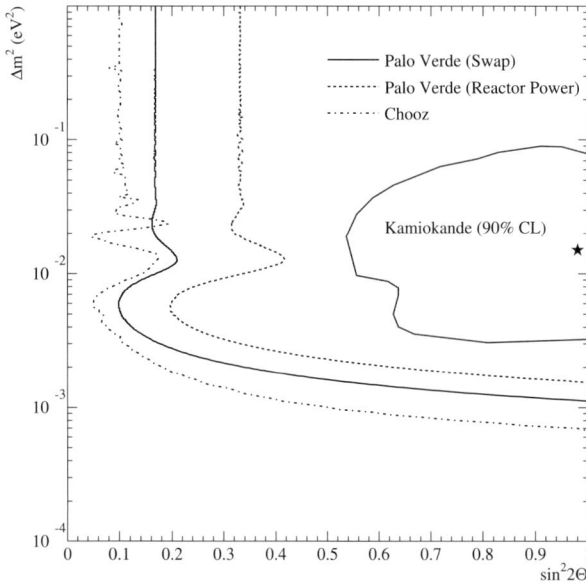

Fig. 4.20. Limit on $\bar{\nu}_e$–$\bar{\nu}_X$ oscillations from the Palo Verde [97] and CHOOZ [96] experiments. The preferred region for the old ν_μ–ν_e interpretation of the atmosheric neutrino anomaly obtained from the Kamiokande experiment is also shown

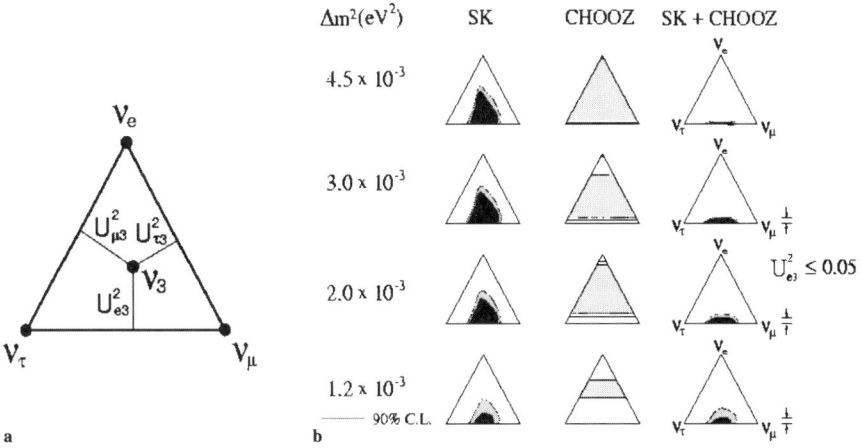

Fig. 4.21. (a) Triangle graph for displaying the possible flavor composition of the mass eigenstate ν_3 in terms of the matrix elements in (4.5). (b) Regions of allowed (U_{e3}^2, $U_{\mu3}^2$, $U_{\tau3}^2$) versus Δm^2, obtained by fitting the Super-Kamiokande and CHOOZ data via three-flavor mixing in the approximation of dominance by one mass scale [98]. First column, allowed by Super-Kamiokande; second column, excluded by CHOOZ; third column, allowed by Super-Kamiokande + CHOOZ (90% and 95% C.L.)

Finally, very long-baseline reactor experiments can test the large-angle MSW solution to the solar neutrino problem. The KamLAND experiment [18, 99] has started to expose 1 kt of liquid scintillator to antineutrinos from several reactors at distances of 140 km or larger, which yields about 450 events per year. The L/E reach is therefore extended by two orders of magnitude with respect to CHOOZ, yielding full sensitivity for $\Delta m^2 > 10^{-5}$ eV^2.

KamLAND occupies the site of the earlier Kamiokande [14] experiment in the Kamioka mine in Japan, which yields a very low cosmic-ray background rate. Here, 1 kt of ultrapure liquid scintillator is contained in a 13 m diameter spherical balloon made of a thin plastic film. A buffer of dodecane and isoparaffin oils between the balloon and an 18 m diameter spherical stainless-steel containment vessel shields the scintillator from external radiation. An array of 1879 photomultiplier tubes, mounted on the steel vessel, completes the inner detector. The containment vessel is surrounded by a 3.3 kt water Cherenkov detector with 225 additional phototubes. This outer detector absorbs gamma rays and neutrons from the surrounding rock and acts as a tag for cosmic-ray muons. Reactor neutrino events are selected by applying fiducial cuts (based on the photomultiplier timing) and energy cuts, and by requiring a space and time correlation between the prompt positron signal and the delayed photon signal from neutron capture.

Figure 4.22 shows the measured prompt energy distribution of the positrons for reactor neutrino events from the first 145 days of running, compared with expectations with and without neutrino oscillations. The absolute event rate,

$$\frac{N_{\mathrm{obs}} - N_{\mathrm{bg}}}{N_{\mathrm{expected}}} = 0.611 \pm 0.085(\mathrm{stat}) \pm 0.041(\mathrm{sys}) \,,$$

indicates a deficit of neutrino events consistent with $\bar{\nu}_e$ oscillations, as illustrated in Fig. 4.23.

Figure 4.24 shows the range of oscillation parameters allowed by KamLAND, obtained both from the rate suppression and from an analysis of the shape of the energy distribution. This allowed range is in very good agreement with the LMA solution for solar neutrino oscillations, while all other solutions (see Sect. 4.3) are excluded. This beautiful confirmation of the solar neutrino oscillation hypothesis by a completely different technique based on artificial reactor neutrinos is really a triumph for both neutrino physics and solar astrophysics.

In addition, this result yields significant constraints on previously allowed nonoscillation interpretations [100], and on the radioactive properties of the earth from the contribution of "geo-neutrinos" [101] tentatively indicated in Fig. 4.22, which originate mainly from uranium and thorium decays in the crust and mantle of the earth.

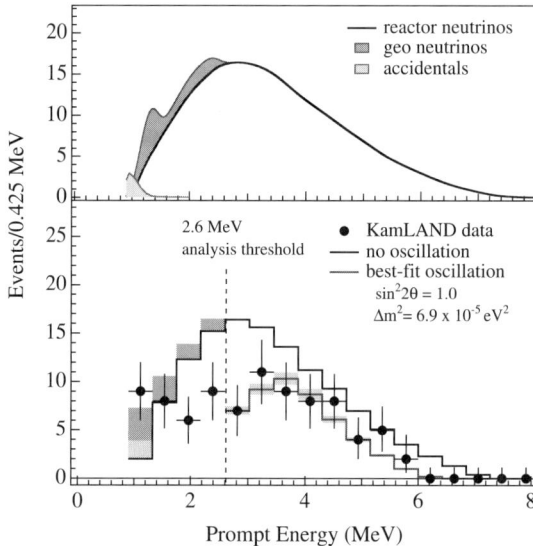

Fig. 4.22. KamLAND positron energy spectrum [18]. *Upper panel*: expected energy spectrum of reactor $\bar{\nu}_e$'s, along with the spectrum of geological neutrinos and the background. *Lower panel*: measured energy spectrum of the observed prompt events, compared with the expectations with and without neutrino oscillations. The *dashed vertical line* indicates the analysis cut used to remove the geological-neutrino and background-neutrino contributions

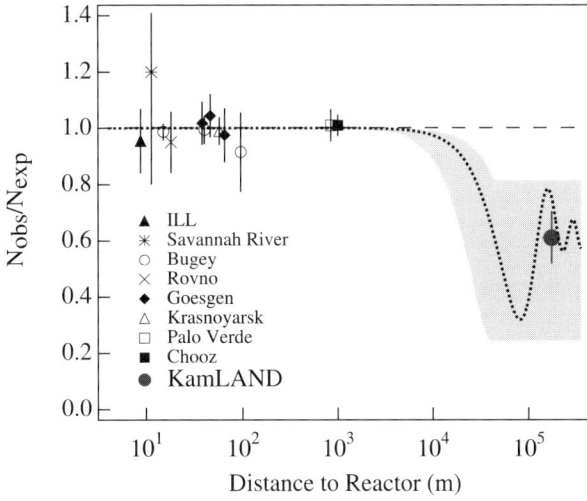

Fig. 4.23. Ratio of measured to expected flux for various reactor neutrino experiments as a function of the average distance between source and detector [18]. The *solid circle* is the KamLAND result. The *shaded region* indicates the range of flux predictions corresponding to the solar LMA neutrino oscillation region (Fig. 4.24). The *dotted curve* corresponds to the best-fit solution of $\sin^2 2\theta = 0.833$ and $\Delta m^2 = 10^5 \, \text{eV}^2$. The *dashed curve* is for no oscillations

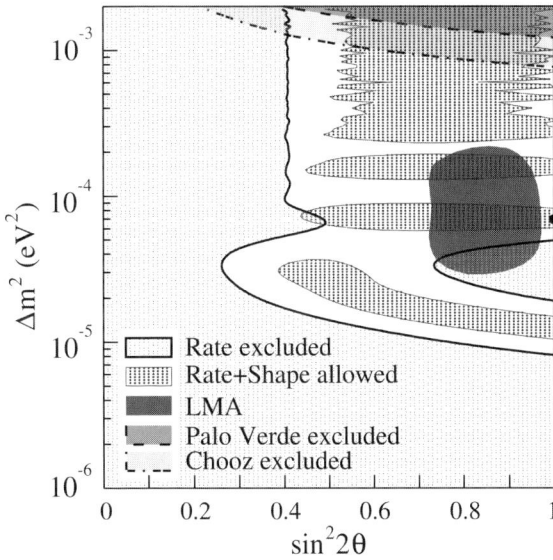

Fig. 4.24. Excluded and allowed regions obtained from KamLAND for rate only and for rate + spectral shape, compared with the solar LMA solution and exclusion regions from other experiments [18]

With somewhat poorer statistics, similar results should soon also be obtained by the Borexino experiment [64] in Gran Sasso, which will measure neutrinos from reactors in France, Switzerland, and the Balkans.

4.6 Accelerator Experiments

If ... the neutrino mass is different from zero, oscillations similar to those in K^0 beams become possible in neutrino beams.

B. Pontecorvo, 1967, referring to his article of 1958 [11]

In the spirit of this idea of Pontecorvo [12] and similar ideas of Maki, Nakagawa, and Sakata [33], essentially all accelerator neutrino experiments have also been used to search for neutrino oscillations [21, 102]. Section 4.6.1 describes the initially very controversial result from LSND and its final status. Negative results from a very similar experiment, KARMEN, are also discussed in this section. On the basis of the idea that at least one of the neutrinos should have a mass of order 10 eV to explain the missing dark matter in the universe [103], two experiments at CERN, CHORUS and NOMAD, were explicitly dedicated to the search for ν_μ–ν_τ oscillations. These experiments started data-taking in 1994/95 and have now been completed. Their most recent results are presented in Sect. 4.6.2.

4.6.1 LSND and KARMEN

LSND (Liquid Scintillator Neutrino Detector) [104] at the Los Alamos Neutron Science Center, formerly the Los Alamos Meson Physics Facility (LAMPF/LANSCE), USA, and KARMEN (KArlsruhe Rutherford Medium Energy Neutrino detector) [105] at the ISIS neutron spallation source at the Rutherford Appleton Laboratory, UK, are experiments optimized for the detection of neutrinos in the energy range 10–100 MeV. A dedicated comparison of the experiments and their results can be found in [106].

In both the LAMPF/LANSCE and the ISIS facilities, neutrinos are produced by stopping 800 MeV protons in a beam stop target. In addition to spallation neutrons, charged pions are produced. The π^- are absorbed by the target nuclei, whereas most of the π^+ are stopped and decay at rest. Muon neutrinos therefore emerge from the decay $\pi^+ \to \mu^+\nu_\mu$. The produced μ^+ are also stopped within the target and decay via $\mu^+ \to e^+\nu_e\overline{\nu}_\mu$. This decay-at-rest (DAR) chain therefore yields fluxes of ν_μ, $\overline{\nu}_\mu$, and ν_e of equal intensities, with an energy distribution up to 52 MeV, as shown in Fig. 4.25. A small fraction of the π^+ decay in flight (DIF) before being stopped, yielding a ν_μ energy spectrum which extends up to 300 MeV (Fig. 4.1). An even smaller fraction of π^- also decay in flight, yielding a small contamination of $\overline{\nu}_e/\nu_e < 6-8 \times 10^{-4}$ from the subsequent μ^- decay. While LANSCE delivers a high-intensity continuous proton beam, the lower-intensity beam at ISIS is

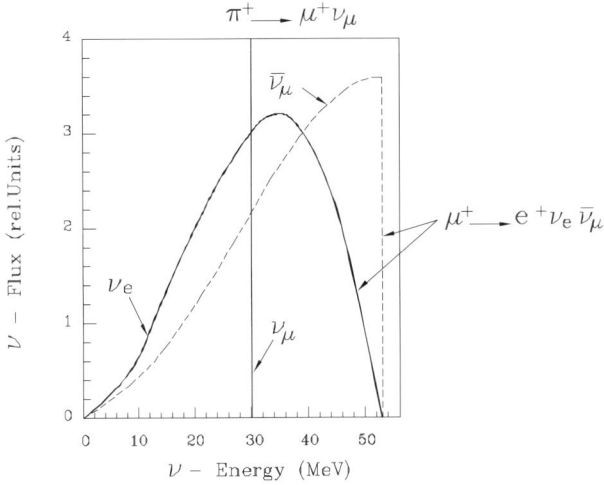

Fig. 4.25. Energy distribution of neutrino fluxes from the pion decay-at-rest chain [106]

pulsed with a repetition frequency of 50 Hz. This offers an additional handle on backgrounds via their different time structure, at the expense of event statistics.

Both experiments are looking for $\bar{\nu}_e$ from $\bar{\nu}_\mu$–$\bar{\nu}_e$ oscillations of DAR neutrinos via the reaction $\bar{\nu}_e + p \rightarrow e^+ + n$, which provides a spatially correlated delayed-coincidence signature of a prompt e^+ and a subsequent neutron capture signal. LSND has also studied ν_μ–ν_e oscillations of DIF neutrinos.

LSND uses a homogeneous detector volume of mineral oil containing a low concentration of scintillator, equipped with 1220 phototubes (Fig. 4.26). Excellent particle identification is obtained by detecting a directional Cherenkov cone and isotropic scintillation light, which lead to a characteristic pattern of photomultipliers that are hit.

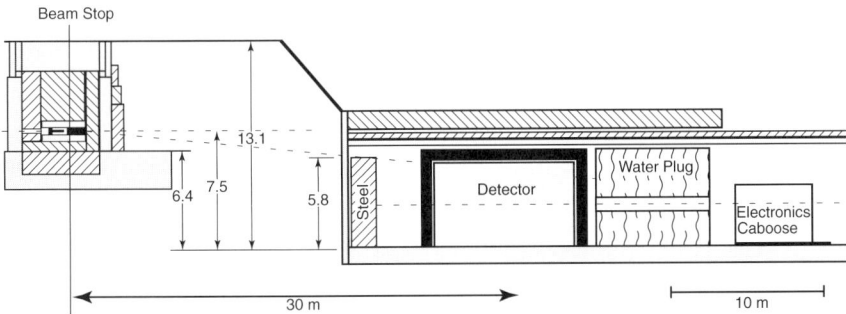

Fig. 4.26. The LSND experimental area [104]

In two data sets taken during the periods 1993–95 and 1996–98, LSND has observed a clear beam-on minus beam-off excess of DAR events with a $\bar{\nu}_e$ signature, i.e. (e$^+$, n) sequences [19]. The initial results from the LSND experiment, based on a fraction of the data set from the first period, were extremely controversial [107]. However, since then the statistics have improved, and no severe systematic problems have been identified. Also, the DAR $\bar{\nu}_\mu$–$\bar{\nu}_e$ signal has been confirmed by an independent (but less significant) $\nu_\mu \rightarrow \nu_e$ signal in the DIF sample [19].

In the decay-at-rest analysis, the oscillation candidate events are selected by constructing a likelihood ratio R_γ from the positron energy and from the time difference and spatial distance between the e$^+$ and n detections. The distribution of this likelihood ratio is shown in Fig. 4.27a. A total excess of about 88 events has been observed, consistent with neutrino oscillations with a combined $\bar{\nu}_\mu$–$\bar{\nu}_e$ oscillation probability of $0.264 \pm 0.067 \pm 0.045\%$. While the rates of the excess are in perfect agreement for the two data-taking periods, there are some differences in the shape of the positron energy distribution [53, 106], yielding somewhat different contours for the preferred oscillation parameters. Since these differences are not statistically significant, the combined "gold-plated" e$^+$ excess spectrum is shown in Fig. 4.27b. The corresponding favored areas for the oscillation parameters are shown in Fig. 4.28.

KARMEN [105] is a segmented liquid scintillation calorimeter, with excellent time and energy resolution obtained by exploiting the distinct time structure of the ISIS neutrino source. Thus, at KARMEN a $\bar{\nu}_e$ excess arising from $\bar{\nu}_\mu \rightarrow \bar{\nu}_e$ would be identified by requiring its time distribution to follow

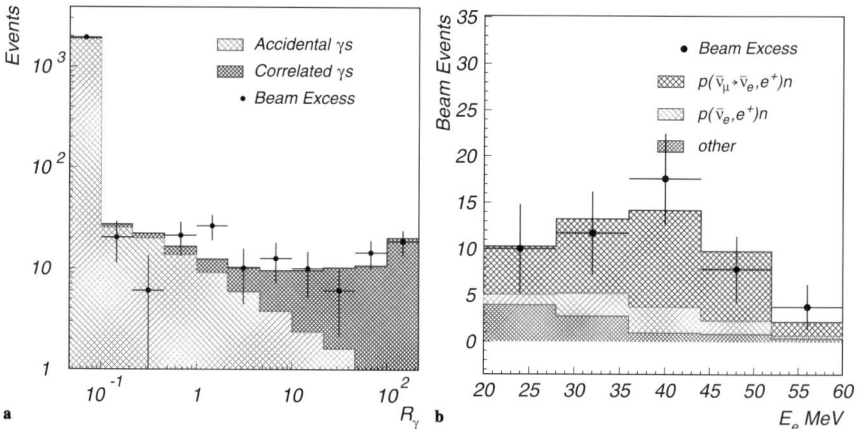

Fig. 4.27. (**a**) LSND likelihood ratio R_γ. The data (*points*) are compared with the best-fit superposition of signal (*dark hatched area*) and background (*light hatched area*) contributions. (**b**) Energy distribution for the LSND 1993–1998 beam excess, applying a cut-off of $R_\gamma > 10$ [19]. The expected background and the shape of the expected signal are also shown

the $2.2\,\mu s$ slope expected from the decay of the parent μ^+. An anomaly in this time distribution found by KARMEN in an unrelated context [108] will not be discussed further here.

KARMEN has not found any excess events above the expected background [109]. For both the potential $\bar{\nu}_e$ signal and the measured background, the energy, time, and spatial distributions for both the prompt and the delayed events are precisely known. Using this spectral information (Fig. 4.29) also leads to no hint of oscillations. Therefore, KARMEN has not confirmed the LSND result.

Figure 4.28 shows the final resulting 90% confidence exclusion limit, compared with the LSND favored region, for $\bar{\nu}_\mu \to \bar{\nu}_e$ oscillations. Despite the negative result, KARMEN cannot fully exclude the low-Δm^2 solution preferred by the LSND result. This is essentially due to the fact that KARMEN is located closer $(18\,m)$ to the neutrino source than LSND is $(30\,m)$, and therefore probes slightly larger values of Δm^2. In view of the importance of the LSND result in the context of overall oscillation scenarios (Sect. 4.7), further checks of this result will therefore be needed.

A dedicated study [106] finds a probability of only 36% that the two results are incompatible. Given this situation, it is instructive to evaluate the allowed region when both experimental results are combined, assuming that both of them are correct. No unique prescription exists for such a combination.

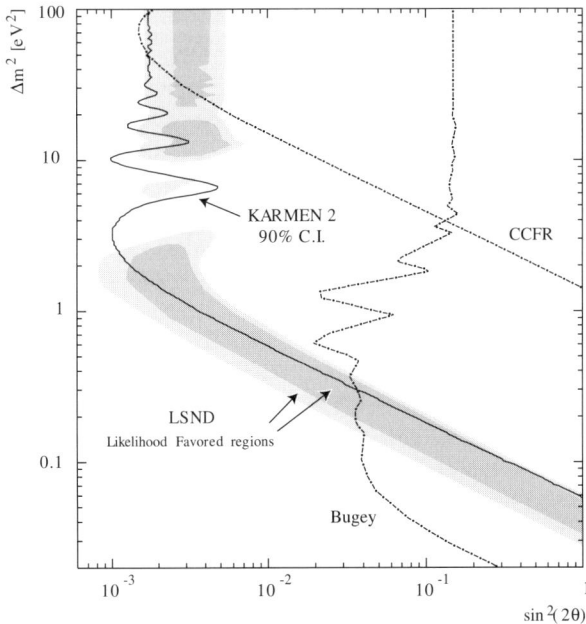

Fig. 4.28. Allowed region obtained from the LSND combined analysis, compared with the final limits from KARMEN2 and other experiments [53, 109]

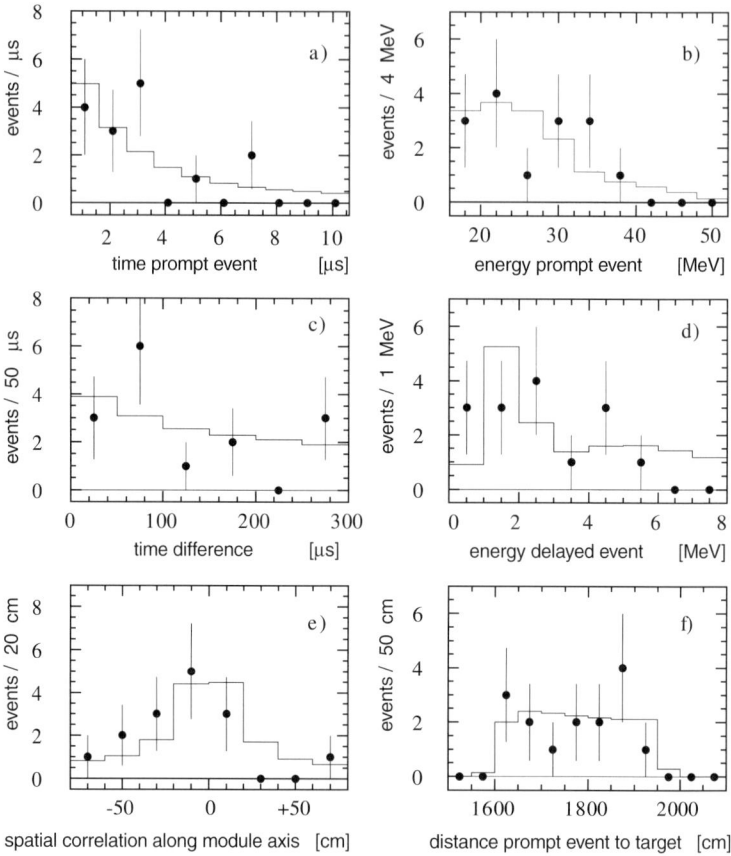

Fig. 4.29. KARMEN events remaining after all cuts [106, 109]. The data (15 events) and the expected background distributions (15.8 ± 0.5 events) are shown

Figure 4.30 shows the result of a combined likelihood fit [106] obtained using the Feldman and Cousins [111] approach. As is to be expected, the combined favored region moves to slightly lower values of $\sin^2 2\theta$, and the large-Δm^2 region is almost fully excluded (at 95% C.L.).

The MiniBooNE experiment [113] reuses part of the LSND equipment for an experiment on a new neutrino beam at the Fermilab booster, which yields an average neutrino energy of 800 MeV. The neutrino detection technique is similar to that of LSND, although at higher neutrino energy and with a much larger event rate. With the current LSND parameters, a signal of about 1000 ν_e events is expected over a background of about 3000 events, mainly originating from the intrinsic ν_e beam contamination. The result is therefore expected to be statistically very significant, but control of the systematics of the beam contamination is crucial. MiniBooNE has started data-taking in October 2002. An upgrade with a second detector at a different distance

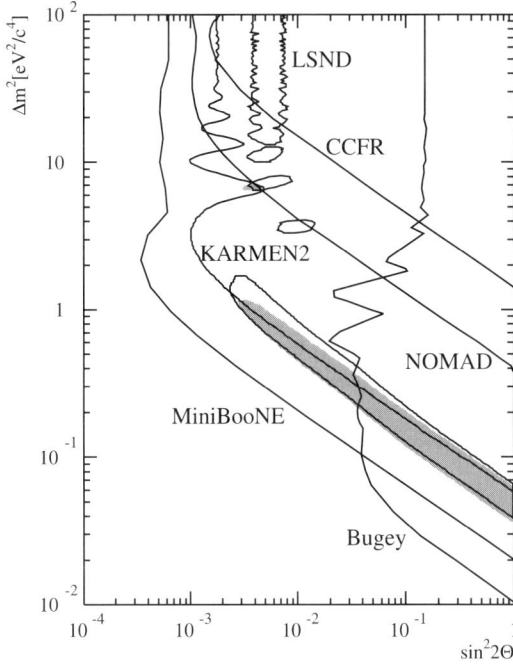

Fig. 4.30. Regions of 90% confidence level (*gray area*) for a combined analysis of the KARMEN and LSND data [106]. The contours obtained from KARMEN and LSND alone, as well as the limits from Bugey [93], CCFR [110], and NOMAD [112], are also shown. The envisaged sensitivity of MiniBooNE [113] is also indicated

(BooNE) is possible at a later stage, in the case of a positive or ambiguous result.

The MiniBooNE experiment is therefore be unique and will be crucial settling the question of whether the LSND claim is right or wrong from the experimental point of view.

4.6.2 CHORUS and NOMAD

NOMAD [114] and CHORUS [115] are experiments on the SPS wideband neutrino beam at the CERN West Area Neutrino Facility (WANF), searching for $\nu_\mu \to \nu_\tau$ oscillations in the cosmologically interesting mass range $\Delta m^2 > 1\,\mathrm{eV}^2/c^4$ [103]. Both experiments use the appearance method [116, 117] and are sensitive to mixing angles about an order of magnitude smaller than the best limit previously set by the Fermilab E531 experiment [118]. More detailed reviews can be found in [21, 102].

The CERN SPS wideband beam is produced by $450\,\mathrm{GeV}/c$ protons impinging on a beryllium target (Fig. 4.2). The resulting pion and kaon yield has been measured by a dedicated experiment [119]. Positive particles are focused into a quasi-parallel beam by a system of magnetic horns and allowed to decay in a 290 m long evacuated tunnel, followed by an earth and iron shield. The neutrinos are produced at an average distance of about 600 m (CHORUS) and 625 m (NOMAD) from the detector. The relative beam composition is dominated by ν_μ, as shown in Fig. 4.3. Direct contamination by ν_τ from the decay of D_s mesons produced at the beryllium target is found to be negligible [120]. Any signal of ν_τ would therefore be evidence for a flavor changing process, presumably neutrino oscillations. Potential ν_τ candidates are identified through their CC interaction $\nu_\tau N \to \tau^- X$. Note that, owing to the dominance of ν over $\bar{\nu}$ in the beam, τ^+ final states are not expected to occur.

The concept of the CHORUS experiment is based on the explicit detection of the τ decay vertex [102]. This requires a high tracking resolution owing to the short lifetime of the τ lepton. Sheets of bulk nuclear emulsion with a total mass of 800 kg supply a three-dimensional resolution of 1 μm and combine the properties of an active detector and a heavy target. The disadvantage of the rather slow offline emulsion analysis is partially overcome by the use of fully automatic scanning procedures. Scintillating fiber trackers give precise tracking close to the emulsion target and define the exit point of the tracks in the emulsion. Figure 4.31 shows a schematic view of this procedure. A hexagonal air-core magnet yields momentum measurements of hadronic tracks. The tracker region is followed by a high-resolution calorimeter. A muon spectrometer measures the charge and momenta of muons.

CHORUS started data-taking in 1994. Preselecting only events with a vertex position (determined from the external trackers) compatible with the emulsion target yields a sample of 713 000 events with a muon identified in

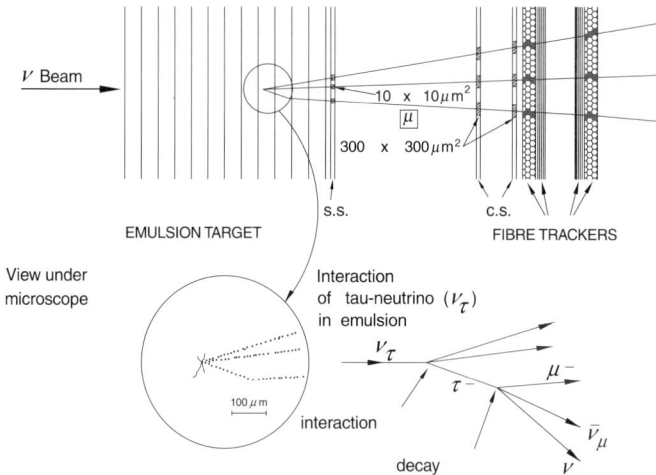

Fig. 4.31. Schematic illustration [102] of the detection of the ν_τ appearance reaction $\nu_\tau N \to \tau^- X$ in CHORUS. "C.S." and "S.S." denote changeable, external emulsion sheets

the final state (1μ), and a sample of 335 000 events without an identified muon (0μ). Electron identification has not been attempted so far. In phase 1 of the analysis, tracks in the momentum and angular range compatible with their being a τ^- decay product (*negative* muons and pions) were scanned back into the emulsion by an automatic procedure to find the primary vertex and possibly a secondary vertex. About 14 000 events with a potential secondary vertex were then submitted to a detailed visual scan [121].

Charm decays can be distinguished from the expected τ signal because they yield predominantly *positive* decay products, and are accompanied by another lepton at the primary vertex. An observed event compatible with the production and decay chain

$$\nu_\mu N \to N' D_s^+ \mu^- \,, \quad D_s^+ \to \tau^+ \nu_\tau \,, \quad \tau^+ \to \mu^+ \nu_\mu \bar{\nu}_\tau \qquad (4.24)$$

is shown in Fig. 4.32 [122], and illustrates the τ detection method of CHORUS.

No event with a secondary vertex compatible with the decay of a primary τ^- was observed. This yields the exclusion curve in the $\sin^2 2\theta$ vs. Δm^2 plane indicated in Fig. 4.33. Currently, phase 2 of the analysis, based on a full scan of the emulsion, independent of the predictions of the outer trackers, is ongoing. This procedure is expected to greatly improve the efficiency. No results are available yet.

While the main sensitivity of CHORUS arises from the explicit detection of the kink in the τ track caused by $\tau \to \mu\nu\nu$ decay, NOMAD is optimized for indirect detection of the τ via its decay products, with particular emphasis on $\tau \to e\nu\nu$ decays.

In general, NOMAD is designed to detect both the leptonic ($e^- \bar{\nu}_e \nu_\tau$ or $\mu^- \bar{\nu}_\mu \nu_\tau$) and hadronic (1-prong or 3-prong) decay modes of the produced τ^-. The ν_τ interactions are recognized through kinematic criteria based on the isolation of the τ decay products from the remainder of the event and on the momentum imbalance in the transverse plane (missing p_T) induced by

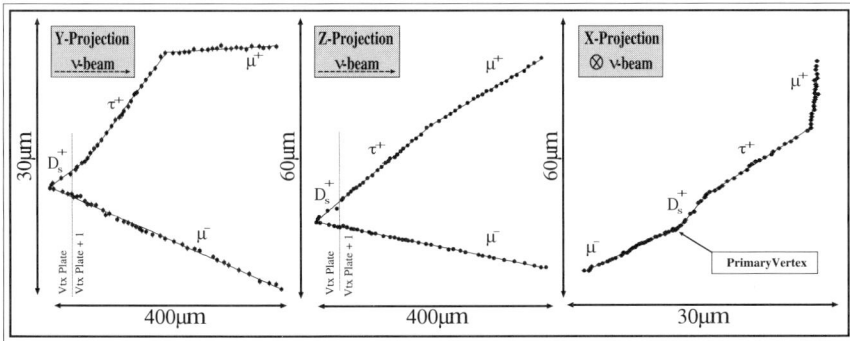

Fig. 4.32. CHORUS candidate for the production and decay chain $\nu_\mu N \to N' D_s^+ \mu^-$, $D_s^+ \to \tau^+ \nu_\tau$, $\tau^+ \to \mu^+ \nu_\mu \bar{\nu}_\tau$ [122]

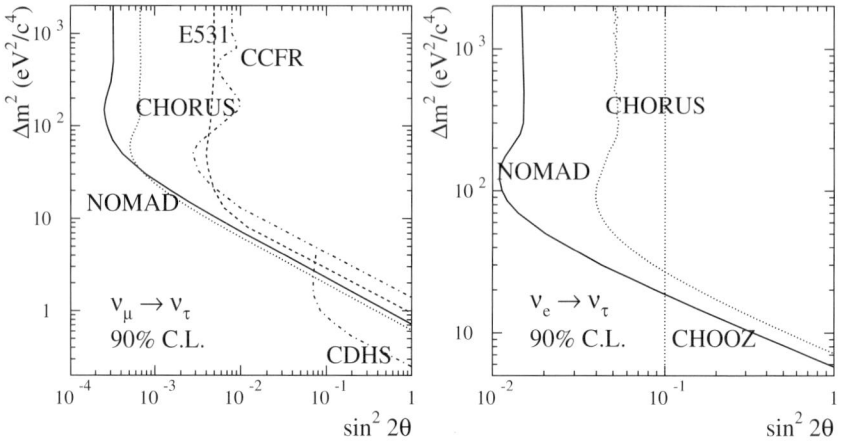

Fig. 4.33. Current best limits on ν_μ–ν_τ (*left*) and ν_e–ν_τ (*right*) oscillations in the high-Δm^2 region in the two-flavor approximation. The ν_τ appearance limits from NOMAD [123], CHORUS [121], CCFR [124], and E531 [118], the ν_μ disappearance limit from CDHS [125], and the ν_e disappearance limit from CHOOZ [96] are shown

the momentum carried away by the final-state neutrino(s) in τ decay. The smallness of the ν_e component makes the search for $\tau^- \to e^- \bar{\nu}_e \nu_\tau$ particularly sensitive, as few ν_e CC background events will be produced.

The central components of the NOMAD detector are located in a large dipole magnet (0.4 T). The target part of the detector was designed to accommodate two conflicting requirements: to be as light as possible in order to allow the precise measurement of charged-particle momenta and to minimize secondary interactions, and to be as heavy as possible in order to produce a significant number of primary neutrino interactions.

This conflict was resolved using an active target (2.7 tons) made of drift chambers, where the target mass is given by the chamber structure, with an average density of $0.1\,\mathrm{g/cm^3}$. The drift chambers are followed by a transition radiation detector, a preshower detector, and an electromagnetic calorimeter. The combination of the transition radiation information with tracking and calorimetry yields excellent electron identification. An iron–scintillator hadronic calorimeter is located just outside the magnet coil and is followed by two muon detection stations consisting of large-area drift chambers.

NOMAD has recorded about 1 350 000 ν_μ CC events and corresponding numbers of $\bar{\nu}_\mu$, ν_e, and NC interactions in the data-taking periods from 1995 to 1998. ν_τ CC event candidates were selected by identifying particles (a muon, an electron, or a hadron or hadrons) consistent with being produced in τ decay. The remaining particles were then combined to form the associated hadronic system. Kinematic variables formed from these two systems were then used to separate the signal from the backgrounds, using a "data simulator" method [116] based on the data themselves.

In order to avoid biases due to knowledge about individual τ candidates in the data, a "blind analysis" approach was adopted [116]. Further cross-checks were made by comparing the results with a (null) search for "wrong-sign" τ^+ candidates in the signal region. Once all efficiencies and performances were understood, the cuts were frozen, and the "box" was opened. Any significant excess over the expected background would then be interpreted as evidence for ν_μ–ν_τ oscillations.

Details of the analyses of various τ decay channels and kinematic regions can be found in [116, 123]. An example of a likelihood distribution obtained from the analysis of the $\tau^- \to e^-\overline{\nu}_e\nu_\tau$ decay channel is shown in Fig. 4.34. The typical expected backgrounds in each analysis range from a fraction of an event to ten events. None of the eight analyses observes a significant excess. The results of the individual analyses have been combined using the statistical approach of Feldman and Cousins [111]. The final limits obtained from this procedure [123] are shown in Fig. 4.33.

So far, it has been implicitly assumed that any observed τ signal would originate from oscillations of the majority ν_μ component of the beam. Alternatively, τ events could also be produced by the minority ν_e component through ν_e–ν_τ oscillations. This would change the energy distribution of the expected τ candidates, but does not affect the predictions of the background.

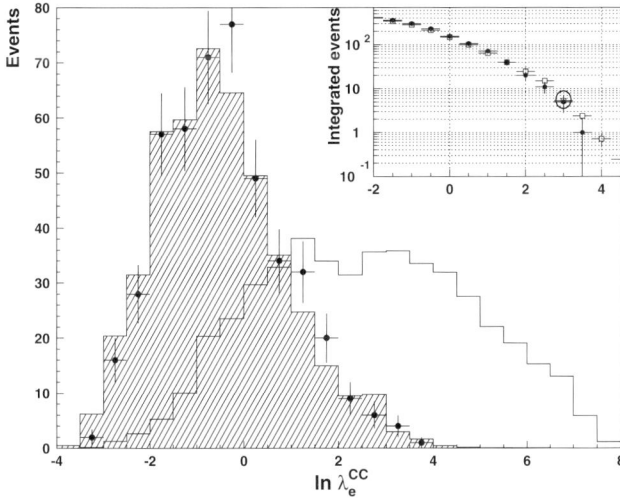

Fig. 4.34. Log-likelihood distribution for $\tau \to e\nu\nu$ candidates passing all analysis cuts except the cut on this variable [116]. The data (*black dots*), the background prediction obtained via the data simulator method (*hatched*), and the shape of the predicted signal distribution are shown. The *inset* shows the integrated distribution for the same data (*black dots*) and the background distributions (*empty squares*). The *circled points* indicate the boundary of the initially "blind" signal region from which the limit was obtained

The nonobservation of τ candidates in the ν_μ–ν_τ search can therefore be translated into a limit on ν_e–ν_τ oscillations [123, 126]. The corresponding current best limits are also shown in Fig. 4.33.

Furthermore, NOMAD has carried out an explicit search for ν_μ–ν_e oscillations, motivated by the LSND result discussed in the previous section. Such oscillations would manifest themselves in an excess of ν_e events in the ν_e/ν_μ ratio at low energies with respect to the predictions of the content of the beam. No such excess has been observed [112]. The corresponding limit is shown in Fig. 4.30.

Both experiments are now completed. However, the CHORUS experiment has not yet reached its final analysis sensitivity, owing to the lengthy emulsion-scanning process, which is still ongoing. Some further improvements can therefore be expected, but the parameter range covered does not and will not overlap with that suggested by the atmospheric neutrino anomaly. If, as is commonly believed, the atmospheric neutrino anomaly is mainly due to $\nu_\mu \rightarrow \nu_\tau$ oscillations, no signal is therefore expected in short-baseline experiments. Should nature have chosen one of the alternative options for this anomaly, then a signal in CHORUS/NOMAD or in a similar more precise experiment would still be possible. Several oscillation scenarios exist which allow this possibility (see e.g. the review in [127]).

Despite the negative results, the CHORUS and NOMAD experiments have firmly established two independent analysis techniques with similar sensitivity for τ appearance searches, which will be used in future long-baseline appearance experiments (see next section). New short-baseline [128] or intermediate-baseline [129] experiments which would extend the parameter range covered are currently not being pursued further, but might be needed at some point to discriminate between competing oscillation scenarios [130], particularly if the LSND result is confirmed.

4.6.3 Long-Baseline Experiments

In order to cover the L/E range probed by atmospheric-neutrino experiments with artificial and therefore controllable neutrino beams, the detector-to-source distance (baseline) needs to be increased to several hundred kilometers. Currently, one such "long-baseline" program is in operation, and two more are under construction. These experiments are the following:

– The K2K experiment in Japan [131, 132]. Here, a low-energy ($\langle E \rangle \sim$ 1.5 GeV) neutrino beam is sent from the KEK laboratory (Fig. 4.35) to the Super-Kamiokande detector, a distance of 250 km away. A near detector is located at a distance of 300 m from the target. The main goal is the measurement of the ν_μ CC disappearance rate from the far/near event ratio. This experiment started data-taking in 1999.

– The NUMI/MINOS project in the USA [133]. Fermilab provides a somewhat higher-energy tunable neutrino beam ($3\,\text{GeV} < \langle E_\nu \rangle < 12\,\text{GeV}$),

Fig. 4.35. The KEK accelerator complex and neutrino beam line

which is sent to the MINOS detector in the Soudan mine, a distance of 730 km away. Again, a near detector monitors the actual neutrino flux. The basic measurements are similar to those performed at K2K, but the much better statistics offer increased sensitivity, including significant measurements of the NC/CC ratio and of potential electron neutrino appearance. The first physics run with a low-energy beam is scheduled for 2005.

– The CNGS project in Europe [134]. A high-energy neutrino beam ($\langle E_\nu \rangle \sim$ 17 GeV) is sent from CERN to the Gran Sasso laboratory in Italy; again the distance is 730 km. Two detectors optimized for the explicit detection of τ appearance by techniques based on those used in CHORUS and NOMAD are planned; one has been approved (OPERA [135]), and the other has been proposed on the basis of an almost complete 600 t prototype (ICARUS [136]). Since the project is primarily focused on τ appearance, a high-energy beam is mandatory. A near detector is neither needed nor foreseen. The beam is approved to start operation in May 2005. Currently, this schedule is being reassessed.

The three projects are complementary in the sense that the K2K experiment has started to confirm the atmospheric neutrino results semiquantitatively, the NUMI/MINOS project will yield a more precise measurement of the oscillation parameters and strong indications of the oscillation mode, and the CNGS project will definitely settle the question of whether the primary oscillation mode involves the τ neutrino. A fourth program at the Japan

Hadron Facility (JHF), sending a K2K-like beam with an intensity two orders of magnitude to Super-Kamiokande from a distance of 290 km [137], is currently being proposed. On an even longer term, a beam from a so-called neutrino factory [25] (neutrinos from a high-energy muon storage ring) could open up the possibility of the measurement of CP violation in the lepton sector. Here, we shall focus on the K2K experiment, which has already collected its first data. The other projects are discussed in more detail in Chap. 7.

K2K started data-taking in spring 1999. The Super-Kamiokande detector is used as the far detector. The near detector consists of a 1 kt water Cherenkov detector (using the same principle as in Super-Kamiokande) and a fine-grained detector consisting of a scintillating fiber tracker, a lead glass counter, and a muon range detector. There is no magnetic field.

The main goal is the observation of a charged-current event deficit in the far detector with respect to the near detector to confirm the atmospheric

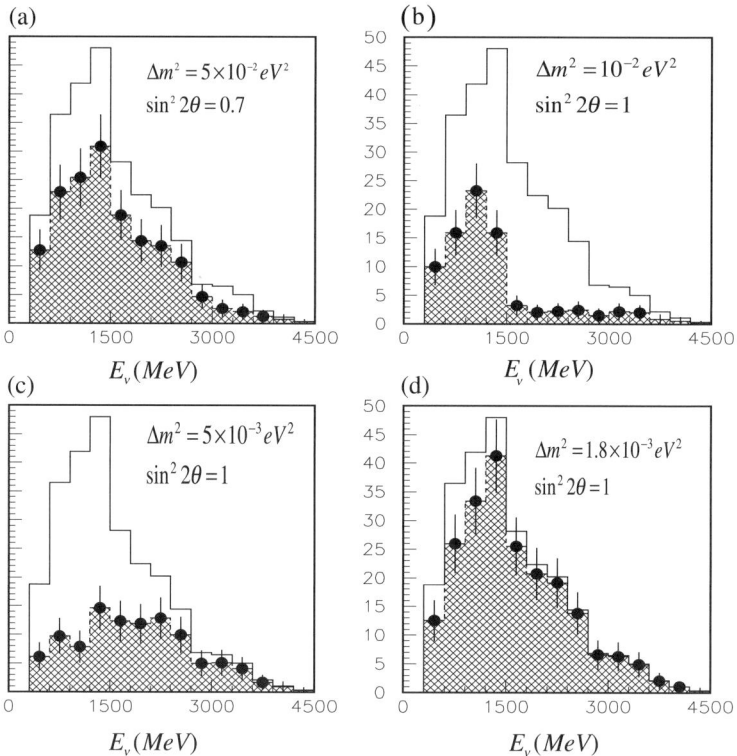

Fig. 4.36. Expected ν_μ energy spectra at the K2K far detector (Super-Kamiokande) for various ν_μ–ν_τ oscillation parameters (*hatched histogram* and *points with error bars*) and for the no-oscillation case (*empty histogram*). The statistics correspond to about three times the rate expected for the first two years [132]

neutrino anomaly and further constrain the parameter space for neutrino oscillations. The experiment is sensitive to $\Delta m^2 \sim 2 \times 10^{-3}\,\text{eV}^2$ or larger. Up to July 2001, K2K had collected several tens of thousands of events in the 1 kt near detector, which provided a successful cross-check of the expected beam properties and fixed the neutrino flux normalization. In the same period, 56 events were collected inside the fiducial volume of the far detector (Super-Kamiokande), the first artificial neutrinos to be detected several hundreds of kilometers away from their production point [132]. This is to be compared with an expectation of 81 ± 6 events for the no-oscillation case, and 54 events with $\Delta m^2 = 2.8 \times 10^{-3}\,\text{eV}^2$ and full mixing. The expected and measured energy spectra for an early subsample of these events are shown in Figs. 4.36 and 4.37. A quantitative analysis of the oscillation parameters based on the rate measurement has also been performed (Fig. 4.38). The results are in excellent agreement with the Super-Kamiokande atmospheric-neutrino result.

These preliminary results therefore confirm qualitatively and quantitatively the atmospheric neutrino deficit. Further data-taking has been delayed until an ongoing repair of the Super-Kamiokande detector is completed. Unless Δm^2 is at the low edge of the currently allowed range, a combination of future more precise measurements from K2K, MINOS, and OPERA/ICARUS should be able to further test the oscillation hypothesis and finally settle the question of the oscillation mode.

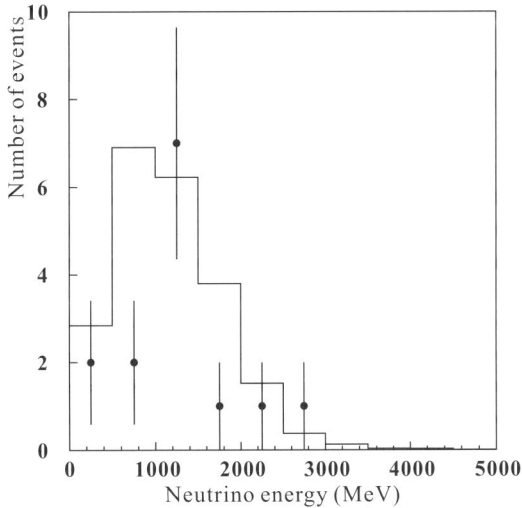

Fig. 4.37. Measured ν_μ energy spectrum for 14 well-measured events [132]

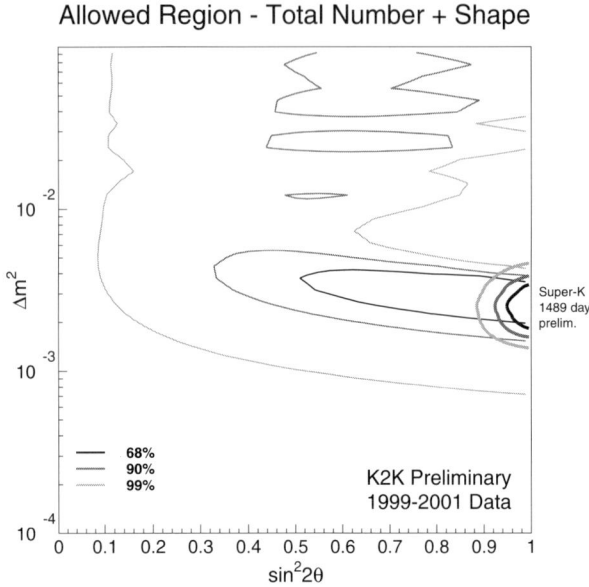

Fig. 4.38. Allowed regions for ν_μ–ν_τ oscillations from an analysis of the K2K normalization and shape [132], compared with the corresponding regions for atmospheric neutrinos obtained by the Super-Kamiokande collaboration

4.7 Summary and Conclusions

... the fascination with neutrinos and the unanswered questions concerning them – such as their masses – are motivating a broad line of research in astrophysics, accelerator physics, and nuclear physics.

J. Steinberger, 1989

Many neutrino-related topics, such as its existence, parity violation in its interactions, and, most recently, its mass and the question of lepton number conservation, were milestones on the path towards understanding the fundamental laws of nature. The current line of neutrino research, referred to by Steinberger above [23], is aiming at a continuation of this tradition. However, as in the past, a lot of patience will be required to see the theoretical concepts tested and finally confirmed or refuted by experiment.

A minirevolution, requiring the revision or at least a minimal extension of the Standard Model, has recently been induced by the accumulation of experimental indications of neutrino oscillations in solar, reactor, accelerator, and atmospheric neutrino experiments (Fig. 4.39). This culminated in the official announcement of evidence for ν_μ oscillations by the Super-Kamiokande experiment in 1998, and most recently by SNO and KamLAND. Even if some of the evidence has to be attributed to nonoscillation effects in the future, it seems increasingly unlikely that *all* indications are wrong or have been misinterpreted. Therefore, the resulting hypothesis of nonzero neutrino mass has become the generally accepted default. The available positive evidence is complemented by increasingly precise negative evidence in other regions of the oscillation parameter space, which helps to narrow down the options for neutrino oscillation models.

More specifically, the combined evidence from seven different experiments for oscillations of solar electron neutrinos into other active neutrinos becomes increasingly compelling, but other (new-physics) interpretations remain possible. Currently, KamLAND and SNO are delivering the keys that may finally settle the question. Similar arguments hold for atmospheric $\nu_\mu \to \nu_\tau$ oscillations, for which the evidence is dominated by the Super-Kamiokande results. The final confirmation of this evidence will rely to a large extent on the long-baseline beam program, which has yielded encouraging initial data, but will take five to ten more years to yield fully conclusive results. The experimental methods pioneered by recent short-baseline experiments are an essential ingredient of this program. Finally, the evidence for ν_μ–ν_e oscillations from LSND has still been neither confirmed nor refuted by another experiment, but will be checked by MiniBooNE very soon.

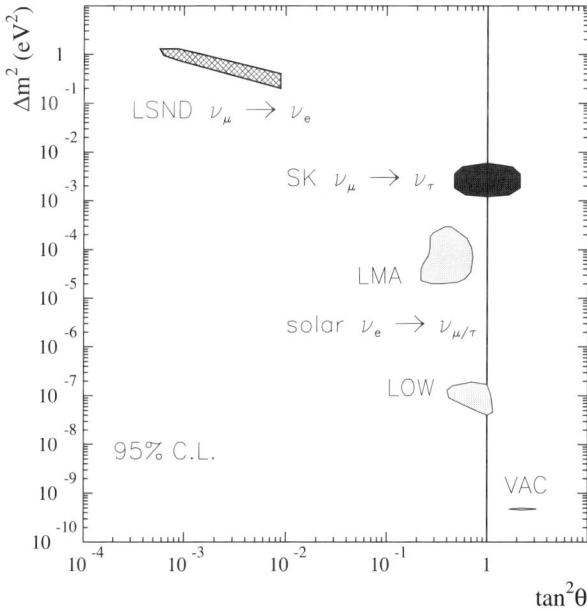

Fig. 4.39. 95% C.L. allowed areas of oscillation parameters for atmospheric neutrinos [68] (*dark shaded area*), solar neutrinos [62] (*light shaded areas*), and the LSND results [106] (*hatched area*). The $\tan^2\theta > 1$ image of the LSND allowed area is not shown explicitly. The LMA solar neutrino solution has recently been confirmed by KamLAND [18], therefore excluding the LOW and VAC options

If all three positive indications are confirmed to be due to neutrino oscillations, the three known active neutrino flavors will not supply a sufficient number of degrees of freedom to describe the data, and contributions from additional sterile neutrinos will be needed (Chap. 5). The future experimental checks mentioned here and described in more detail in Chap. 7 are therefore of very fundamental importance.

In summary, an exciting new chapter in neutrino physics has just been opened, which might yield further subjects for experimental exploration and theoretical debate for the next 30 years before being fully understood.

Acknowledgments

The material presented in this chapter was collected from the publicly available sources cited and I owe thanks to the authors for making figures and sources available for further processing. I am grateful to Isabell Pellmann and Oliver Gutsche for their substantial contributions to the proofreading of the manuscript. Finally, I would like to thank the editors for the opportunity to contribute this chapter.

References

1. W. Pauli, in *Aufsätze und Vorträge über Physik und Erkenntnistheorie*, ed. W. Westphal (Vieweg, Braunschweig, 1961), p. 156.
2. F. Reines and C.L. Cowan, Nature **178**, 445 (1956); Nature **178**, 523 (1956) (erratum); C.L. Cowan et al., Science **124**, 103 (1956); F. Reines, Rev. Mod. Phys. **68**, 317 (1996).
3. See e.g. K. Winter (ed.), *Neutrino Physics*, Cambridge Monographs on Particle Physics, Nuclear Physics, and Cosmology, Vol. 14, 2nd edition (Cambridge University Press, Cambridge, 2000) 539.7215 N3993; M. Riordan, in *Current Aspects of Neutrino Physics*, ed. D. Caldwell (Springer, Berlin, Heidelberg, 2001), p. 6.
4. M. Goldhaber et al., Phys. Rev. **109**, 1015 (1958).
5. For a recent review, see e.g. J.M. Conrad, M.H. Shaevitz, and T. Bolton, Rev. Mod. Phys. **70**, 1341 (1998).
6. L.M. Lederman, Rev. Mod. Phys. **61**, 547 (1989).
7. D. Karlen, Rev. Part. Phys., Phys. Rev. **D66**, 387 (2002).
8. K. Kodama et al. (DONUT Collaboration), Phys. Lett. B **504**, 218 (2001).
9. Particle Data Group (PDG), Rev. Part. Phys., Phys. Rev. **D66**, 1 (2002).
10. E. Majorana, Nuovo Cim. **14**, 171 (1937); G. Racah, Nuovo Cim. **14**, 322 (1937).
11. B. Pontecorvo, Zh. Eksp. Teor. Fiz. **34**, 247 (1958); Sov. Phys. JETP **6**, 429 (1958); Sov. Phys. JETP **7**, 172 (1958).
12. B. Pontecorvo, Zh. Eksp. Teor. Fiz. **53**, 1717 (1967); Sov. Phys. JETP **26**, 984 (1968).
13. R. Davis, Prog. Part. Nucl. Phys. **32**, 13 (1994), and references therein.

14. K.S. Hirata et al. (Kamiokande Collaboration), Phys. Lett. B **205**, 416 (1988); Phys. Lett. B **280**, 146 (1992); Y. Fukuda et al. (Kamiokande Collaboration), Phys. Lett. B **335**, 237 (1994).

15. Y. Fukuda et al. (Super-Kamiokande Collaboration), Phys. Rev. Lett. **81**, 1562 (1998); Phys. Lett. B **436**, 33 (1998); Phys. Lett. B **433**, 9 (1998).

16. Y. Fukuda et al. (Super-Kamiokande Collaboration), Phys. Rev. Lett. **81**, 1158 (1998); Phys. Rev. Lett. **81**, 4279 (1998) (erratum); Phys. Rev. Lett. **82**, 2430 (1999); Phys. Rev. Lett. **82**, 1810 (1999); Phys. Rev. Lett. **86**, 5656 (2001).

17. Q.R. Ahmad et al. (SNO Collaboration), Phys. Rev. Lett. **89**, 011301 (2002); Phys. Rev. Lett. **89**, 011302 (2002).

18. E. Eguchi et al. (KamLAND Collaboration), Phys. Rev. Lett. **90**, 021802 (2003).

19. C. Athanassopoulos et al. (LSND Collaboration), Phys. Rev. Lett. **75**, 2650 (1995); Phys. Rev. Lett. **77**, 3082 (1996); Phys. Rev. Lett. **81**, 1774 (1998); Phys. Rev. C **54**, 2685 (1996); Phys. Rev. C **58**, 2489 (1998); A. Aguilar et al. (LSND Collaboration), Phys. Rev. D **64**, 112007 (2001).

20. See e.g. J. Primack and A. Klypin, in *Proceedings of the 2nd Symposium on Critique of the Sources of Dark Matter in the Universe*, Santa Monica, CA, USA, 14–16 Feb. 1996, Nucl. Phys. (Proc. Suppl.) **51**, 30 (1996).

21. A. Geiser, Rep. Prog. Phys. **63**, 1779 (2000).

22. M. Schwartz, Rev. Mod. Phys. **61**, 527 (1989).

23. J. Steinberger, Rev. Mod. Phys. **61**, 533 (1989).

24. G. Collazuol et al., presented at NOW98 Workshop, Amsterdam, 7–9 Sept. 1998, CERN preprint OPEN-98-032.

25. C. Albright et al., *Physics at a Neutrino Factory*, Fermilab-FN-692 (May 2000).

26. See e.g. S. Sarkar, Rep. Prog. Phys. **59**, 1493 (1996).

27. F. Reines et al., Phys. Rev. Lett. **15**, 429 (1965); Phys. Rev. D **4**, 80 (1971); H. Achar et al., Phys. Lett. **15**, 196 (1965); H.R. Krishnaswamy et al., Proc. Phys. Soc. Lond. A **323**, 489 (1971).

28. See e.g. W. Hillebrand, Prog. Part. Nucl. Phys. **32**, 75 (1994), and references therein.

29. T.K. Gaisser, F. Halzen, and T. Stanev, Phys. Rep. **258**, 173 (1995); A. Roberts, Rev. Mod. Phys. **64**, 259 (1992); F. Halzen, Phys. Rep. **307**, 243 (1998).

30. C. Spiering, in *Proceedings of the 19th International Conference on Neutrino Physics and Astrophysics (Neutrino 2000)*, Sudbury, Canada, 16–21 June 2000, eds. J. Law, R.W. Ollerhead, and J.J. Simpson, Nucl. Phys. B (Proc. Suppl.) **91**, 445 (2001).

31. M. Gell-Mann, P. Ramond, and R. Slansky, in *Supergravity*, eds. P. van Nieuwenhuizen and D.Z. Freedman (North-Holland, Amsterdam, 1979), p. 315; T. Yanagida, in *Proceedings of the Workshop on the Unified Theory and Baryon Number in the Universe*, eds. O. Sawada and A. Sugamoto (KEK, Tsukuba 1979), p. 95; R.N. Mohapatra and G. Senjanović, Phys. Rev. Lett. **44**, 912 (1980).

32. See e.g. G. Gelmini and R. Roulet, Rep. Prog. Phys. **58**, 1207 (1995), and references therein; N. Schmitz, *Neutrinophysik* (Teubner, Stuttgart, 1997), and references therein; S.M. Bilenky and S.T. Petcov, Rev. Mod. Phys. **59**, 671 (1987).

33. Z. Maki, M. Nakagawa, and S. Sakata, Prog. Theor. Phys. **28**, 870 (1962).
34. See e.g. F.J. Gilman, K. Kleinknecht, and B. Renk, Rev. Part. Phys., Eur. J. Phys. C **3**, 103 (1998), and references therein.
35. T.K. Kuo, Phys. Rep. **242**, 242 (1994); T.K. Kuo and J. Pantaleone, Rev. Mod. Phys. **61**, 937 (1989).
36. L. Wolfenstein, Phys. Rev. D **17**, 2369 (1978); Phys. Rev. D **20**, 2634 (1979); S.P. Mikheyev and A.Y. Smirnov, Sov. J. Nucl. Phys. **42**, 913 (1985); Nuovo Cim. C **9**, 17 (1986).
37. J.N. Bahcall, SLAC Beam Line **31**(1), 2 (2001).
38. J.N. Bahcall, P.I. Krastev, and A.Yu. Smirnov, Phys. Rev. D **58**, 096016 (1998); N. Hata and P. Langacker, Phys. Rev. D **56**, 6107 (1997); G.L. Fogli, E. Lisi, and D. Montanino, Astropart. Phys. **9**, 119 (1998); M.F. Altmann, R.L. Mössbauer, and L.J.N. Oberauer, Rep. Prog. Phys. **64**, 97 (2001).
39. A. de Santo, Int. J. Mod. Phys. A **16**, 4085 (2001).
40. S. Basu et al., Mon. Not. R. Astron. Soc. **292**, 1402 (1997).
41. S.M. Bilenky et al., Prog. Part. Nucl. Phys. **43**, 1 (1999).
42. J.N. Bahcall, M.H. Pinsonneault, and S. Basu, Astrophys. J. **555**, 990 (2001).
43. J.N. Bahcall's home page, http://www.sns.ias.edu/~jnb.
44. B.T. Cleveland et al., Astrophys. J. **496**, 505 (1998); K. Lande et al., in *Proceedings of the 19th Internatinal Conference on Neutrino Physics and Astrophysics (Neutrino 2000)*, Sudbury, Canada, 16–21 June 2000, eds. J. Law, R.W. Ollerhead, and J.J. Simpson, Nucl. Phys. B (Proc. Suppl.) **91**, 50 (2001).
45. P. Anselmann et al. (GALLEX Collaboration), Phys. Lett. B **327**, 377 (1994); Phys. Lett. B **342**, 440 (1995); Phys. Lett. B **357**, 237 (1995); Phys. Lett. B **361**, 235 (1996) (erratum); W. Hampel et al. (GALLEX Collaboration), Phys. Lett. B **388**, 384 (1996); Phys. Lett. B **420**, 114 (1998); Phys. Lett. B **436**, 158 (1998); Phys. Lett. B **447**, 127 (1999).
46. W. Hampel et al. (GALLEX Collaboration), Phys. Lett. B **420**, 114 (1998); Phys. Lett. B **436**, 158 (1998).
47. M. Altmann et al. (GNO Collaboration), Phys. Lett. B **490**, 16 (2000); E. Bellotti et al. (GNO Collaboration), in *Proceedings of the 19th Internatinal Conference on Neutrino Physics and Astrophysics (Neutrino 2000)*, Sudbury, Canada, 16–21 June 2000, eds. J. Law, R.W. Ollerhead, and J.J. Simpson, Nucl. Phys. B (Proc. Suppl.) **91**, 44 (2001).
48. J.N. Abdurashitov et al. (SAGE Collaboration), Phys. Lett. B **328**, 234 (1994); Nucl. Phys. B **38**, 60 (1995); Phys. Rev. Lett. **77**, 4708 (1996); Phys. Rev. Lett. **83**, 4686 (1999); V. Gavrin et al. (SAGE Collaboration), in *Proceedings of the 19th Internatinal Conference on Neutrino Physics and Astrophysics (Neutrino 2000)*, Sudbury, Canada, 16–21 June 2000, eds. J. Law, R.W. Ollerhead, and J.J. Simpson, Nucl. Phys. B (Proc. Suppl.) **91**, 36 (2001).
49. D.N. Abdurashnitov et al. (SAGE Collaboration), Phys. Rev. Lett. **77**, 4708 (1996).
50. Y. Fukuda et al. (Kamiokande Collaboration), Phys. Rev. Lett. **77**, 1683 (1996); K.S. Hirata et al. (Kamiokande Collaboration), Phys. Rev. D **44**, 2241 (1991); Phys. Rev. D **45**, 2170 (1992) (erratum); Phys. Rev. Lett. **66**, 9 (1991); Phys. Rev. Lett. **65**, 1297 (1990); Phys. Rev. Lett. **65**, 1301 (1990).
51. Super-Kamiokande home page, http://www-sk.icrr.u-tokyo.ac.jp/doc/sk/index.html.

52. K.S. Hirata et al. (Kamiokande Collaboration), Phys. Lett. B **220**, 308 (1989).

53. M. Nakahata et al. (Super-Kamiokande Collaboration), Nucl. Instrum. Meth. A **421**, 113 (1999).

54. E. Blaufuss et al. (Super-Kamiokande Collaboration), Nucl. Instrum. Meth. A **458**, 638 (2001).

55. Q.R. Ahmad et al. (SNO Collaboration), Phys. Rev. Lett. **87**, 071301 (2001); J.R. Klein (SNO Collaboration), in *Proceedings of Lepton Photon 2001*, Rome, 23–28 July 2001, http://www.lp01.infn.it/ and arXiv:hep-ex/0111040.

56. J.N. Bahcall, Phys. Rev. C **65**, 015802 (2002).

57. J.N. Bahcall, P.I. Krastev, and A.Y. Smirnov, Phys. Rev. D **62**, 093004 (2000).

58. Y. Suzuki (Super-Kamiokande Collaboration), in *Proceedings of 18th International Conference on Neutrino Physics and Astrophysics (Neutrino 98)*, Takayama, Japan, 4–9 June 1998, eds. Y. Suzuki and Y. Tosuka (Elsevier, Amsterdam, 1999), p. 35.

59. G.L. Fogli et al., Phys. Rev. D **64**, 093007 (2001); M.V. Garzelli and C. Giunti, J. High Energy Phys. **0112**, 017 (2001); A. Bandyopadhyay et al., Phys. Lett. B **519**, 83 (2001).

60. J.N. Bahcall, M.C. Gonzalez-Garcia, and C. Pena-Garay, J. High Energy Phys. **0108**, 014 (2001); P.I. Krastev and A.Y. Smirnov, Phys. Rev. D **65**, 073022 (2002); M.C. Gonzalez-Garcia, M. Maltoni, and C. Pena-Garay, arXiv:hep-ph/0108073.

61. P.C. de Holanda and A.Y. Smirnov, Phys. Rev. D **66**, 113005 (2002); V. Barger et al., Phys. Lett. B **357**, 179 (2002).

62. See e.g. J.N. Bahcall, M.C. Gonzalez-Garcia, and C. Pena-Garay, J. High Energy Phys. **0204**, 007 (2002).

63. See e.g. A.M. Gago et al., Phys. Rev. D **65**, 073012 (2002).

64. C. Arpesella at al., Borexino proposal, internal report, INFN Milan (1992); G. Ranucci et al. (Borexino Collaboration), in *Proceedings of the 19th International Conference on Neutrino Physics and Astrophysics (Neutrino 2000)*, Sudbury, Canada, 16–21 June 2000, eds. J. Law, R.W. Ollerhead, and J.J. Simpson, Nucl. Phys. B (Proc. Suppl.) **91**, 58 (2001).

65. T. Kajita (Super-Kamiokande Collaboration), in *Proceedings of 18th International Conference on Neutrino Physics and Astrophysics (Neutrino 98)*, Takayama, Japan, 4–9 June 1998, Nucl. Phys. (Proc. Suppl.) **77**, 123 (1999).

66. G.L. Fogli et al., Phys. Rev. D **55**, 4385 (1997); Phys. Rev. D **57**, 5893 (1998); E.W. Beier et al., Phys. Lett. B **283**, 446 (1992).

67. W.A. Mann, Int. J. Mod. Phys. **A15**(S1), 229 (2000); G. Giacomelli, M. Giorgini, and M. Spurio, lectures at the 6th School on Non-Accelerator Astroparticle Physics, Trieste, Italy, 9–20 July 2001, arXiv:hep-ex/0201032.

68. J.G. Learned, in *Current Aspects of Neutrino Physics*, ed. D.O. Caldwell (Springer, Berlin, Heidelberg, 2001), p. 89; arXiv:hep-ex/0007056; J. Goodman, in *Proceedings of Lepton Photon 2001*, Rome, 23–28 July 2001, http://www.lp01.infn.it/.

69. D. Ayres et al., Phys. Rev. D **29**, 902 (1984); P. Fisher, B. Kayser, and K.S. McFarland, Annu. Rev. Nucl. Part. Sci. **49**, 481 (1999).

70. T. Futagami et al. (Super-Kamiokande Collaboration), Phys. Rev. Lett. **82**, 5194 (1999); K. Munakata et al. (Kamiokande Collaboration), Phys. Rev. D **56**, 23 (1997).

71. D. Casper et al. (IMB Collaboration), Phys. Rev. Lett. **66**, 2561 (1991); R. Becker-Szendy et al., Phys. Rev. D **46**, 3720 (1992); M. Aglietta et al. (NUSEX Collaboration), Europhys. Lett. **8**, 611 (1989); C. Berger et al. (Frejus Collaboration), Phys. Lett. B **227**, 489 (1989); Phys. Lett. B **245**, 305 (1990); K. Daum et al., Z. Phys. C **66**, 417 (1995).

72. S. Ahlen et al. (MACRO Collaboration), Nucl. Instrum. Meth. A **324**, 337 (1993).

73. W.W.M. Allison et al. (Soudan 2 Collaboration), Nucl. Instrum. Meth. A **376**, 36 (1996); Nucl. Instrum. Meth. A **381**, 385 (1996); W.P. Oliver et al., Nucl. Instrum. Meth. A **275**, 371 (1989).

74. V.J. Stenger, in *Proceedings of Snowmass '94, Particle and Nuclear Astrophysics and Cosmology in the Next Millenium*, Snowmass, Colorado, 29 June–14 July 1994, p. 167.

75. M.C. Gonzalez-Garcia et al., Phys. Rev. D **58**, 033004 (1998); J.W. Flanagan, J.G. Learned, and S. Pakvasa, Phys. Rev. D **57**, R2649 (1998).

76. W.W.M. Allison et al. (Soudan 2 Collaboration), Phys. Lett. B **391**, 491 (1997); Phys. Lett. B **449**, 137 (1999); W.A. Mann (Soudan 2 Collaboration), in *Proceedings of the 19th International Conference on Neutrino Physics and Astrophysics (Neutrino 2000)*, Sudbury, Canada, 16–21 June 2000, eds. J. Law, R.W. Ollerhead, and J.J. Simpson, Nucl. Phys. B (Proc. Suppl.) **91**, 134 (2001).

77. R. Foot, R.R. Volkas, and O. Yasuda, Phys. Rev. D **58**, 013006 (1998); Phys. Rev. D **57**, R1345 (1998); P. Lipari and M. Lusignoli, Phys. Rev. D **58**, 073005 (1998); M.C. Gonzalez-Garcia et al., Nucl. Phys. B **543**, 3 (1999).

78. Y. Fukuda et al. (Super-Kamiokande Collaboration), Phys. Rev. Lett. **82**, 2644 (1999); Phys. Lett. B **467**, 185 (1999).

79. S. Hatekayama et al. (Kamiokande Collaboration), Phys. Rev. Lett. **81**, 2016 (1998); T. Gaisser (Baksan Collaboration), in *Neutrino '96, Proceedings of the 17th International Conference on Neutrino Physics and Astrophysics*, Helsinki, 13 June 1996, eds. K. Enqvist, K. Huitu and J. Maalampi (World Scientific, Singapore, 1997), p. 211; R. Becker-Szendy et al. (IMB Collaboration), Phys. Rev. D **46**, 3720 (1992).

80. M. Ambrosio et al. (MACRO Collaboration), Phys. Lett. B **434**, 451 (1998); Phys. Lett. B **517**, 59 (2001); S. Ahlen et al., Phys. Lett. B **357**, 481 (1995).

81. V. Barger et al., Phys. Lett. B **462**, 109 (1999); Phys. Rev. Lett. **82**, 2640 (1999).

82. See e.g. S. Pakvasa, in *Proceedings of the 8th International Workshop on Neutrino Telescopes*, Venice, 23–26 Feb. 1999, p. 283; arXiv:hep-ph/9905426; S. Pakvasa, Pramana **54**, 65 (2000), and references therein. Some explicit examples can be found in E. Ma and P. Roy, Phys. Rev. Lett. **80**, 4637 (1998); L.M. Johnson and D.W. McKay, Phys. Lett. B **433**, 355 (1998); R. Foot, C.N. Leung, and O. Yasuda, Phys. Lett. B **443**, 185 (1998); M.C. Gonzalez-Garcia et al., Phys. Rev. Lett. **82**, 3202 (1999).

83. MONOLITH Proposal, LNGS P26/2000, CERN/SPSC 2000-031 (Aug. 2000); P. Antonioli (MONOLITH Collaboration), presented at Europhysics Neutrino Oscillation Workshop (NOW 2001), Conca Spechiulla, Otranto, Italy, 9–12 Sept. 2000, arXiv:hep-ex/0101040.

84. S. Wojcicki, in *Proceedings of 18th International Conference on Neutrino Physics and Astrophysics (Neutrino 98)*, Takayama, Japan, 4–9 June 1998, eds. Y. Suzuki and Y. Tosuka (Elsevier, Amsterdam, 1999), p. 182.

85. J.W. Flanagan, J.G. Learned, and S. Pakvasa, Phys. Rev. D **57**, 2649 (1998); J.G. Learned, S. Pakvasa, and J.L. Stone, Phys. Lett. B **435**, 131 (1998).
86. L.J. Hall and H. Murayama, Phys. Lett. B **436**, 323 (1998).
87. F. Vissani and A.Yu. Smirnov, Phys. Lett. B **432**, 376 (1998).
88. P. Lipari and M. Lusignoli, Phys. Rev. D **58**, 073005 (1998); Q.Y. Liu, S.P. Mikheyev, and A.Y. Smirnov, Phys. Lett. B **440**, 319 (1998).
89. Y. Fukuda et al. (Super-Kamiokande Collaboration), Phys. Rev. Lett. **85**, 3999 (2000).
90. A. Geiser (MONOLITH Collaboration), Phys. Scr. T **93**, 86 (2001).
91. B. Pontecorvo, reprinted report PD-205 of the National Research Council of Canada, Division of Atomic Energy, Chalk River, Ontario (13 Nov. 1946), in *Neutrino Physics*, ed. K. Winter, 2nd edn. (Cambridge University Press, Cambridge, 2000), p. 23.
92. F. Reines and C.L. Cowan, Phys. Rev. **92**, 830 (1953).
93. B. Achkar et al. (Bugey Collaboration), Nucl. Phys. B **434**, 503 (1995).
94. G. Zacek et al. (Goesgen Collaboration), Phys. Rev. D **34**, 2621 (1986).
95. G.S. Vidyakin et al. (Krasnoyarsk Collaboration), JETP Lett. **59**, 237 (1994).
96. M. Apollonio et al. (CHOOZ Collaboration), **420**, 397 (1998); Phys. Lett. B **466**, 415 (1999).
97. F. Boehm et al. (Palo Verde proposal), Californian Institute of Technology (1994); F. Boehm et al. (Palo Verde Collaboration), Phys. Rev. Lett. **84**, 3764 (2000).
98. G.L. Fogli et al., Phys. Rev. D **59**, 033001 (1998); O. Yasuda, Phys. Rev. D **58**, 091301 (1998); V. Barger and K. Whisnant, Phys. Rev. D **59**, 093007 (1999); T. Sakai and T. Teshima, Prog. Theor. Phys. **102**, 629 (1999); O.L.G. Peres and A.Yu. Smirnov, Phys. Lett. B **456**, 204 (1999).
99. A. Piepke (KamLAND Collaboration), in *Proceedings of the 19th International Conference on Neutrino Physics and Astrophysics (Neutrino 2000)*, Sudbury, Canada, 16–21 June 2000, eds. J. Law, R.W. Ollerhead, and J.J. Simpson, Nucl. Phys. B (Proc. Suppl.) **91**, 99 (2001).
100. S. Pakvasa and J.W.F. Valle, arXiv:hep-ph/0301061.
101. S. Mohanty, arXiv:hep-ph/0302060.
102. J. Brunner, Fortschr. Phys. **45**, 343 (1997).
103. H. Harari, Phys. Lett. B **216**, 413 (1989); J. Ellis, J.L. Lopez, and D.V. Nanopoulos, Phys. Lett. B **292**, 189 (1992).
104. C. Athanassopoulos et al. (LSND Collaboration), Nucl. Instrum. Meth. A **388**, 149 (1997).
105. G. Drexlin et al. (KARMEN Collaboration), Nucl. Instrum. Meth. A **289**, 490 (1990).
106. K. Eitel, New J. Phys. **2**, 1 (2000); W. Church et al., Phys. Rev. D **66**, 013001 (2002).
107. J.E. Hill, Phys. Rev. Lett. **75**, 2654 (1995).
108. R. Bilger et al. (KARMEN collaboration), Phys. Lett. B **348**, 19 (1995).
109. B. Zeitnitz et al. (KARMEN Collaboration), Prog. Part. Nucl. Physics **40**, 169 (1998); B. Armbruster et al. (KARMEN Collaboration), Phys. Rev. D **65**, 112001 (2002).
110. A. Romosan et al. (CCFR Collaboration), Phys. Rev. Lett. **78**, 2912 (1997).
111. G.J. Feldman and R.D. Cousins, Phys. Rev. D **57**, 3873 (1998).

112. V. Valuev, *Proceedings of International Europhysics Conference on High Energy Physics (HEP 2001)*, Budapest, Hungary, 12–18 Jul 2001, PRHEP-hep2001/190.

113. E. Church et al., BooNE letter of intent, arXiv:nucl-ex/9706011; E. Church et al., BooNE proposal, FERMILAB-P-0898 (1997); A. Bazarko (MiniBooNE Collaboration), in *Proceedings of the 19th International Conference on Neutrino Physics and Astrophysics (Neutrino 2000)*, Sudbury, Canada, 16–21 June 2000, eds. J. Law, R.W. Ollerhead, and J.J. Simpson, Nucl. Phys. B (Proc. Suppl.) **91**, 210 (2001).

114. J. Altegoer et al. (NOMAD Collaboration), Nucl. Instrum. and Meth. A **404**, 96 (1998).

115. E. Eskut et al. (CHORUS Collaboration), Nucl. Instrum. Meth. A **401**, 7 (1997).

116. J. Altegoer et al. (NOMAD Collaboration), Phys. Lett. B **431**, 219 (1998); P. Astier et al., Phys. Lett. B **453**, 169 (1999).

117. E. Eskut et al. (CHORUS Collaboration), Phys. Lett. B **434**, 205 (1998); Phys. Lett. B **424**, 202 (1998).

118. N. Ushida et al. (E531 Collaboration), Phys. Rev. Lett. **57**, 2897 (1986).

119. G. Ambrosini et al. (NA56/SPY Collaboration), Eur. Phys. J. C **10**, 605 (1999); Phys. Lett. B **425**, 208 (1998).

120. M.C. Gonzales-Garcia and J.J. Gomez-Cadenas, Phys. Rev. D **55**, 1297 (1997); B. Van de Vijver and P. Zucchelli, Nucl. Instrum. Meth. A **385**, 91 (1997).

121. L. Ludovici (CHORUS collaboration), in *Proceedings of the 19th International Conference on Neutrino Physics and Astrophysics (Neutrino 2000)*, Sudbury, Canada, 16–21 June 2000, eds. J. Law, R.W. Ollerhead, and J.J. Simpson, Nucl. Phys. B (Proc. Suppl.) **91**, 177 (2001).

122. P. Annis et al. (CHORUS Collaboration), Phys. Lett. B **435**, 458 (1998).

123. P. Astier et al. (NOMAD Collaboration), Nucl. Phys. B **611**, 3 (2001).

124. K.S. McFarland et al. (CCFR Collaboration), Phys. Rev. Lett. **75**, 3993 (1995).

125. F. Dydak et al. (CDHS Collaboration), Phys. Lett. B **134**, 281 (1984).

126. P. Astier et al. (NOMAD Collaboration), Phys. Lett. B **471**, 406 (2000).

127. S.M. Bilenky et al., arXiv:hep-ph/9906251.

128. A.S. Ayan et al., TOSCA letter of intent, CERN-SPSC/97-5, SPSC/I 213 (1997); addendum to letter of intent, CERN-SPSC/98-20, SPSC/I 213 add. (May 1998); R.A. Sidwell (COSMOS Collaboration), in *Neutrino '96, Proceedings of the 17th International Conference on Neutrino Physics and Astrophysics*, Helsinki, 13 June 1996, eds. K. Enqvist, K. Huitu, and J. Maalampi (World Scientific, Singapore, 1997), p. 152.

129. A. Rubbia et al., proposal for ICARUS in the Jura, CERN/SPSLC 96-58, SPSLC/P 304 (1996); D. Autiero et al., Letter of Interest for Jura experiment, CERN/SPSC/97-23, SPSC/I 217 (Oct. 1997).

130. A. Geiser, Eur. Phys. J. C **7**, 437 (1999).

131. M.H. Ahn et al. (K2K Collaboration), Phys. Rev. Lett. **90**, 041801 (2003).

132. K. Nishikawa (K2K Collaboration), *Proceedings of XXth International Conference on High Energy Physics and Astrophysics*, Munich, Germany, 25–30 May 2002, p. 129; S.H. Ahn et al. (K2K Collaboration), Phys. Lett.

B **511**, 178 (2001); Y. Oyama (E362 (K2K) Collaboration), in *Proceedings of 35th Rencontres de Moriond "Electroweak Interactions and Unified Theories"*, Les Arcs, Savoie, France, 11–18 March 2000, arXiv: hep-ex/0004015, and references therein; K. Nakamura (K2K Collaboration), in *Proceedings of the 19th International Conference on Neutrino Physics and Astrophysics (Neutrino 2000)*, Sudbury, Canada, 16–21 June 2000, eds. J. Law, R.W. Ollerhead, and J.J. Simpson, Nucl. Phys. B (Proc. Suppl.) **91**, 203 (2001); C.K. Jung, in *Proceedings of Lepton Photon 2001*, Rome, 23–28 July 2001, http://www.lp01.infn.it/.

133. MINOS Collaboration, Fermilab proposal P875 (1995); MINOS Collaboration, *Neutrino Oscillation Physics at Fermilab: The NuMI-MINOS Project*, NuMI-L375 (May 1998); P. Adamson et al., MINOS technical design report, NUMI-L-337 (Oct. 1998).

134. G. Aquistapace et al., CERN 98-02, INFN/AE-98/05 (1998); CERN-SL/99-034 (1999); A. Ball et al., CERN-SL/2000-063 (2000).

135. M. Guler et al., OPERA proposal, CERN-SPSC/2000-028, LNGS-P25/2000 (2000); M. Guler et al., OPERA status report, CERN-SPSC/2001-025, LNGS-EXP-30-2001-ADD-1 (2001); A. Gocco, Nucl. Phys. Proc. Suppl. **85**, 125 (2000).

136. P. Benetti et al. (ICARUS Collaboration), Nucl. Instrum. Meth. A **327**, 327 (1993); Nucl. Instrum. Meth. A **332**, 332 (1993); P. Cennini et al., Nucl. Instrum. Meth. A **333**, 567 (1993); Nucl. Instrum. Meth. A **345**, 230 (1994); Nucl. Instrum. Meth. A **355**, 355 (1995); F. Arneodo et al. (ICARUS Collaboration), Nucl. Instrum. Meth. A **471**, 272 (2000); arXiv:hep-ex/0103008.

137. Y. Itow et al., arXiv:hep-ex/0106019.

5 Theoretical Interpretation of Current Neutrino Oscillation Data

Gianluigi Fogli and Eligio Lisi

We discuss the theoretical interpretation of neutrino oscillation data in terms of 3ν and 4ν mixing. Two-neutrino oscillations, often used to describe experimental results in a first approximation, are briefly recalled (Sect. 5.1). The main focus of our review is 3ν mixing (Sect. 5.2), which accommodates both the negative results of oscillation searches at reactors (Sect. 5.3) and the evidence for flavor transitions obtained from atmospheric and solar neutrino data (Sects. 5.4 and 5.5). The status and problems of 4ν scenarios embedding the additional LSND signal are also discussed (Sect. 5.7). Finally, we outline the impact of the very latest data (Sect. 5.8). Standard electroweak neutrino interactions are assumed in all cases; scenarios with nonstandard dynamics are beyond the scope of this review.

5.1 Two-Neutrino Oscillations

Since both the 3ν and the 4ν scenarios admit effective 2ν limits of phenomenological interest, we start by recalling the basics of 2ν oscillations. The mixing between two flavor eigenstates (say ν_α and ν_β) and two mass eigenstates (ν_1 and ν_2), separated by a gap $\Delta m^2 = m_2^2 - m_1^2$, can be described through

$$\begin{pmatrix} \nu_\alpha \\ \nu_\beta \end{pmatrix} = \begin{pmatrix} \cos\theta & \sin\theta \\ -\sin\theta & \cos\theta \end{pmatrix} \begin{pmatrix} \nu_1 \\ \nu_2 \end{pmatrix} = U(\theta) \begin{pmatrix} \nu_1 \\ \nu_2 \end{pmatrix}, \qquad (5.1)$$

where $\theta \in [0, \pi/2]$ is the mixing angle and $U(\theta)U^\dagger(\theta) = I$.

The 2ν evolution equation in the flavor basis,

$$i\frac{d}{dx}\begin{pmatrix} \nu_\alpha \\ \nu_\beta \end{pmatrix} = H \begin{pmatrix} \nu_\alpha \\ \nu_\beta \end{pmatrix}, \qquad (5.2)$$

is governed, in general, by a Hamiltonian containing both a kinetic ("vacuum") term [1] and an interaction ("matter") term [2],

$$H = H_{\text{kin}} + H_{\text{int}} = \frac{k}{2}U(\theta)\begin{pmatrix} -1 & 0 \\ 0 & +1 \end{pmatrix}U^\dagger(\theta) + \frac{V}{2}\begin{pmatrix} +1 & 0 \\ 0 & -1 \end{pmatrix}, \qquad (5.3)$$

where irrelevant terms, proportional to the unit matrix I, have been traced away. In (5.3), $k = \Delta m^2/2E$ is the wave number of the oscillation, and

$V = V_\alpha - V_\beta$ is the difference between the interaction energies of ν_α and ν_β in matter, which is typically proportional to a background fermion density N_f times G_F. Solutions of (5.2) have been studied in a variety of contexts [3, 4].

5.2 Three-Neutrino Oscillations

In this section, we consider the combined interpretation of solar and atmospheric neutrino data within a 3ν oscillation framework (implicitly excluding the LSND result). It will be shown that the emerging 3ν scenario for the squared-mass gaps and mixing angles of the neutrinos exhibits a nontrivial self-consistency, which justifies its wide use in both theoretical model-building and proposals for experimental studies. We start with some order-of-magnitude estimates, which are useful for understanding the main features of the current 3ν phenomenology.

High-statistics observations of atmospheric neutrino events show no unexpected features in the ν_e-induced electrons, but indicate a strong disappearance of ν_μ-induced muons [5]. By exclusion, the disappearing ν_μ's must, basically, convert into (unobservable) ν_τ's, implying an oscillation dynamics dominated by two states (ν_μ, ν_τ). In the leading approximation, we also have $H_{int} = 0$, since ν_μ and ν_τ have the same interaction energies in the matter of the earth $(V_\mu - V_\tau = 0$, see (5.3)). Denoting by (m^2, ψ) the oscillation parameters $(\Delta m^2, \theta)$ of the (ν_μ, ν_τ) subspace, the corresponding transition probability is $P_{\mu\tau} = \sin^2 2\psi \sin^2(m^2 L/4E_\nu)$. Since the onset of the disappearance of atmospheric ν_μ's is best observed for E_ν of the order of a few times GeV (multi-GeV events) and for a path length $L \sim 10^3$ km, the observation of a large effect $(m^2 L/4E_\nu \sim \pi/2)$ implies that m^2 is of the order of a few 10^{-3} eV^2. For such values of m^2, the results of the CHOOZ reactor experiment forbid significant ν_e mixing [6], which is consistent with the effective decoupling of ν_e inferred from atmospheric electron events. Concerning the mixing angle ψ, the $\sim 50\%$ disappearance observed for upgoing multi-GeV muon events (characterized by $L \sim 2R_\oplus$ and thus by $\langle \sin^2(m^2 L/4E_\nu)\rangle \sim 1/2$) implies $P_{\mu\tau} \sim 1/2$ and thus $\sin^2 2\psi \sim 1$, namely, $\psi \sim \pi/4$.

All observations of solar neutrino events so far [7–11] indicate a significant (1/2 to 1/3) suppression of the expected ν_e flux from the sun [12]. In terms of 2ν oscillations, with the solar-neutrino mass-mixing parameters denoted by $(\delta m^2, \omega)$, the earth–sun distance sets a lower limit on the oscillation wave number and thus on δm^2 $(\delta m^2 \gtrsim 10^{-12}$ eV$^2)$. An upper limit on δm^2 is not unambiguously set by solar neutrino data alone, since energy-independent scenarios (corresponding to $\delta m^2 \to \infty$), although globally disfavored, are not ruled out by any single observation. A more solid upper limit $(\delta m^2 \lesssim 7 \times 10^{-4}$ eV$^2)$ is provided by CHOOZ [6], forbidding large ν_e disappearance above that range. An important consequence is that $\delta m^2 < m^2$ (and, more often than not, $\delta m^2 \ll m^2$). Concerning ω, both upper and lower limits are found on the allowed range of δm^2 (details are given in Sect. 5.5).

In conclusion, both upper and lower bounds can currently be set on two independent squared-mass differences (δm^2 and m^2), and on two independent mixing angles (ψ and ω). A third mixing angle, φ, which parameterizes subleading 3ν effects (neglected in a first approximation) in *both* atmospheric *and* solar neutrino oscillations, is limited only from above (by reactor data). This qualitative understanding of the 3ν oscillation parameters ($\delta m^2, m^2, \omega, \varphi, \psi$) will be quantitatively refined in the following sections, starting from the precise definition of those parameters.

5.2.1 3ν Mass Spectra

Current ν oscillation phenomenology favors a "solar" squared-mass difference δm^2 significantly smaller than the "atmospheric" squared-mass difference m^2, i.e.

$$\delta m^2 \ll m^2 . \tag{5.4}$$

Figure 5.1 shows the two possible spectra that satisfy the above condition, characterized by a solar neutrino doublet either lighter (Fig. 5.1a) or heavier (Fig. 5.1b) than the third, isolated mass eigenstate. The cases illustrated in Fig. 5.1a,b are usually referred to as the "normal" and "inverted" hierarchies, respectively, of the squared-mass differences.

In order to uniquely define the mixing angles, labels $(1, 2, 3)$ have to be attached to the mass eigenstates. In our convention, we do so by defining the squared-mass matrix $\mathcal{M}^2 = \mathrm{diag}(m_1^2, m_2^2, m_3^2)$ in the (ν_1, ν_2, ν_3) basis as

$$\mathcal{M}^2 = \mathrm{diag}(-\delta m^2/2, +\delta m^2/2, +m^2) \quad \text{for the normal hierarchy ,} \tag{5.5}$$
$$\mathcal{M}^2 = \mathrm{diag}(-\delta m^2/2, +\delta m^2/2, -m^2) \quad \text{for the inverted hierarchy ,} \tag{5.6}$$

Reference mass spectrum

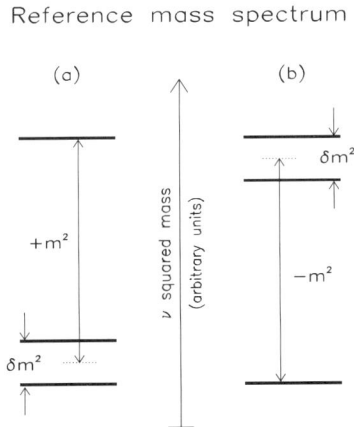

Fig. 5.1. Three-neutrino spectra of squared-mass differences that satisfy (5.4): (**a**) normal hierarchy, (**b**) inverted hierarchy. The zero of the vertical axis (i.e. the absolute squared neutrino mass) is undetermined in oscillation searches

up to an overall additive term (unobservable in oscillation searches, as is the case for any absolute squared mass of a neutrino). With the above notation, the two spectra of Fig. 5.1a,b can be interchanged by simply flipping the sign of m^2. The current neutrino oscillation phenomenology is, however, very weakly sensitive to $\text{sign}(m^2)$. Unless otherwise noted, we shall consider the case $+m^2$ (normal hierarchy) as the default.

5.2.2 3ν Mixing

The unitary mixing matrix U connecting flavor states ν_α and mass states ν_i,

$$\nu_\alpha = U\nu_i , \tag{5.7}$$

can be parameterized in a number of ways. We adopt the standard parameterization in terms of three Euler rotations [3], ordered as

$$U = U_{23}(\psi)U_{13}(\varphi)U_{12}(\omega) \tag{5.8}$$

$$= \begin{pmatrix} 1 & 0 & 0 \\ 0 & c_\psi & s_\psi \\ 0 & -s_\psi & c_\psi \end{pmatrix} \begin{pmatrix} c_\varphi & 0 & s_\varphi \\ 0 & 1 & 0 \\ -s_\varphi & 0 & c_\varphi \end{pmatrix} \begin{pmatrix} c_\omega & s_\omega & 0 \\ -s_\omega & c_\omega & 0 \\ 0 & 0 & 1 \end{pmatrix} \tag{5.9}$$

$$= \begin{pmatrix} c_\varphi c_\omega & c_\varphi s_\omega & s_\varphi \\ -c_\psi s_\omega - s_\varphi s_\psi s_\omega & c_\psi c_\omega - s_\varphi s_\psi s_\omega & c_\varphi s_\psi \\ s_\psi s_\omega - s_\varphi c_\psi c_\omega & -s_\psi c_\omega - s_\varphi c_\psi s_\omega & c_\varphi c_\psi \end{pmatrix} , \tag{5.10}$$

where $c = \cos$ and $s = \sin$, and we have neglected a possible CP-violating phase, to which the current oscillation phenomenology is basically insensitive.
    The angles (ω, φ, ψ) are also often written as

$$(\omega, \varphi, \psi) \equiv (\theta_{12}, \theta_{13}, \theta_{23}) , \tag{5.11}$$

and take values in the first quadrant $[0, \pi/2]$. The two octants $[0, \pi/4]$ and $[\pi/4, \pi/2]$ embed, in general, different oscillation physics [13–15]. Therefore, the associated (and time-honored) variable $\sin^2 2\theta_{ij}$, which maps the first octant only, is being currently replaced by the more appropriate two-octant variables $\sin^2 \theta_{ij}$ (on a linear scale) or $\tan^2 \theta_{ij}$ (on a logarithmic scale) [13, 14].

5.2.3 3ν Dynamics

In the flavor basis $\nu = (\nu_e, \nu_\mu, \nu_\tau)^{\text{T}}$, the 3ν evolution equation reads, in general,

$$i\frac{d\nu}{dx} = \left(U\frac{\mathcal{M}^2}{2E}U^\dagger + H_{\text{int}} \right)\nu , \tag{5.12}$$

where the first (kinetic) term is defined through (5.5)–(5.8), while the second (interaction) term is given by $H_{\text{int}} = \text{diag}(V(x), 0, 0)$, where [2]

$$V(x) = \sqrt{2}G_{\text{F}}N_{\text{e}}(x) \, . \tag{5.13}$$

Here G_{F} is the Fermi constant, and $N_{\text{e}}(x)$ is the electron density at the position x (for neutrinos traveling in matter). For antineutrinos, $V \to -V$.

Solutions of the above equation involve, in general, all the mass-mixing parameters $(\delta m^2, m^2, \omega, \varphi, \psi)$, as well as a (sometimes complicated) functional dependence on the profile of $N_{\text{e}}(x)$ in matter. As we shall see, phenomenological considerations allow us to reduce the 3ν dynamics to an effective 2ν dynamics in the two classes of "solar" and "atmospheric" experiments. We start, however, from an exception to this reduction rule, provided by the CHOOZ reactor experiment.

5.3 Reactor Neutrinos in a 3ν Interpretation

The CHOOZ experiment has not observed disappearance of reactor ν_{e}'s[1] over a baseline $L \simeq 1\,\text{km}$ [6]. In terms of 2ν mixing, the CHOOZ bounds on $(\Delta m^2, \theta)$ are shown in Fig. 5.2. At maximal mixing ($\theta \simeq \pi/4$), the upper limit on Δm^2 is $\sim 7 \times 10^{-3}\,\text{eV}^2$. At large Δm^2, the upper limit on $\sin^2 2\theta$ is ~ 0.1, corresponding to θ close to either 0 or $\pi/2$. Somewhat weaker bounds have been obtained at Palo Verde [16].

In the case of 3ν oscillations, the survival probability of reactor ν_{e}'s obtained by CHOOZ for the normal hierarchy reads [17–19]

$$P_{\text{ee}}^{\text{reac}} = 1 - c_{\varphi}^4 s_{2\omega}^2 \sin^2\left(\frac{\delta m^2}{4E}x\right)$$
$$- s_{2\varphi}^2 \left[s_{\omega}^2 \sin^2\left(\frac{m^2 - \delta m^2/2}{4E}x\right) + c_{\omega}^2 \sin^2\left(\frac{m^2 + \delta m^2/2}{4E}x\right) \right] \, . \tag{5.14}$$

The case of an inverted hierarchy is obtained through the replacement $+m^2 \to -m^2$. Notice that the m^2 sign flip can be absorbed by swapping the octants for the angle ω ($s_{\omega} \to c_{\omega}$). Therefore, reactor experiments *cannot* distinguish a normal from an inverted hierarchy, unless ω is *constrained to be in one octant* by other (e.g. solar) neutrino experiments. This remark is relevant in the context of recent studies of the possibility of hierarchy discrimination at future reactor experiments [20, 21]. Notice also that $P_{\text{ee}}^{\text{reac}}$ does

---

[1] We speak loosely about reactor ν_{e}'s rather than $\bar{\nu}_{\text{e}}$'s, since matter effects are negligible in this context, and thus the oscillation properties of ν and $\bar{\nu}$ are formally equivalent.

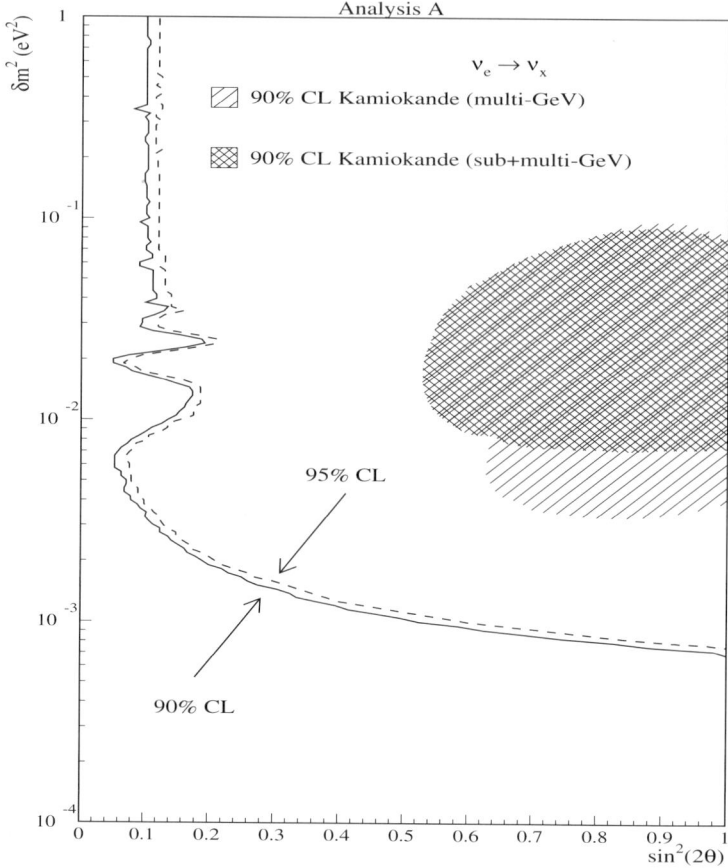

Fig. 5.2. Region excluded by CHOOZ in the two-family parameter space $(\Delta m^2, \theta)$, reported in [6]

not depend on the angle $\psi = \theta_{23}$, as a result of the ignorance of our final flavor state, ν_μ or ν_τ, in reactor experiments.

Equation (5.14) admits two interesting 2ν subcases. The first is obtained for δm^2 well below the CHOOZ sensitivity (namely, δm^2 much less than a few times 10^{-3} eV^2), as favored by current solutions to the solar neutrino problem (see Sect. 5.5). In this case one can effectively take $\delta m^2 = 0$, so that

$$\delta m^2 = 0 \quad \Longrightarrow \quad P_{ee}^{\text{reac}} = 1 - s_{2\varphi}^2 \sin^2\left(\frac{m^2}{4E}x\right) , \qquad (5.15)$$

which is equivalent to the two-family probability, up to the identifications $(m^2, \varphi) \equiv (\Delta m^2, \theta)$, which allows one to apply the same limits as in Fig. 5.2. Equation (5.15) is often used to combine the CHOOZ results with atmospheric neutrino data. Since these data constrain m^2 to be close to

$\sim 3 \times 10^{-3}$ eV^2, and forbid the case $s_\varphi^2 \sim 1$, the CHOOZ + atmospheric combination constrains s_φ^2 to be smaller than a few times 10^{-2} [17, 22, 23] (see Sect. 5.4.4).

The second 2ν subcase of (5.14) is obtained for $\varphi = 0$:

$$\varphi = 0 \quad \Longrightarrow \quad P_{ee}^{reac} = 1 - s_{2\omega}^2 \sin^2 \left(\frac{\delta m^2}{4E} x \right) , \qquad (5.16)$$

which leads to the identification $(\delta m^2, \omega) \equiv (\Delta m^2, \theta)$ in Fig. 5.2. Equation (5.16) is often used to combine the CHOOZ results with solar neutrino data. Since these data constrain $\sin^2 \omega$ to be close to $\sim 1/2$ if δm^2 approaches the CHOOZ sensitivity limits, the CHOOZ + solar combination constrains δm^2 to be smaller than $\sim 7 \times 10^{-4}$ eV^2 (see Sect. 5.5).

However, when *both δm^2 and s_φ^2* approach their upper limits (δm^2 of the order of a few times 10^{-4} eV^2 and s_φ^2 of the order of a few percent), neither of the two subcases in (5.15) and (5.16) applies, and the full 3ν expression in (5.14) must be used for P_{ee}^{reac}. In this case, the 2ν limits in Fig. 5.2 cannot be directly applied, and a statistical reanalysis of the CHOOZ spectral data is needed for any given set of $(\delta m^2, m^2, \omega, \varphi)$ values [17, 18, 24]. Such a reanalysis has been performed in recent CHOOZ + solar combined 3ν fits [17, 25, 26], and its implications will be briefly illustrated in Sect. 5.5.2.

5.4 Atmospheric Neutrinos in a 3ν Interpretation

In the 3ν analysis of atmospheric neutrino data, the zeroth-order approximation in the small parameter $\delta m^2/m^2$ (sometimes called the one-dominant-mass-scale approximation) is often used [17, 22]. The results discussed here have been obtained within such an approximation. Perturbations due to $\delta m^2 \neq 0$ have rather small effects in the current phenomenology [24, 27–30].

5.4.1 Parameter Space

In the analysis of atmospheric (+ reactor) neutrino oscillations, the assumption $\delta m^2 = 0$ implies an effective degeneracy of the "solar" doublet (ν_1, ν_2), which makes the rotation $U_{12}(\omega)$ (5.8) irrelevant in the 3ν evolution Hamiltonian (5.12). Therefore, to zeroth order in $\delta m^2/m^2$, the subspace of terrestrial (atmospheric + reactor) neutrino oscillations reduces to

$$\text{terrestrial } \nu \text{ parameter space} \simeq (m^2, \varphi, \psi) . \qquad (5.17)$$

The only observable parameters are those associated with the "lone" state ν_3, i.e. the largest mass gap m^2, and the mixing-matrix elements $U_{\alpha 3}$,

$$(m^2, U_{e3}, U_{\mu 3}, U_{\tau 3}) \equiv (m^2, s_\varphi, c_\varphi s_\psi, c_\varphi c_\psi) , \qquad (5.18)$$

consistently with (5.17).

In the vacuum case ($N_e = 0$), the evolution equation (5.12) is easily solved, giving the following probabilities of flavor appearance ($\alpha \neq \beta$) and disappearance ($\alpha = \beta$), which are useful in interpreting accelerator and reactor oscillation searches:

$$P_{\alpha\beta}^{\mathrm{vac}} = 4U_{\alpha3}^2 U_{\beta3}^2 \sin^2(m^2 x/4E) , \qquad (5.19)$$

$$P_{\alpha\alpha}^{\mathrm{vac}} = 1 - 4U_{e3}^2(1 - U_{e3}^2) \sin^2(m^2 x/4E) . \qquad (5.20)$$

Such probabilities are formally equivalent to those in the two-family case, modulo the identifications $\Delta m^2 \leftrightarrow m^2$ and $\sin^2 2\theta \leftrightarrow 4U_{\alpha3}^2 U_{\beta3}^2$ or $\sin^2 2\theta \leftrightarrow 4U_{\alpha3}^2(1 - U_{\alpha3}^2)$, which allow one to map published 2ν limits onto the 3ν terrestrial parameter space in (5.17) [13, 31]. Notice that the above equations are invariant under the replacement $+m^2 \to -m^2$, i.e.

$$P_{\alpha\beta}^{\mathrm{vac}}(+m^2) = P_{\alpha\beta}^{\mathrm{vac}}(-m^2) , \qquad (5.21)$$

and thus do not distinguish the normal from the inverted hierarchy.

However, the vacuum limits in (5.19) and (5.20) are not applicable to atmospheric neutrino oscillations, which largely occur within the matter of the earth. For $N_e \neq 0$, it turns out that 3ν oscillations do not reduce to 2ν subcases, since the degeneracy of the solar neutrino doublet is lifted in matter [32], and genuine 3ν effects take place. A relevant consequence in matter is that

$$P_{\alpha\beta}^{\mathrm{mat}}(+m^2) \neq P_{\alpha\beta}^{\mathrm{mat}}(-m^2) , \qquad (5.22)$$

which provides a handle for hierarchy discrimination. Unfortunately, the differences in (5.22) vanish as $\varphi \to 0$, and are currently undetectable within the stringent CHOOZ upper limits on φ.

5.4.2 Data

Concerning atmospheric neutrinos, our statistical analysis [23] refers to the Super-Kamiokande (SK) data with 79.5 kton/year (kTy) exposure [33], as shown in Fig. 5.3. The data in the first two panels represent the observed rates of e-like events (induced by ν_e or $\bar{\nu}_e$) at sub-GeV (SGe) or multi-GeV (MGe) energies, as a function of the lepton zenith angle ϑ (where $\cos\theta = -1$, 0, and $+1$ corresponds to upgoing, horizontal, and downgoing leptons). The remaining four panels represent the observed rates of μ-like events (induced by ν_μ or $\bar{\nu}_\mu$) at sub-GeV and multi-GeV energies (SGμ and MGμ), and also at the higher energies that characterize the sample of upgoing muons originating in the rock below the detector, which can either stop in (USμ) or pass through (UTμ) the detector. In the last panel, the overwhelming μ shower background from above restricts the event selection to zenith angles below the horizon, i.e. $\cos\vartheta \in [-1, 0]$. The error bars are statistical and correspond to $\pm 1\sigma$.

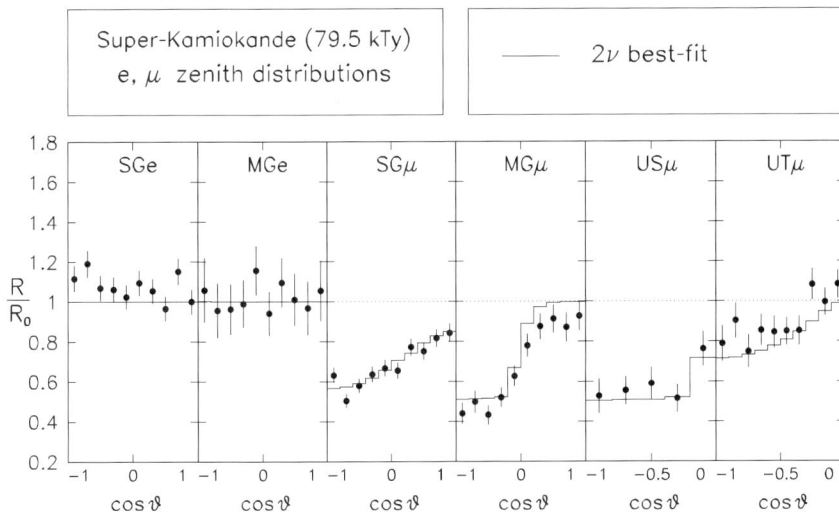

Fig. 5.3. Reference atmospheric neutrino data from the Super-Kamiokande experiment, together with the 2ν best-fit predictions in the $\nu_\mu \to \nu_\tau$ channel [23]

The systematic errors (not shown, but included in the analysis) are large (20%–30%) and highly correlated [22].

In each panel of Fig. 5.3, the lepton rates R are normalized to the expectations R_0 in the absence of oscillations, so that any deviation from the horizontal line at $R/R_0 = 1$ signals a possible flavor oscillation effect. It can be seen that the electron samples are roughly in agreement with the no-oscillation hypothesis, the small overall excess in the SGe sample being smaller than the systematic (normalization) uncertainties. In contrast, strong distortions of the zenith distributions are visible in all μ samples; none of these distortions can be fully absorbed by normalization effects. The striking evidence for ν_μ disappearance is well fitted, in all its zenith-energy spectral features, by the simple hypothesis of pure $\nu_\mu \to \nu_\tau$ oscillations (solid lines), shown here for our current best-fit point $(m^2/\text{eV}^2, \sin^2 2\psi) = (0.003, 0.98)$ [23].

5.4.3 Graphical Representations

We shall use two kinds of graphical representation to show the results of the SK (+ CHOOZ) 3ν oscillation analysis: a triangle plot and a bi-logarithmic plot. Both representations map the full quadrant $[0, \pi/2]$ spanned by the mixing angles φ and ψ. Similar plots will also be used for the 4ν analysis (Sect. 5.7).

Figure 5.4 shows the triangle plot, which embeds the unitarity condition on ν_3, $U_{e3}^2 + U_{\mu3}^2 + U_{\tau3}^2 = 1$, through the geometrical property $h_1 + h_2 + h_3 = 1$, which links the three heights h_i of any point, measured perpendicular to the sides, inside an equilateral triangle of unit height. The mapping in terms of

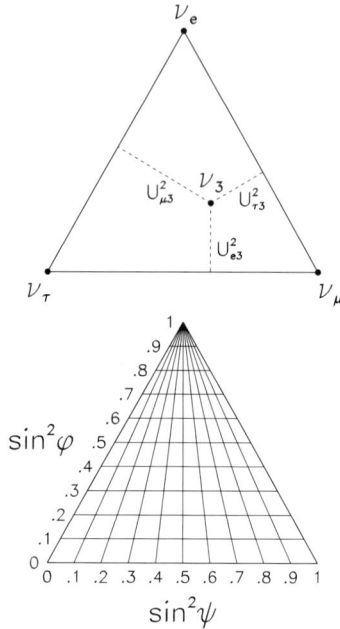

Fig. 5.4. Triangular representation of the terrestrial 3ν parameter space defined in (5.17) and (5.18) [22], embedding the unitarity constraint $U_{e3}^2 + U_{\mu3}^2 + U_{\tau3}^2 = 1$

$\sin^2 \varphi$ and $\sin^2 \psi$ [22, 34] is also shown. The inner points, the sides, and the corners of the triangle correspond to 3ν, 2ν, and no oscillations, respectively. In particular, the bottom and right sides represent pure $\nu_\mu \to \nu_\tau$ and pure $\nu_\mu \to \nu_e$ oscillations, respectively; between the bottom and the right sides, genuine 3ν oscillations $\nu_\mu \to \nu_{\tau,e}$ occur.

The second graphical representation uses sections (or projections) of allowed regions in the following log-scale rectangular coordinates:

$$\text{coordinates} = (m^2, \tan^2 \varphi, \tan^2 \psi) . \qquad (5.23)$$

The choice of $\tan^2 \theta_{ij}$ on a log scale is motivated by the property $\log \tan^2(\theta_{ij}) = -\log \tan^2(\pi/2 - \theta_{ij})$, which preserves graphically the octant symmetry of the oscillation probability, when applicable (e.g. for the 2ν subcases in vacuum).

5.4.4 Constraints on 3ν Parameters

Constraints on the mass-mixing oscillation parameters can be obtained through a statistical χ^2 comparison of experimental data and theoretical predictions [17, 22]. Figure 5.5 shows χ^2 from SK atmospheric neutrino data as a function of the squared-mass difference m^2 [35], for unconstrained values

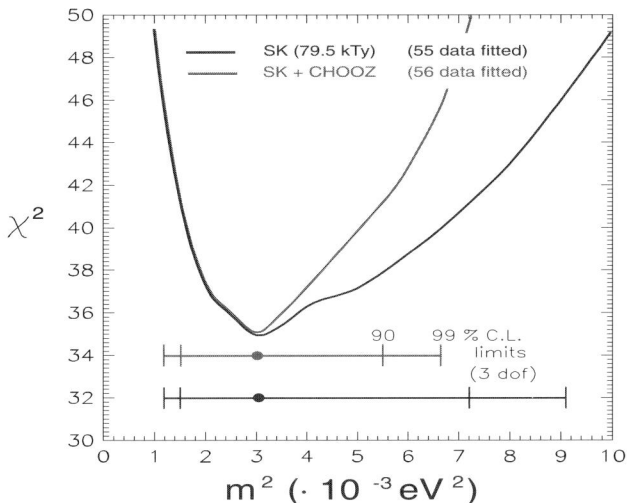

Fig. 5.5. 3ν analysis of terrestrial neutrino data: χ^2 as a function of m^2, for unconstrained mixing angles φ and ψ, from SK and SK + CHOOZ data [35]

of φ and ψ. The best fit is located at $m^2 = 3 \times 10^{-3}$ eV^2, a value that appears to be rather stable (the differences in m^2 between independent state-of-the-art analyses being smaller than $\pm 0.5 \times 10^{-3}$ eV^2). The inclusion of CHOOZ data in Fig. 5.5 improves the upper bound on m^2, without shifting its best-fit value.

The favored m^2 range is within a factor of ~ 2 of the central value (the precise interval depending on the confidence level chosen). Therefore, we have taken six representative values of m^2 from 1.5 to 6×10^{-3} eV^2 in order to illustrate the complementary constraints on 3ν mixing, in the triangle plots in Fig. 5.6. In the left column of triangles, the regions allowed by the SK data always touch the lower side, corresponding to pure $\nu_\mu \to \nu_\tau$ oscillations (not surprisingly, given the excellent quality of the $\nu_\mu \to \nu_\tau$ fit in Fig. 5.3). Such regions extend considerably within the triangle, to areas where genuine 3ν oscillations occur and the additional $\nu_\mu \to \nu_e$ channel is open. However, such a channel cannot be dominant: the SK-allowed regions are always far from pure $\nu_\mu \to \nu_e$ oscillations (right side of the triangle). The middle column in Fig. 5.6 shows the CHOOZ constraints, which exclude wide horizontal bands. Indeed, the CHOOZ limit $\sin^2 2\varphi \lesssim 0.1$ allows only narrow strips in the triangle at $s_\varphi^2 \sim 0$ (lower side) or $s_\varphi^2 \sim 1$ (upper corner), the latter being excluded by SK.

The combination of the SK + CHOOZ data is shown in the right column of Fig. 5.6. The allowed region is squeezed into a small strip close to the lower side. The height of this strip is just U_{e3}^2 and is smaller than a few times 10^{-2}. The center of the strip corresponds to "maximal" $\nu_\mu \to \nu_\tau$ mixing

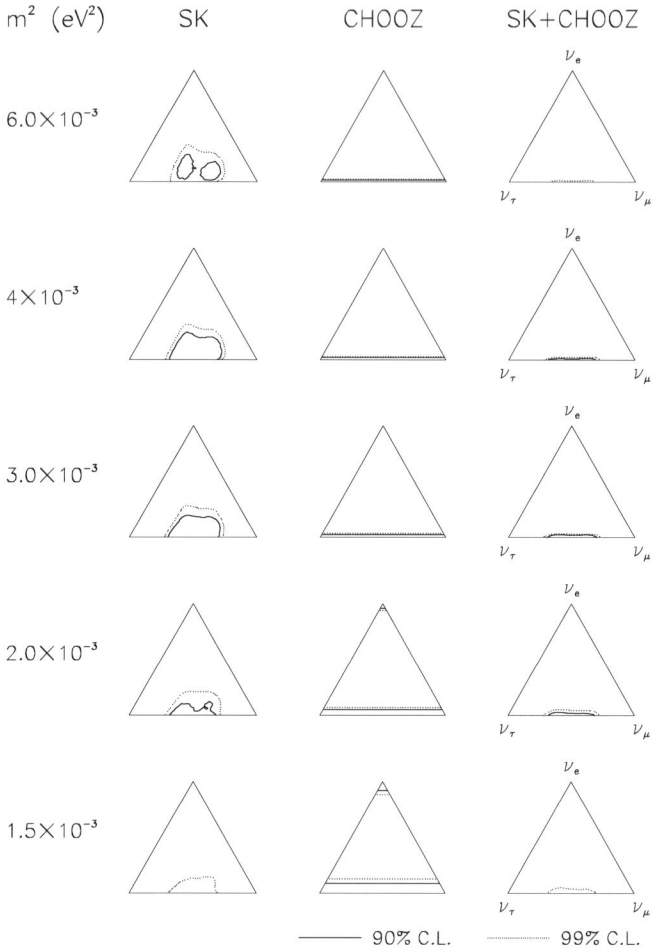

Fig. 5.6. Triangle plot of SK, CHOOZ, and SK + CHOOZ constraints on the 3ν mass-mixing oscillation parameters. The *curves* represent sections of the region allowed at 90% and 99% C.L. (for $N_{DF} = 3$) by a χ^2 analysis of the data, for six representative values of m^2. See also [36]

$(U_{\mu3}^2/U_{\tau3}^2 = 1)$. Deviations from maximal mixing are confined to a range of approximately $1/2 \lesssim U_{\mu3}^2/U_{\tau3}^2 \lesssim 2$. The CHOOZ data appear to have a tremendous impact on the mixing parameters, leaving room only for a small admixture of ν_e on top of dominant $\nu_\mu \to \nu_\tau$ oscillations.

Since the SK + CHOOZ allowed region is rather "squeezed" in the plots of Fig. 5.6, a log-scale enhancement appears useful. Figure 5.7 show the projections of the 3ν volume allowed by the SK data at 90% and 99% C.L. ($\Delta\chi^2 = 6.25$ and 11.36, respectively, for $N_{DF} = 3$) in the $(m^2, \tan^2\psi, \tan^2\varphi)$ parameter space. The best fit is reached at $(m^2, \tan^2\psi, \tan^2\varphi) = (3 \times 10^{-3}\ \text{eV}^2, 0.9,$

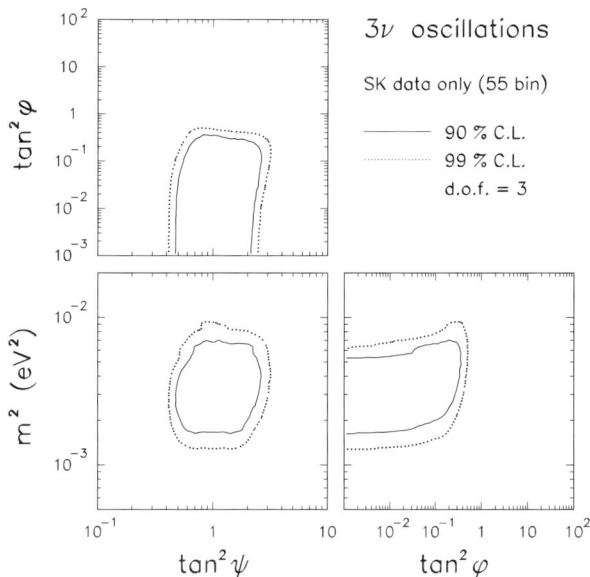

Fig. 5.7. Projections onto the coordinate planes of the allowed regions in the 3ν parameter space $(m^2, \tan^2\psi, \tan^2\varphi)$ at 90% and 99% C.L. for $N_{\rm DF} = 3$. The fit includes SK data only (79.5 kTy) [23]

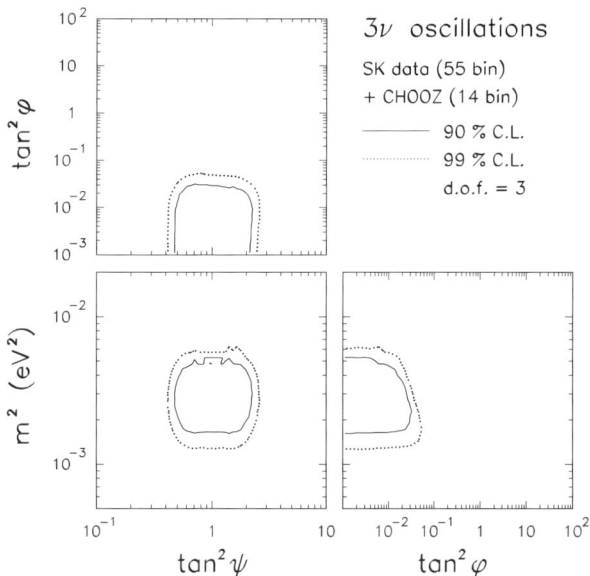

Fig. 5.8. Projections onto the coordinate planes of the allowed regions in the 3ν parameter space $(m^2, \tan^2\psi, \tan^2\varphi)$ at 90% and 99% C.L. for $N_{\rm DF} = 3$. The fit includes SK and CHOOZ data [23]

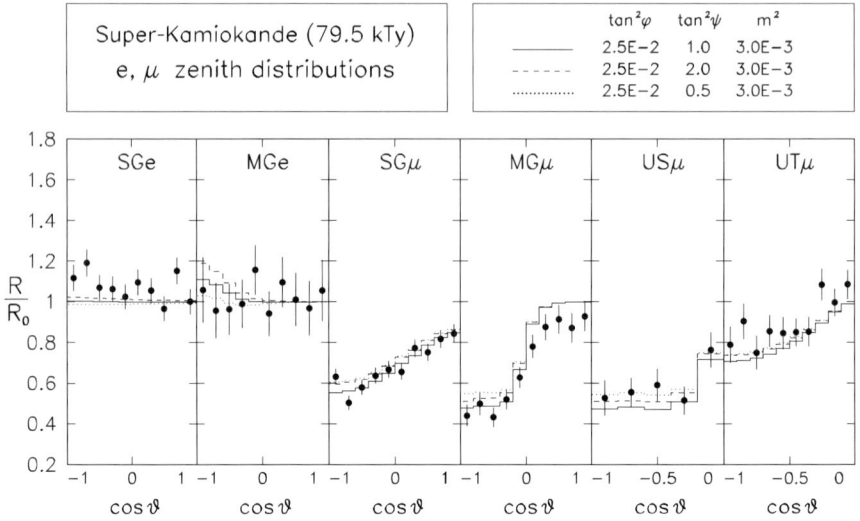

		tan²φ	tan²ψ	m²
———		2.5E−2	1.0	3.0E−3
- - - - -		2.5E−2	2.0	3.0E−3
··········		2.5E−2	0.5	3.0E−3

Super-Kamiokande (79.5 kTy)

e, μ zenith distributions

Fig. 5.9. SK zenith distributions for three representative 3ν cases with $\tan^2 \varphi = 2.5 \times 10^{-2}$, allowed at 90% C.L. by SK + CHOOZ

0.01). The slight deviation of the best-fit mixing from the pure $\nu_\mu \leftrightarrow \nu_\tau$ maximal mixing $((\tan^2 \psi, \tan^2 \varphi) = (1, 0))$, although intriguing, is – unfortunately – not statistically significant $(\Delta\chi^2 \lesssim 1)$. This also implies that there is no significant indication of possible matter effects related to ν_e mixing $(\tan^2 \varphi > 0)$ in the SK data. On the other hand, the SK preference for small values of φ represents a very important and nontrivial consistency check of the 3ν oscillation scenario, which is independently confirmed by solar neutrino data (see Sect. 5.5) and, as we have just seen, by CHOOZ. Notice that the bounds on $\tan^2 \psi$ in Fig. 5.7 are octant-symmetric only in the 2ν limit $\tan^2 \varphi \to 0$ (as should be the case), and show a slight preference for ψ in the second octant when $\tan^2 \varphi > 0$. Correspondingly, slightly higher values of m^2 are preferred. The (weak) positive correlation between $\tan^2 \varphi$ and $\tan^2 \psi$ or m^2, however, is largely suppressed by the CHOOZ data, as we now discuss.

Figure 5.8 shows the projections of the $(m^2, \tan^2 \psi, \tan^2 \varphi)$ volume allowed by SK + CHOOZ. The (degenerate) best-fit points are now at $(m^2, \tan^{\pm 2} \psi, \tan^2 \varphi) = (3 \times 10^{-3} \text{ eV}^2, 0.76, 0)$. By comparing Fig. 5.8 with Fig. 5.7, we can see that CHOOZ improves the upper bound on $\tan^2 \varphi$ by an order of magnitude. As expected, at the small values of $\tan^2 \varphi$ allowed by SK + CHOOZ, both the octant asymmetry in ψ and the upper limit on m^2 are reduced. This also explains the improvement of the upper limit in Fig. 5.5 for the SK + CHOOZ case. We have repeated the fit for the case $m^2 < 0$ (not shown), corresponding to Fig. 5.1b. For negative m^2, we obtain somewhat weaker bounds on $\tan^2 \varphi$ ($\lesssim 0.5$ at 90% C.L.) in the fit to the SK data only,

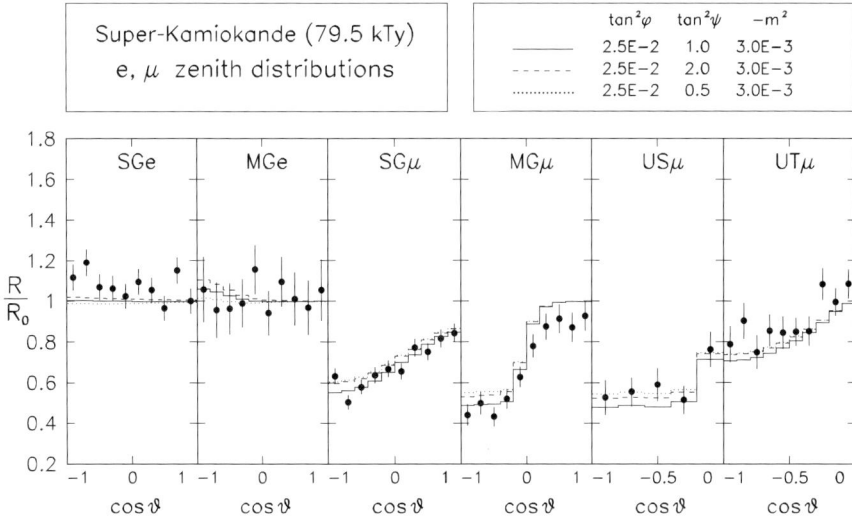

	$\tan^2\varphi$	$\tan^2\psi$	$-m^2$
————	2.5E−2	1.0	3.0E−3
- - - - - -	2.5E−2	2.0	3.0E−3
··············	2.5E−2	0.5	3.0E−3

Super-Kamiokande (79.5 kTy)

e, μ zenith distributions

Fig. 5.10. As in Fig. 5.9, but for negative m^2 (inverted-hierarchy case)

while the fit to the SK + CHOOZ data gives results almost identical to those in Fig. 5.8. This fact shows that current atmospheric + reactor data are, basically, unable to discriminate between the two signs of m^2 in 3ν scenarios.

In conclusion, we learn from the SK data that $\nu_\mu \to \nu_\tau$ oscillations are favored, and from CHOOZ that any additional ν_e mixing must be very small. Is the ν_e mixing (i.e. $U_{e3}^2 = \sin^2\varphi$) nonzero? This is one of the most important questions in neutrino physics, since the case $\varphi \neq 0$ opens the door – at least in principle – to observations of effects of the matter in the earth [37, 38] and of $\text{sign}(m^2)$ [32, 34].

Is there any residual chance to infer effects of $\varphi \neq 0$ from atmospheric neutrino data? Figures 5.9 and 5.10 show some typical predicted effects of $\varphi \neq 0$ on the SK zenith distributions, for the normal [23] and the inverted hierarchy, respectively. It appears that the best chances of observing an effect are confined to the MGe distribution, through an increase in the upgoing electron rate. However, it will be hard to reach enough statistical significance. If an effect is seen, it will also be difficult to assess the underlying hierarchy (the zenith distortions are not too different in the two figures). Therefore, the detection of effects of nonzero φ appears to be essentially a task for future, higher-statistics experiments.

5.4.5 2ν Subcase: Impact of the First K2K Data

We complete the discussion of atmospheric neutrino oscillations with the recent (supporting) indications obtained from the KEK-to-Kamioka (K2K) accelerator neutrino experiment [39], characterized by a value of E of the order of a few GeV and $L \simeq 250$ km. The recent K2K result (44 observed neutrino events vs. 63.9 expected [39]) is already inconsistent with no oscillations at $\sim 97\%$ C.L., and suggests instead ν_μ disappearance. The K2K collaboration is being understandably conservative about the oscillation interpretation, pending higher statistics and a reduction of various uncertainties. However, it is tempting to use the first K2K result for a preliminary analysis, at least in the simplest 2ν subcase [40].

Figure 5.11 shows the results in the $\nu_\mu \to \nu_\tau$ parameter space $(m^2, \tan^2 \psi)$. The upper left panel corresponds to the SK data only, with degenerate best-fit points (stars) at $m^2 = 3 \times 10^{-3}$ eV^2 and at octant-symmetric mixing values $\tan^{\pm 2} \psi = 0.76$. The upper right panel shows the (much weaker) constraints from the K2K data only. The lower left panel shows the combination of the SK and K2K results. The bounds on $\tan^2 \psi$ are not significantly modified, and the best-fit value of m^2 is only slightly lowered ($m^2 = 2.9 \times 10^{-3}$ eV^2). However, the 90% and 99% C.L. ranges of m^2 are appreciably reduced, both from below and (more strongly) from above. Therefore, the K2K experiment is already having a nonnegligible impact in the determination of m^2 parameter relevant to the $\nu_\mu \to \nu_\tau$ channel. The lower right panel shows the further (prospective) m^2 range reduction that would be achieved if the K2K errors were halved. In conclusion, the first K2K results corroborate the atmospheric evidence of ν_μ oscillation.

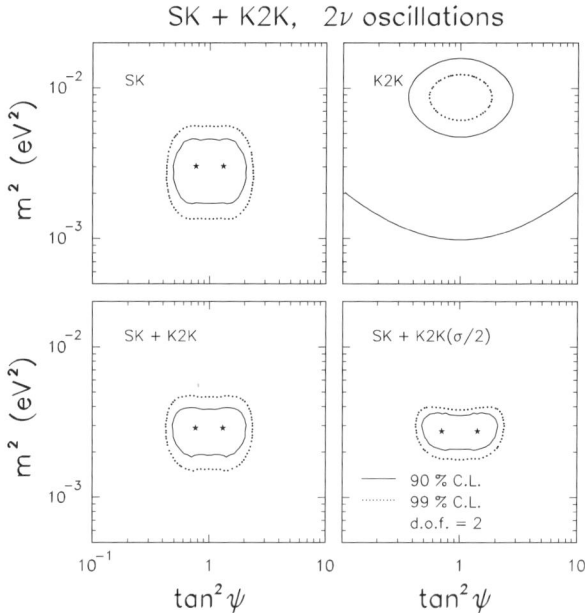

Fig. 5.11. 2ν analyses of SK and K2K data, and of their combination (including the result that would be obtained if the K2K errors were halved) [40]

5.5 Solar Neutrinos in a 3ν Interpretation

The zeroth-order approximation in $\delta m^2/m^2$ (introduced in Sect. 5.4 for atmospheric neutrinos) turns out to be a very good approximation also for solar neutrinos [3, 14, 31]. In this case, the approximation implies the averaging out of all oscillating terms associated with m^2 ($m^2 \to \infty$). Corrections for the finite value of $\delta m^2/m^2$ are currently negligible [18, 26]. Moreover, at solar neutrino energies, the ν_μ's and ν_τ's cannot be distinguished in the final state. Therefore, not only m^2 but also $\psi = \theta_{23}$ decouples from the solar 3ν parameter space [3], which reduces to

$$\text{solar } \nu \text{ parameter space} \simeq (\delta m^2, \varphi, \omega) \,. \tag{5.24}$$

The 2ν limit, involving only $\nu_1 \to \nu_2$ oscillations with no ν_3 mixing, is recovered for $\varphi = 0$. For $\varphi \neq 0$, it turns out that the 3ν and 2ν survival probabilities are linked by the simple relation (see [41] and references therein)

$$P_{ee}^{3\nu}(\delta m^2, \omega, \varphi) = s_\varphi^4 + c_\varphi^4 P_{ee}^{2\nu}(\delta m^2, \omega)\Big|_{N_e \to c_\varphi^2 N_e} \,, \tag{5.25}$$

which shows that the solar 3ν and 2ν cases differ by terms of $O(s_\varphi^2)$. We consider, in succession, the cases $\varphi = 0$, $\varphi \neq 0$ (but small), and φ unconstrained.

5.5.1 2ν Analysis ($\varphi = 0$)

The main results of a recent solar 2ν analysis [42] including the first SNO charged-current (CC) data [11] are shown in Figs. 5.12 and 5.13, in the plane $(\delta m^2, \tan^2 \omega)$. The asymmetry of the allowed regions with respect to maximal mixing ($\tan^2 \omega = 1$) is due to matter effects.

Figure 5.12 includes the total event rates obtained from the chlorine [7] and gallium [8] experiments, and from the SK [10] and SNO CC [11] data. It also includes the CHOOZ data, which limit δm^2 from above (see (5.16)). The solutions are usually referred to as follows: small-mixing-angle (SMA), at $\tan^2 \omega \sim 10^{-3}$; large-mixing-angle (LMA), at $\tan^2 \omega \sim O(1)$ and $\delta m^2 \gtrsim 10^{-5}$ eV^2; and low-δm^2 (LOW), at $\delta m^2 \sim O(10^{-7})$ eV^2, which is continuously connected to the vacuum (VAC) solution at $\sim O(10^{-10})$ eV^2 through quasi-vacuum oscillations (QVO). The fit in Fig. 5.12 favors the SMA and LMA solutions over the others. The different roles of the kinetic and interaction terms in (5.3) in the various solutions have been widely studied [3, 4, 41]. Recent developments include an accurate description of interaction terms in the QVO range [43]. Notice that maximal mixing ($\tan^2 \omega = 1$) is increasingly allowed as δm^2 decreases. Scenarios with $\tan^2 \omega \simeq 1$ *and* $\tan^2 \psi = 1$ (suggested by atmospheric ν data) are usually dubbed "bimaximal-mixing" cases [44–46].

Figure 5.13 shows a global 2ν analysis of solar + CHOOZ data [42], including the SK day–night spectrum (see also [47]). The absence of time–energy spectral distortions [48] excludes both the SMA and (most of) the

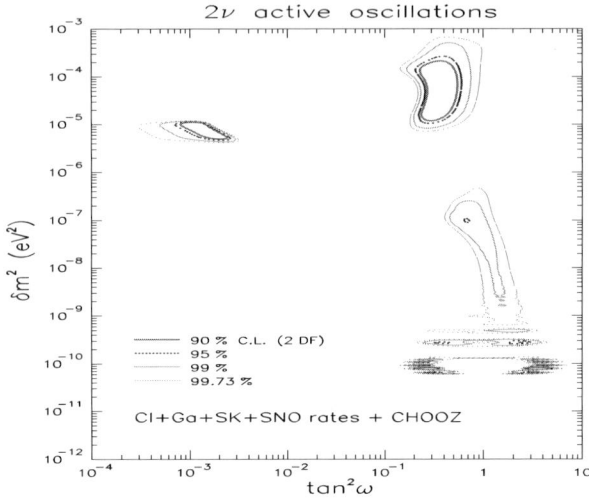

Fig. 5.12. 2ν fit of total rates from the Cl, Ga, SK, and SNO CC experiments [42]

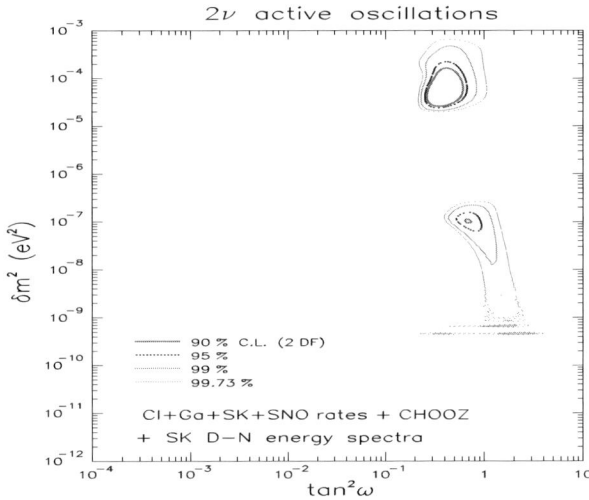

Fig. 5.13. As in Fig. 5.12, but including the SK day–night energy spectrum [42]

VAC regions, leaving only those fractions of the LMA, LOW, and QVO solutions which predict no or only mild spectral deviations. The best fit is reached within the LMA parameter region, which should be accessible to the current long-baseline reactor experiment KamLAND [49]. The (less favored) LOW solution is still in good shape, and should be tested through day–night earth matter effects in the Borexino experiment [50] or, with less sensitivity, through winter–summer matter effects in GNO [50]. Notice that the LOW solution extends down to the QVO range, which might be probed in Borexino by pushing its time-variation sensitivity close to its upper limits [51].

5.5.2 3ν Analysis ($\varphi \neq 0$, Within CHOOZ Limits)

Equation 5.25 suggests that the solutions shown in Fig. 5.13 should be only mildly modified by small-φ 3ν corrections. This is confirmed by quantitative analyses of solar + CHOOZ data [25, 26, 52], as shown in Fig. 5.14 for a representative value of m^2. In Fig. 5.14, the CHOOZ data play a double role, constraining both δm^2 and φ from above [19, 26], through the full dependence on the 3ν parameters in (5.14). This leads to an anticorrelation between the

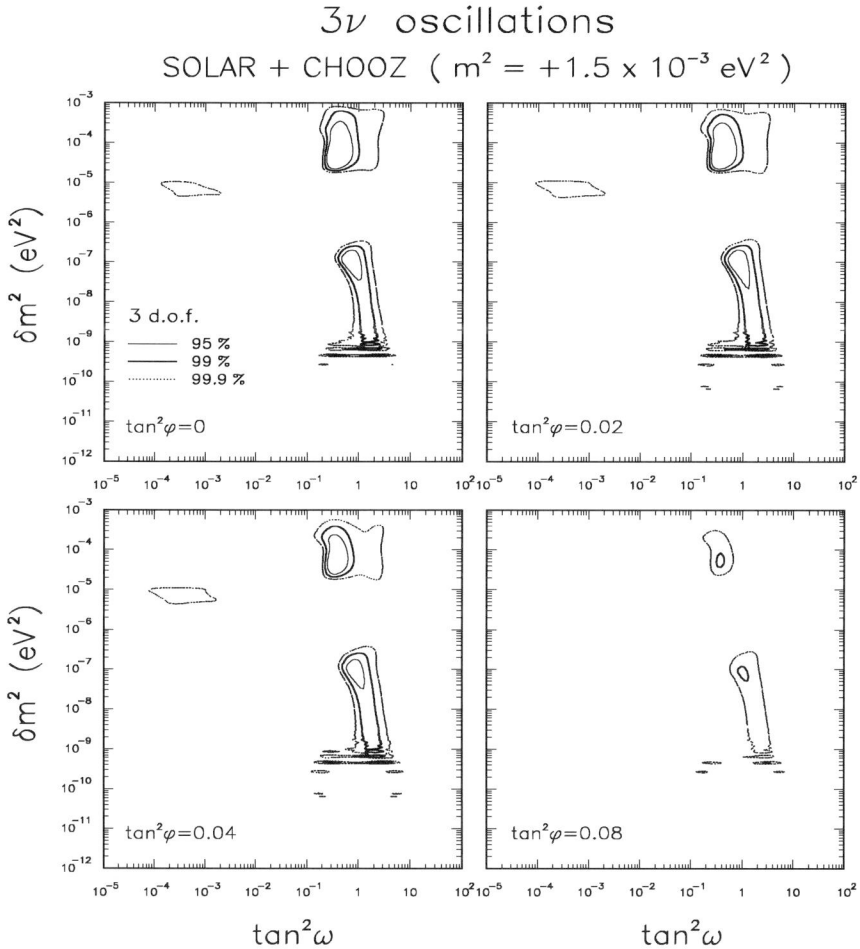

Fig. 5.14. Global 3ν analysis of solar + CHOOZ data, for a fixed representative value of m^2 ($+1.5 \times 10^{-3}$ eV^2), and for $\tan^2 \varphi = 0$, 0.02, 0.04, and 0.08 [26]

maximum allowed values of δm^2 and φ: as φ increases in Fig. 5.14, not only do the solutions vanish, but the highest value of δm^2 in the LMA region decreases [26]. This fact should be taken into account in the evaluation of CP-violation effects in future experiments: CP effects increase with φ and δm^2, but Fig. 5.14 shows that these two parameters cannot both be maximized.

5.5.3 3ν Analysis (φ Unconstrained)

It is relevant to check whether solar neutrinos favor, by themselves, small values of φ, consistently with the independent indications from SK + CHOOZ.

Both pre-SNO CC [41] and post-SNO CC [25, 52] solar-neutrino analyses show that this is the case. In particular, Fig. 5.15 [52] shows that the upper limits on $U_{e3}^2 = \sin^2 \varphi$ from solar data alone are competitive with those placed on φ by atmospheric neutrino data alone, although they are much weaker, of course, than those placed on it by CHOOZ. The preference of all data subsets for small φ is a nontrivial consistency check of the 3ν oscillation scenario [17, 22].

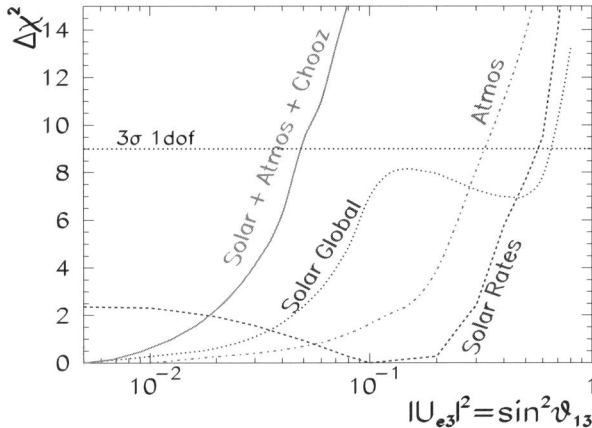

Fig. 5.15. 3ν limits on $\sin^2 \varphi = U_{e3}^2$ from all ν oscillation data, from [52]

5.6 Summary of 3ν Mixing

The currently explorable 3ν parameter space $(\delta m^2, m^2, \omega, \varphi, \psi)$ is constrained by oscillation data in the following way. Atmospheric neutrinos alone provide upper and lower limits on the dominant parameters (m^2, ψ), and a weak upper limit on φ. Solar neutrinos set upper and lower limits on the dominant parameters $(\delta m^2, \omega)$, and a weak upper limit on φ. Reactor (CHOOZ) neutrinos probe $(\delta m^2, m^2, \omega, \varphi)$, placing the strongest upper limit on φ (in combination with atmospheric neutrino data) and improving the upper limit on δm^2 (in combination with solar neutrino data). The preference of all data subsets for small φ is impressive and nontrivial. We anticipate that, conversely, 4ν mixing will show a tension between different data subsets which, although not strong enough to rule it out, makes 4ν mixing phenomenologically more "fragile."

Although the 3ν scenario emerging from current solar, atmospheric, and reactor data is rather satisfactory, one should not forget that many important questions remain open to future investigations. Can the LSND signal really be excluded? Is $\varphi \neq 0$? Is the hierarchy normal or inverted? How can we make matter effects emerge? Is mixing (bi)maximal? Which is the true solution of the solar neutrino problem? How can we see effects beyond the zeroth order in $\delta m^2/m^2$, including possible CP violations – for which there is currently no indication?

5.7 Four-Neutrino Oscillations in "2+2" Scenarios

Four-neutrino models, involving $(\nu_e, \nu_\mu, \nu_\tau)$ plus one hypothetical sterile state (ν_s), are meant to accommodate not only the solar and atmospheric data, but also the controversial LSND signal for small-amplitude $(\sim 3 \times 10^{-3})$ oscillations in the $\nu_\mu \to \nu_e$ channel [53]. Using a fourth ν state, a third (LSND) squared-mass difference $M^2 \sim O(1)$ eV^2 can be constructed (with $M^2 \gg m^2 > \delta m^2$).

There are two possible options [4, 54]: "3+1" spectra, with a lone state separated by a gap M^2 from a solar + atmospheric triplet (with subgaps δm^2 and m^2); and "2+2" spectra, with two (solar and atmospheric) doublets separated by the largest gap, M^2. We discuss in more detail the 2+2 case, and comment on the 3+1 case at the end of this section.

5.7.1 4ν Mass Spectrum, Mixing, and Dynamics

Figure 5.16 shows our reference 2+2 mass spectrum, with a lower "solar neutrino doublet" (ν_1, ν_2), and an upper "atmospheric neutrino doublet" (ν_3, ν_4), separated by a relatively large LSND mass gap. Other phenomenologically allowed 2+2 spectra can be obtained by interchanging the two doublets $(M^2 \to -M^2)$ and/or the two states in a doublet $(\delta m^2 \to -\delta m^2$ or $m^2 \to -m^2$, or both). The oscillation physics for such alternative 2+2 spectra is hardly distinguishable from that for the reference spectrum, within current oscillation phenomenology.

Concerning the 4ν mixing matrix U, we arrange the neutrino flavor and mass eigenstates in column vectors as $\nu_\alpha = (\nu_e, \nu_s, \nu_\mu, \nu_\tau)^T$ and $\nu_i = (\nu_1, \nu_2, \nu_3, \nu_4)^T$, respectively, where U is defined by $\nu_\alpha = U_{\alpha i} \nu_i$.

Reference 4ν mass spectrum

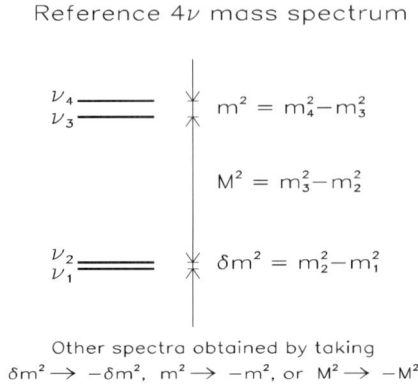

$$\nu_4 \underline{}$$
$$\nu_3 \underline{}$$

$$m^2 = m_4^2 - m_3^2$$

$$M^2 = m_3^2 - m_2^2$$

$$\nu_2 \underline{}$$
$$\nu_1 \underline{}$$

$$\delta m^2 = m_2^2 - m_1^2$$

Other spectra obtained by taking
$\delta m^2 \to -\delta m^2$, $m^2 \to -m^2$, or $M^2 \to -M^2$

Fig. 5.16. Reference four-neutrino spectrum of squared-mass differences in the 2+2 scenario. Spectra obtained by inverting any mass gap are also allowed

It turns out that the mixing of ν_μ with the solar doublet (ν_1, ν_2) must be small, and the mixing of ν_e with the atmospheric doublet (ν_3, ν_4) must also be small, otherwise (a) large (unobserved) ν_μ and ν_e disappearance effects would have occurred in the CDHSW and Bugey experiments, respectively, and (b) the LSND oscillation amplitude would be unacceptably large [4, 54]. Conversely, the relative ν_s components in the atmospheric and solar neutrino oscillations ($|\nu_s|_{\mathrm{atm}}$ and $|\nu_s|_{\mathrm{solar}}$) are, in principle, constrained only by unitarity:

$$|\nu_s|^2_{\mathrm{atm}} + |\nu_s|^2_{\mathrm{solar}} = 1 \, . \tag{5.26}$$

The above constraints on $\nu_{\mu,e,s}$ mixing can be be embedded by use of a mixing matrix having the following approximate structure [55, 56]:

$$U \simeq \begin{pmatrix} U_{e1} & U_{e2} & 0 & 0 \\ U_{s1} & U_{s2} & U_{s3} & U_{s4} \\ 0 & 0 & U_{\mu3} & U_{\mu4} \\ U_{\tau1} & U_{\tau2} & U_{\tau3} & U_{\tau4} \end{pmatrix} = \begin{pmatrix} c_\omega & s_\omega & 0 & 0 \\ -s_\omega c_\xi & c_\omega c_\xi & -s_\psi s_\xi & c_\psi s_\xi \\ 0 & 0 & c_\psi & s_\psi \\ s_\omega s_\xi & -c_\omega s_\xi & -s_\psi c_\xi & c_\psi c_\xi \end{pmatrix} \, . \tag{5.27}$$

This leads to $\langle \nu_\mu | \nu_{1,2} \rangle \simeq 0 \simeq \langle \nu_e | \nu_{3,4} \rangle$, as desired. Such a structure for the 4ν mixing matrix U describes atmospheric and solar neutrino oscillations in the channels

$$\nu_\mu \to \nu_+ \quad \text{(atmospheric)} \, , \tag{5.28}$$
$$\nu_e \to \nu_- \quad \text{(solar)} \, , \tag{5.29}$$

where the states $\nu_\pm$ represent linear (orthogonal) combinations of ν_τ and ν_s through a specific mixing angle ξ [55]:

$$\nu_+ = +\cos\xi \, \nu_\tau + \sin\xi \, \nu_s \, , \tag{5.30}$$
$$\nu_- = -\sin\xi \, \nu_\tau + \cos\xi \, \nu_s \, . \tag{5.31}$$

The angle ξ modulates the sterile-neutrino admixture shared by the atmospheric neutrino doublet ($\propto \sin^2 \xi$) and by the solar neutrino doublet ($\propto \cos^2 \xi$), so that the sum rule in (5.26) can be interpreted in terms of the identity

$$\sin^2 \xi + \cos^2 \xi = 1 . \tag{5.32}$$

The angles ψ and ω continue to parameterize (ν_μ, ν_τ) and (ν_1, ν_2) mixing in atmospheric and solar neutrino oscillations, respectively.

Finally, because of (5.27) and the fact that $\delta m^2 \ll m^2 \ll M^2$, it turns out that atmospheric and solar 4ν dynamics decouple into two effective 2ν evolution equations [55, 56], which are the basis of the phenomenological analysis:

$$i \frac{d}{dx} \begin{pmatrix} \nu_\mu \\ \nu_+ \end{pmatrix} \simeq \left[\frac{m^2}{4E} \begin{pmatrix} c_{2\psi} & s_{2\psi} \\ s_{2\psi} & -c_{2\psi} \end{pmatrix} + \sqrt{2} \, G_F \begin{pmatrix} 0 & 0 \\ 0 & N_+ \end{pmatrix} \right] \begin{pmatrix} \nu_\mu \\ \nu_+ \end{pmatrix} , \tag{5.33}$$

$$i \frac{d}{dx} \begin{pmatrix} \nu_e \\ \nu_- \end{pmatrix} \simeq \left[\frac{\delta m^2}{4E} \begin{pmatrix} c_{2\omega} & s_{2\omega} \\ s_{2\omega} & -c_{2\omega} \end{pmatrix} + \sqrt{2} \, G_F \begin{pmatrix} N_- & 0 \\ 0 & 0 \end{pmatrix} \right] \begin{pmatrix} \nu_e \\ \nu_- \end{pmatrix} , \tag{5.34}$$

where $N_+ = s_\xi^2 N_n/2$ and $N_- = N_e - c_\xi^2 N_n/2$, and N_e and N_n are, respectively, the electron and neutron densities along the neutrino trajectory.

5.7.2 Atmospheric Neutrinos

Within the approximations discussed in the previous section, the atmospheric-neutrino parameter space is spanned by (m^2, ψ, ξ). Equivalently, for a fixed value of m^2, atmospheric neutrino oscillations are described by the ν_4 (or ν_3) component in terms of $(\nu_\mu, \nu_\tau, \nu_s)$, namely, $\nu_4 = s_\psi \nu_\mu + c_\psi (c_\xi \nu_\tau + s_\xi \nu_s)$. The corresponding unitarity condition $U_{\mu 4}^2 + U_{\tau 4}^2 + U_{s 4}^2 = 1$ can thus be embedded (in analogy with the 3ν case) in a triangle plot, whose corners are $(\nu_\mu, \nu_\tau, \nu_s)$, as shown in Fig. 5.17. The left and right sides correspond to pure $\nu_\mu \rightarrow \nu_\tau$ and pure $\nu_\mu \rightarrow \nu_s$ oscillations, respectively.

Figure 5.18 shows the results of a 4ν analysis of the SK atmospheric neutrino data for separate and combined data sets [57]. It can be seen that, in the combination, the case of pure $\nu_\mu \rightarrow \nu_\tau$ oscillations (left side) is allowed, while the case of pure $\nu_\mu \rightarrow \nu_s$ oscillations (right side) is significantly disfavored. There are also intermediate solutions for $\sin^2 \xi \neq 0$, with a significant admixture of ν_s. Such results and constraints emerge from the interplay of low-energy data (which are more sensitive to m^2) and high-energy data (which are more sensitive to the ν_s component through matter effects, scaling as $s_\xi^2 N_n/2$ (5.33)).

The marked preference for $\nu_\mu \rightarrow \nu_\tau$ as compared with $\nu_\mu \rightarrow \nu_s$ is confirmed by all recent analyses [57–60], and is consistent with the ∼99% rejection of the pure $\nu_\mu \rightarrow \nu_s$ case discussed by the SK [61] and MACRO [62] collaborations. This preference is also consistent with the SK statistical indication of ν_τ appearance and the nonobservation of neutral-current event depletion [33]. In summary, the ν_s admixture in atmospheric neutrino oscillation is practically

Atmospheric ν parameter space

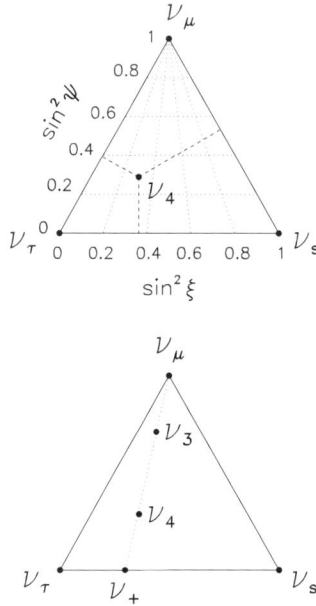

Fig. 5.17. Triangle plot for 4v mixing of atmospheric neutrinos, embedding the unitarity constraint $U_{\mu 4}^2 + U_{\tau 4}^2 + U_{s4}^2 = 1$

zero at best fit, and is bounded from above. According to a recent global analysis [52],

$$|\nu_s|_{atm}^2 = \sin^2 \xi \lesssim 0.5 \ (99\% \ \text{C.L.}) \ . \tag{5.35}$$

5.7.3 Solar Neutrinos

Solar neutrinos, like atmospheric neutrinos, show no indication of a ν_s admixture. This feature, already present in earlier 4v analyses [56], has been strengthened after the SNO evidence in favor of active $\nu_e \to \nu_{\mu,\tau}$ transitions [47, 63]. Any possible sterile-neutrino component [64] must be subdominant [63]. Figure 5.19 shows the results of a recent global analysis of solar neutrino data [52]: pure active oscillations ($\cos^2 \xi = 0$) provide the best fit, while the pure sterile case ($\cos^2 \xi = 1$) is strongly disfavored. Therefore, an upper bound can be placed on the ν_s admixture in solar neutrino oscillations:

$$|\nu_s|_{solar}^2 = \cos^2 \xi \lesssim 0.5 \ (99\% \ \text{C.L.}) \ . \tag{5.36}$$

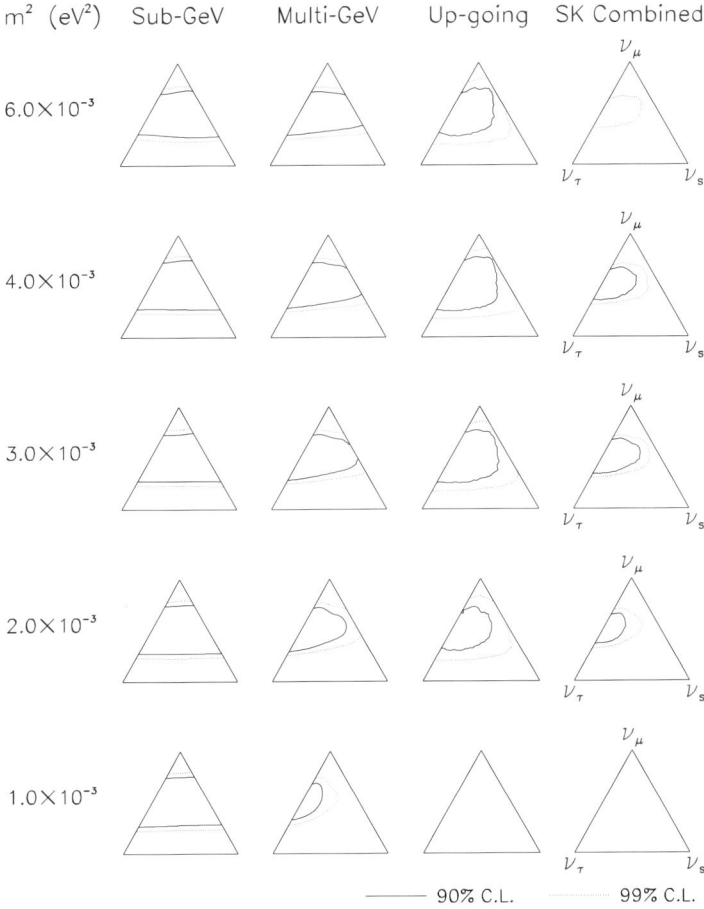

Fig. 5.18. 4ν analysis in a $(\nu_\mu, \nu_s, \nu_\tau)$ triangle plot, for five representative values of m^2. *First three columns*: separate analyses of SGe + SGµ, MGe + MGµ, and USµ + UTµ data. *Right column*: all SK data (70.5 kTy). The allowed regions typically include pure $\nu_\mu \leftrightarrow \nu_\tau$ oscillations (*left* side of the triangle) and disfavor pure $\nu_\mu \rightarrow \nu_s$ oscillations (*right* side of the triangle) [57]

5.7.4 Global "2+2" Interpretation

The bounds in (5.35) and (5.36) are clearly incompatible, at 99% C.L., with the sum rule in (5.26) and (5.32). In other words, 2+2 scenarios exhibit a strong tension between the solar and atmospheric data sets: none of those scenarios leaves enough room for the ν_s, and a conflict arises over the associated mixing angle ξ. This situation should be contrasted with the 3ν scenario, where, conversely, consistent indications about the mixing angle φ (associated with ν_e mixing) are found in the solar, atmospheric, and reactor data subsets (see Sect. 5.6).

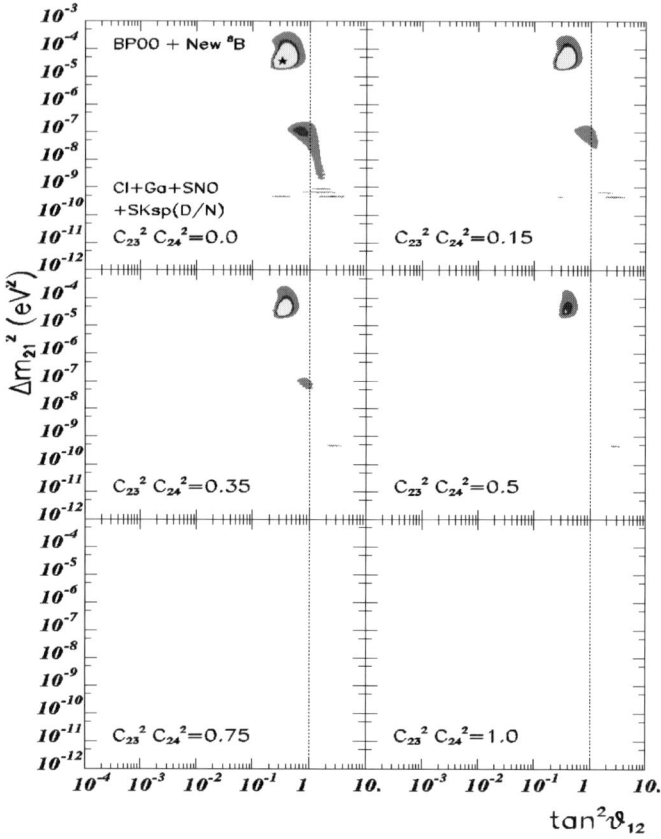

Fig. 5.19. 4v analysis of solar neutrinos (from [52]) in the plane (δm^2, $\tan^2 \omega$), for six representative values of $\cos^2 \xi$ (= $c_{23}^2 c_{24}^2$ in the notation of [52]). The case of pure sterile oscillations ($\cos^2 \xi = 1$, *bottom right panel*) is strongly disfavored with respect to the case of pure active oscillations ($\cos^2 \xi = 0$, *top left panel*)

The 99% C.L. tension between the solar and atmospheric data does not suffice, in itself, to rule out the 2+2 scenario, which is still being actively investigated [60]. However, in our opinion, this tension prevents a meaningful global fit to the 4v parameters, in particular to $\sin^2 \xi$. In any case, the real problems of the 2+2 scenario do not involve χ^2 fits, but crucial experimental data: in order for this scenario to survive, some (so far missing) evidence for nonzero $|v_s|_{atm}$ or $|v_s|_{solar}$ has to be found and, most importantly, the LSND signal (not confirmed by KARMEN [65]) needs to be verified by Mini-BooNE [66].

5.7.5 Comments on "3+1" Mass Spectra

Scenarios with 3+1 mass spectra have also been considered in the literature (see [4, 67–69] and references therein). It turns out that the "lone" state ν_4 must basically coincide with ν_s, implying an effective 3ν phenomenology for the solar + atmospheric triplet (ν_1, ν_2, ν_3) [68, 69]. In this case, instead of the sum rule of the 2+2 case (5.26), there is a "product rule" linking the LSND signal to the survival probabilities in the CDHSW (accelerator) and Bugey (reactor) disappearance experiments [4]. Symbolically,

$$[P_{\mu\mu}]_{\mathrm{CDHSW}} \times [P_{ee}]_{\mathrm{Bugey}} \propto [P_{\mu e}]_{\mathrm{LSND}} . \tag{5.37}$$

Since both $[P_{\mu\mu}]_{\mathrm{CDHSW}}$ and $[P_{ee}]_{\mathrm{Bugey}}$ are experimentally ~ 0, the predicted value of $[P_{\mu e}]_{\mathrm{LSND}}$ turns out to be doubly suppressed. Quantitative studies [68, 70] show that such suppression induces an $\sim 99\%$ C.L. conflict with the experimental LSND signal, although the most recent (a little lower) LSND + KARMEN combined estimate for $P_{\mu e}$ [71] might make this argument slightly less compelling. In any case, a strong tension between data sets arises in the 3+1 case also, at a significance level comparable to that of the 2+2 case.

5.8 Addendum: Impact of the Very Latest Neutrino Data

After this review had been essentially completed, new relevant neutrino data appeared, including the SNO neutral-current (NC) measurement [72, 73], the K2K spectral analysis [74], updated SK solar [75] and atmospheric [76] neutrino data, and updated Ga solar neutrino data [77, 78]. At the time of writing this addendum, the impact of these data on the global 3ν and 4ν scenarios has not yet been quantitatively explored at the same level of detail as for earlier data. However, the main qualitative consequences of the new data can be easily outlined as follows.

5.8.1 Impact of New Solar Neutrino Data

The recent SNO NC (and day–night) data [72, 73] have enhanced the model-independent evidence for solar $\nu_e \to \nu_{\mu,\tau}$ transitions at $\gtrsim 5\sigma$ [72], and have also strengthened the bounds on the oscillation parameters [73, 75, 79, 80].

In the context of 2ν active oscillations, and with respect to the previous bounds shown in Fig. 5.13, it turns out that the latest data strongly favor the LMA solution, as shown in Fig. 5.20 (taken from [80]). However, it is not yet possible to rule out the LOW and QVO solutions at 3σ. The KamLAND experiment [49] will be decisive in this respect.

Concerning 3ν mixing, small values of φ can be expected to induce only mild modifications in the preferred solutions shown in Fig. 5.20. Therefore, the global 3ν picture discussed in Sect. 5.6 remains essentially unchanged.

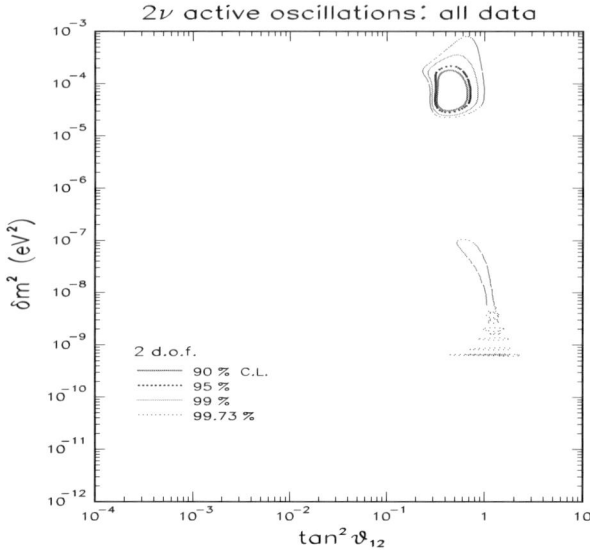

Fig. 5.20. Limits on 2ν mixing from the latest Cl, Ga, SK, and SNO solar ν data, from [80] (see also [73, 75, 79]). The LMA solution is strongly favored, but the LOW and QVO regions are still allowed at 3σ

Concerning 4ν mixing, the SNO CC+NC data definitely rule out the case of pure $\nu_e \to \nu_s$ oscillations, but still allow a sizable sterile-neutrino fraction $|\nu_s|^2_{\rm solar}$ [81, 82]. Indeed, the earlier bound in (5.36) remains basically valid [81], a combination of solar and (future) KamLAND data being needed to significantly limit the solar ν_s fraction [81].

5.8.2 Impact of New Atmospheric and Accelerator Data

The updated SK atmospheric neutrino data [76] and the recent K2K spectral data [74] show an impressive agreement within the $\nu_\mu \to \nu_\tau$ scenario for $m^2 \simeq 2.6 \times 10^{-3}$ eV^2 and $\sin^2 \psi \simeq 1/2$ [74]. This agreement corroborates the 3ν mixing scenario with small φ summarized in Sect. 5.6.

Concerning 4ν mixing, the SK collaboration has presented a preliminary analysis [76], which seems to lower by a factor of $1/2$ the previous upper bound on the atmospheric ν_s fraction in (5.35). Taking such results at face value, the unitarity condition in (5.26) and (5.32) could not be fulfilled in a 2+2 scenario. However, pending a detailed description by the SK collaboration of these preliminary results, the current status of the 2+2 scenario may still fluctuate – according to different viewpoints – from "dead" (see the "autopsy of 2+2 oscillations" in [83]) to "still alive" [84]. In any case, after the latest neutrino data, the 2+2 case appears to be definitely "borderline" (although maybe not ruled out yet), while the 3+1 scenario remains as marginally acceptable as before.

5.9 General Summary

We have reviewed the theoretical interpretation of the current neutrino data in terms of 3ν and 4ν oscillations, and have discussed the phenomenological constraints on the squared-mass differences and mixing angles of the neutrinos.

The interpretation of solar, atmospheric, and reactor neutrino data in a three-flavor oscillation framework appears to be self-consistent and robust. In particular, upper and lower bounds can be put on the two subsets of squared-mass differences and mixing angles, (m^2, ψ) and $(\delta m^2, \omega)$, that dominate atmospheric and solar neutrino oscillations, respectively. Upper bounds are set as $\varphi = \theta_{13}$ by reactor neutrino data. Open issues in the 3ν scenario include the lower bound on φ (if any), the spectrum hierarchy, and the CP-violating phase.

The inclusion of the LSND oscillation signal through a fourth (sterile) neutrino state has also been discussed, mainly in the 2+2 scenario. A tension is seen to arise between the solar and atmospheric data sets, since both disfavor significant ν_s mixing. In the alternative 3+1 case, a similar tension arises between the results of reactor and accelerator flavor disappearance searches. Such discrepancies are not strong enough to definitely rule out 4ν mixing, although the very latest (ν 2002 conference) data seem to jeopardize at least the 2+2 case. Clearly, an independent experimental confirmation or nonconfirmation of the LNSD signal appears to be crucial for the survival of 4ν scenarios.

Note added in proof

The KamLAND experiment [49] has recently found strong evidence in favour of reactor $\bar{\nu}_e$ disappearance over long baselines [85]. These results have a tremendous impact on the solutions to the solar neutrino problem shown in Fig. 5.20: the LMA region is confirmed and further constrained, while the LOW and QVO regions are ruled out [85]. For details, we refer the reader to our recent work [86], where the KamLAND data are combined with other (solar and terrestrial) experimental results in both the 2ν and 3ν oscillation scenarios.

References

1. Z. Maki, M. Nakagawa, and S. Sakata, Prog. Theor. Phys. **28**, 870 (1962); B. Pontecorvo, Zh. Eksp. Teor. Fiz. **53**, 1717 (1968) [Sov. Phys. JETP **26**, 984 (1968)].
2. L. Wolfenstein, Phys. Rev. D **17**, 2369 (1978); S.P. Mikheev and A.Yu. Smirnov, Yad. Fiz. **42**, 1441 (1985) [Sov. J. Nucl. Phys. **42**, 913 (1985)].
3. T.K. Kuo and J. Pantaleone, Rev. Mod. Phys. **61**, 937 (1989).

4. S.M. Bilenkii, C. Giunti, and W. Grimus, Prog. Part. Nucl. Phys. **43**, 1 (1999).
5. T. Kajita and Y. Totsuka, Rev. Mod. Phys. **73**, 85 (2001).
6. M. Apollonio et al. (CHOOZ Collaboration), Phys. Lett. B **466**, 415 (1999).
7. K. Lande, T. Daily, R. Davis, J.R. Distel, B.T. Cleveland, C.K. Lee, P.S. Wildenhain, and J. Ullman, Astrophys. J. **496**, 505 (1998); K. Lande, P. Wildenhain, R. Corey, M. Foygel, and J. Distel, in *Proceedings of Neutrino 2000, 19th International Conference on Neutrino Physics and Astrophysics*, Sudbury, Canada, 2000, Nucl. Phys. B (Proc. Suppl.) **91**, 50 (2001).
8. V. Gavrin (SAGE Collaboration), in *Proceedings of Neutrino 2000, 19th International Conference on Neutrino Physics and Astrophysics*, Sudbury, Canada, 2000, Nucl. Phys. B (Proc. Suppl.) **91**, 36 (2001); E. Bellotti (GALLEX and GNO Collaborations), in *Proceedings of Neutrino 2000, 19th International Conference on Neutrino Physics and Astrophysics*, Sudbury, Canada, 2000, Nucl. Phys. B (Proc. Suppl.) **91**, 44 (2001).
9. Y. Fukuda et al. (Kamiokande Collaboration), Phys. Rev. Lett. **77**, 1683 (1996).
10. S. Fukuda et al. (Super-Kamiokande Collaboration), Phys. Rev. Lett. **86**, 5651 (2001).
11. Q.R. Ahmad et al. (SNO Collaboration), Phys. Rev. Lett. **87**, 071301 (2001).
12. J.N. Bahcall, M.H. Pinsonneault, and S. Basu, Astrophys. J. **555**, 990 (2001).
13. G.L. Fogli, E. Lisi, and G. Scioscia, Phys. Rev. D **52**, 5334 (1995).
14. G.L. Fogli, E. Lisi, and D. Montanino, Phys. Rev. D **54**, 2048 (1996).
15. A. de Gouvea, A. Friedland, and H. Murayama, Phys. Lett. B **490**, 125 (2000).
16. F. Boehm et al. (Palo Verde Collaboration), Phys. Rev. D **64**, 112001 (2001).
17. M.C. Gonzalez-Garcia, M. Maltoni, C. Pena-Garay, and J.W. Valle, Phys. Rev. D **63**, 033005 (2001).
18. G.L. Fogli, E. Lisi, and A. Palazzo, Phys. Rev. D **65**, 073019 (2002).
19. S.M. Bilenky, D. Nicolo, and S.T. Petcov, Phys. Lett. B **538**, 77 (2002).
20. S.T. Petcov and M. Piai, Phys. Lett. B **533**, 94 (2002).
21. S. Schoenert, T. Lasserre and L. Oberauer, Astropart. Phys. **18**, 565 (2003).
22. G.L. Fogli, E. Lisi, A. Marrone, and G. Scioscia, Phys. Rev. D **59**, 033001 (1999).
23. G.L. Fogli, E. Lisi, and A. Marrone, Phys. Rev. D **64**, 093005 (2001).
24. M.C. Gonzalez-Garcia and M. Maltoni, Eur. Phys. J. C **26**, 417 (2003).
25. A. Bandyopadhyay, S. Choubey, S. Goswami, and K. Kar, Phys. Rev. D **65**, 073031 (2002).
26. G.L. Fogli, E. Lisi, D. Montanino, and A. Palazzo, in *Proceedings of NO-VE, International Workshop on Neutrino Oscillations in Venice*, ed. by M. Baldo-Ceolin (University of Padua Press, Padua, 2001), p. 17; A. Palazzo, in *Proceedings of the 37th Rencontres de Moriond*, Les Arcs, France, 2002, to appear.
27. A. Marrone, in *Proceedings of the 3rd Workshop on Neutrino Oscillations and their Origin*, Kashiwa, Japan, 2002, to appear.
28. G.L. Fogli, E. Lisi, and D. Montanino, Astropart. Phys. **4**, 177 (1995).
29. T. Sakai and T. Teshima, Prog. Theor. Phys. **102**, 629 (1999).
30. O.L. Peres and A.Y. Smirnov, Phys. Lett. B **456**, 204 (1999).
31. G.L. Fogli, E. Lisi, and D. Montanino, Phys. Rev. D **49**, 3626 (1994).
32. G.L. Fogli, E. Lisi, A. Marrone, and D. Montanino, Phys. Lett. B **425**, 341 (1998); A. De Rujula, M.B. Gavela, and P. Hernandez, Phys. Rev. D **63**, 033001 (2001).

33. C. McGrew (Super-Kamiokande Collaboration), in *Proceedings of the 9th International Workshop on Neutrino Telescopes*, Venice, 2001, ed. by M. Baldo-Ceolin (University of Padua Press, Padua, 2001), Vol. 1, p. 93.
34. G.L. Fogli, E. Lisi, D. Montanino, and G. Scioscia, Phys. Rev. D **55**, 4385 (1997).
35. G.L. Fogli, E. Lisi, A. Marrone, D. Montanino, and A. Palazzo, in *Proceedings of the 9th International Workshop on Neutrino Telescopes*, Venice, 2001, ed. by M. Baldo-Ceolin (University of Padua press, Padua, 2001), Vol. 1, p. 105.
36. G.L. Fogli, E. Lisi, A. Marrone, D. Montanino, and A. Palazzo, arXiv:hep-ph/0104221.
37. V.D. Barger, K. Whisnant, S. Pakvasa, and R.J. Phillips, Phys. Rev. D **22**, 2718 (1980).
38. M.V. Chizhov and S.T. Petcov, Phys. Rev. Lett. **83**, 1096 (1999); E.K. Akhmedov, A. Dighe, P. Lipari, and A.Y. Smirnov, Nucl. Phys. B **542**, 3 (1999).
39. J.E. Hill (K2K Collaboration), arXiv:hep-ex/0110034.
40. G.L. Fogli, E. Lisi, and A. Marrone, Phys. Rev. D **65**, 073028 (2002).
41. G.L. Fogli, E. Lisi, D. Montanino, and A. Palazzo, Phys. Rev. D **62**, 013002 (2000).
42. G.L. Fogli, E. Lisi, D. Montanino, and A. Palazzo, Phys. Rev. D **64**, 093007 (2001).
43. A. Friedland, Phys. Rev. Lett. **85**, 936 (2000); E. Lisi, A. Marrone, D. Montanino, A. Palazzo, and S.T. Petcov, Phys. Rev. D **63**, 093002 (2001).
44. V.D. Barger, S. Pakvasa, T.J. Weiler, and K. Whisnant, Phys. Lett. B **437**, 107 (1998).
45. A.J. Baltz, A.S. Goldhaber, and M. Goldhaber, Phys. Rev. Lett. **81**, 5730 (1998).
46. R. Barbieri, L.J. Hall, D.R. Smith, A. Strumia, and N. Weiner, J. High Energy Phys. **9812**, 017 (1998).
47. J.N. Bahcall, M.C. Gonzalez-Garcia, and C. Pena-Garay, J. High Energy Phys. **0108**, 014 (2001); A. Bandyopadhyay, S. Choubey, S. Goswami, and K. Kar, Phys. Lett. B **519**, 83 (2001); P.I. Krastev and A.Y. Smirnov, Phys. Rev. D **65**, 073022 (2002); M.V. Garzelli and C. Giunti, J. High Energy Phys. **0112**, 017 (2001); A. Strumia and F. Vissani, J. High Energy Phys. **0111**, 048 (2001); A.M. Gago, M.M. Guzzo, P.C. de Holanda, H. Nunokawa, O.L. Peres, V. Pleitez, and R. Zukanovich Funchal, Phys. Rev. D **65**, 073012 (2002).
48. S. Fukuda et al. (Super-Kamiokande Collaboration), Phys. Rev. Lett. **86**, 5656 (2001).
49. A. Piepke (KamLAND Collaboration), in *Proceedings of Neutrino 2000, 19th International Conference on Neutrino Physics and Astrophysics*, Sudbury, Canada, 2000, Nucl. Phys. B (Proc. Suppl.) **91**, 99 (2001).
50. G.L. Fogli, E. Lisi, D. Montanino, and A. Palazzo, Phys. Rev. D **61**, 073009 (2000).
51. A. de Gouvea, A. Friedland, and H. Murayama, J. High Energy Phys. **0103**, 009 (2001).
52. M.C. Gonzalez-Garcia and Y. Nir, Rev. Mod. Phys. **75**, 345 (2003).
53. A. Aguilar et al. (LSND Collaboration), Phys. Rev. D **64**, 112007 (2001).
54. S.M. Bilenkii, C. Giunti, and W. Grimus, Eur. Phys. J. C **1**, 247 (1998).
55. G.L. Fogli, E. Lisi, and A. Marrone, Phys. Rev. D **63**, 053008 (2001).
56. C. Giunti, M.C. Gonzalez-Garcia, and C. Pena-Garay, Phys. Rev. D **62**, 013005 (2000).

57. G.L. Fogli, E. Lisi, A. Marrone, and D. Montanino, in *Proceedings of Neutrino 2000, 19th International Conference on Neutrino Physics and Astrophysics*, Sudbury, Canada, 2000, Nucl. Phys. B (Proc. Suppl.) **91**, 167 (2001).

58. O. Yasuda, Nucl. Instrum. Meth. A **472**, 343 (2000).

59. M.C. Gonzalez-Garcia, M. Maltoni, and C. Pena-Garay, Phys. Rev. D **64**, 093001 (2001).

60. M. Maltoni, T. Schwetz, and J.W. Valle, Phys. Rev. D **65**, 093004 (2002).

61. S. Fukuda et al. (Super-Kamiokande Collaboration), Phys. Rev. Lett. **85**, 3999 (2000).

62. M. Ambrosio et al. (MACRO Collaboration), Phys. Lett. B **517**, 59 (2001).

63. J.N. Bahcall, M.C. Gonzalez-Garcia, and C. Pena-Garay, J. High Energy Phys. **0204**, 007 (2002).

64. V.D. Barger, D. Marfatia, and K. Whisnant, Phys. Rev. Lett. **88**, 011302 (2002).

65. B. Armbruster et al. (KARMEN Collaboration), Phys. Rev. C **57**, 3414 (1998).

66. A. Bazarko (MiniBooNE Collaboration), in *Proceedings of Neutrino 2000, 19th International Conference on Neutrino Physics and Astrophysics*, Sudbury, Canada, 2000, Nucl. Phys. B (Proc. Suppl.) **91**, 210 (2001).

67. V.D. Barger, B. Kayser, J. Learned, T. Weiler, and K. Whisnant, Phys. Lett. B **489**, 345 (2000).

68. O.L. Peres and A.Y. Smirnov, Nucl. Phys. B **599**, 3 (2001).

69. C. Giunti, Nucl. Phys. B (Proc. Suppl.) **100**, 244 (2001).

70. M. Maltoni, T. Schwetz, and J.W. Valle, Phys. Lett. B **518**, 252 (2001).

71. E.D. Church, K. Eitel, G.B. Mills, and M. Steidl, Phys. Rev. D **66**, 013001 (2002).

72. Q.R. Ahmad et al. (SNO Collaboration), Phys. Rev. Lett. **89**, 011301 (2002).

73. Q.R. Ahmad et al. (SNO Collaboration), Phys. Rev. Lett. **89**, 011302 (2002).

74. J. Shirai (K2K Collaboration), in *Proceedings of Neutrino 2002, 20th International Conference on Neutrino Physics and Astrophysics*, ed. by F. von Feilitzsch and N. Schmitz, Munich, Germany, 2002, Nucl. Phys. B (Proc. Suppl.) **118**, 15 (2003).

75. S. Fukuda et al. (Super-Kamiokande Collaboration), arXiv:hep-ex/0205075.

76. M. Shiozawa (Super-Kamiokande Collaboration), in *Proceedings of Neutrino 2002, 20th International Conference on Neutrino Physics and Astrophysics*, Munich, Germany, 2002 (unpublished contribution).

77. J.N. Abdurashitov et al. (SAGE Collaboration), arXiv:astro-ph/0204245.

78. T. Kirsten et al. (GALLEX/GNO Collaboration), in *Proceedings of Neutrino 2002, 20th International Conference on Neutrino Physics and Astrophysics*, ed. by F. von Feilitzsch and N. Schmitz, Munich, Germany, 2002, Nucl. Phys. B (Proc. Suppl.) **118**, 33 (2003).

79. V. Barger, D. Marfatia, K. Whisnant, and B.P. Wood, Phys. Lett. B **537**, 179 (2002); P. Creminelli, G. Signorelli, and A. Strumia, J. High Energy Phys. **0105**, 052 (2001); arXiv:hep-ph/0102234 v3; A. Bandyopadhyay, S. Choubey, S. Goswami, and D.P. Roy, Phys. Lett. B **540**, 14 (2002); J.N. Bahcall, M.C. Gonzalez-Garcia, and C. Pena-Garay, J. High Energy Phys. **0207**, 054 (2002); P.C. de Holanda and A.Y. Smirnov, Phys. Rev. D **66**, 113005 (2002); A. Strumia, C. Cattadori, N. Ferrari, and F. Vissani, Phys. Lett. B **541**, 327 (2002).

80. G.L. Fogli, E. Lisi, A. Marrone, D. Montanino, and A. Palazzo, Phys. Rev. D **66**, 053010 (2002).

81. J.N. Bahcall, M.C. Gonzalez-Garcia, and C. Pena-Garay, Phys. Rev. C **66**, 0350802 (2002).
82. V. Barger, D. Marfatia, K. Whisnant, and B.P. Wood, Phys. Lett. B **537**, 179 (2002); P. Creminelli, G. Signorelli, and A. Strumia, J. High Energy Phys. **0105**, 052 (2001).
83. P. Creminelli, G. Signorelli, and A. Strumia, J. High Energy Phys. **0105**, 052 (2001).
84. J.W.F. Valle, in *Proceedings of Neutrino 2002, 20th International Conference on Neutrino Physics and Astrophysics*, ed. by F. von Feilitzsch and N. Schmitz, Munich, Germany, 2002.Nucl. Phys. B (Proc. Suppl.) **118**, 255 (2003).
85. K. Eguchi et al. (KamLAND Collaboration), Phys. Rev. Lett. **90**, 021802 (2003).
86. G.L. Fogli, E. Lisi, A. Marrone, D. Montanino, A. Palazzo, and A.M. Rotunno, Phys. Rev. D **67**, 073002 (2003).

6 Theoretical Models of Neutrino Masses and Mixings

Guido Altarelli and Ferruccio Feruglio

We review the theoretical ideas, problems, and implications of several different models for neutrino masses and mixing angles. We give a general discussion of schemes with three or more light neutrinos. Several specific examples are analyzed in some detail, particularly those that can be embedded into grand unified theories.

6.1 Introduction

There is now convincing evidence, from the experimental study of atmospheric and solar neutrinos [1, 2], for the existence of at least two distinct frequencies of neutrino oscillation. This, in turn, implies nonvanishing neutrino masses and a mixing matrix, in analogy with the quark sector and the CKM matrix. So, a priori, the study of masses and mixings in the lepton sector should be considered at least as important as that for the quark sector. However, there are a number of features that make neutrinos especially interesting. The smallness of neutrino masses is probably related to the fact that neutrinos are completely neutral (i.e. they carry no charge, which is exactly conserved) and are Majorana particles with masses inversely proportional to the large scale where lepton number (L) conservation is violated. Majorana masses can arise from the seesaw mechanism [3], in which case there is some relation to Dirac masses, or from higher-dimensional nonrenormalizable operators which come from a sector of the Lagrangian density different from any other fermion mass terms. The relation to L nonconservation and the fact that the observed neutrino oscillation frequencies are well compatible with a large scale for L nonconservation points to a tantalizing connection with grand unified theories (GUTs). So neutrino masses and mixings can represent a probe into physics at GUT energy scales and offer a different perspective on the problem of flavor and the origin of fermion masses. There are also direct connections with important issues in astrophysics and cosmology, for example baryogenesis through leptogenesis [4] and the possibly nonnegligible contribution of neutrinos to hot dark matter in the universe.

At present there are many alternative models of neutrino masses. This variety is mostly due to the considerable experimental ambiguities that still exist. The most crucial questions to be clarified by experiment are whether the

LSND signal [5] will be confirmed or excluded, and which solar neutrino solution will eventually be established. If the LSND result is right, we probably
need at least four light neutrinos; if not, we can manage with only the three
known ones. The answer to the question of which solar solution is correct fixes
the corresponding mass-squared difference and the associated mixing angle.
Another crucial unknown is the absolute scale of neutrino masses. This, in
turn is related to physical questions as diverse as the possible cosmological
relevance of neutrinos as hot dark matter and the rate of neutrinoless double
beta decay ($0\nu\beta\beta$). If neutrinos are an important fraction of the cosmological
density, say $\Omega_\nu \sim 0.1$, then the average neutrino mass must be considerably
heavier than the splittings that are indicated by the observed atmospheric
and solar oscillation frequencies. For example, for three light neutrinos, only
models with almost degenerate neutrinos, with a common mass $|m_\nu| \approx 1\,\mathrm{eV}$,
are compatible with a large hot-dark-matter component, but in this case the
existing bounds on $0\nu\beta\beta$ decay represent an important constraint. In contrast,
hierarchical three-neutrino models (with both signs of Δm_{23}^2) have the largest
neutrino mass fixed by $|m| \approx \sqrt{\Delta m_{\mathrm{atm}}^2} \approx 0.05\,\mathrm{eV}$. In view of all these impor-
tant questions still pending, it is no wonder that many different theoretical
avenues are open and have been explored in the vast literature on the subject.

Here, we shall briefly summarize the main categories of neutrino mass
models, discuss their respective advantages and difficulties, and give a number
of examples. We shall illustrate how forthcoming experiments will be able to
discriminate among the various alternatives. We shall devote special attention
to the most constrained set of models, those with only three widely split
neutrinos, with masses dominated by the seesaw mechanism and inversely
proportional to a large mass close to the grand unification scale M_{GUT}. In
this case one can aim at a comprehensive discussion of all fermion masses in
the framework of a GUT. This is possible to some extent in models based on
$\mathrm{SU}(5) \times \mathrm{U}(1)_\mathrm{F}$ on $\mathrm{SO}(10)$ (we consider only supersymmetric (SUSY) GUTs).

6.2 Neutrino Masses
and Lepton Number Nonconservation

Neutrino oscillations imply neutrino masses, which in turn demand either
the existence of right-handed (RH) neutrinos (Dirac masses) or violation of
the conservation of lepton number L (Majorana masses), or both. Given that
neutrino masses are certainly extremely small, it is really difficult from the
theoretical point of view to avoid the conclusion that L conservation must
be violated. In fact, in terms of lepton number nonconservation, the small
neutrino masses can be explained as being inversely proportional to the very
large scale where conservation of L is violated, of order M_{GUT} or even M_{Pl},
the Planck mass.

Once we accept L nonconservation, we gain an elegant explanation for
the smallness of neutrino masses. If L is not conserved, even in the absence

of heavy RH neutrinos, Majorana masses can be generated for neutrinos by dimension-5 operators [6] of the form

$$O_5 = \frac{(Hl)_i^{\mathrm{T}} \lambda_{ij} (Hl)_j}{\Lambda} + \text{h.c.} , \qquad (6.1)$$

where H is the ordinary Higgs doublet, l_i the SU(2) lepton doublets, λ a matrix in flavor space, and Λ a large scale of mass, of order M_{GUT} or M_{Pl}. The neutrino masses generated by O_5 are of the order $m_\nu \approx v^2/\Lambda$ for $\lambda_{ij} \approx O(1)$, where $v \sim O(100\,\mathrm{GeV})$ is the vacuum expectation value of the ordinary Higgs.

We consider that the existence of RH neutrinos ν^c is quite plausible, because all GUT groups larger than SU(5) require them. In particular the fact that ν^c completes the representation 16 of SO(10), i.e. $16=\bar{5}+10+1$, so that all fermions of each family are contained in a single representation of the unifying group, is too impressive not to be significant. At least as a classification group, SO(10) must be of some relevance. Thus, in the following we assume that there is nonconservation of both ν^c and L. With these assumptions, the seesaw mechanism [3] is possible. To fix the notation, we recall that in its simplest form, the seesaw mechanism arises as follows. Consider the SU(3) $\times$ SU(2) $\times$ U(1) invariant Lagrangian giving rise to Dirac masses and to Majorana masses of ν^c (for the time being, we consider the Majorana mass terms of ν as comparatively negligible):

$$\mathcal{L} = -\nu^{c\,\mathrm{T}} y_\nu (Hl) + \frac{1}{2} \nu^{c\,\mathrm{T}} M \nu^c + \text{h.c.} \qquad (6.2)$$

The Dirac mass matrix $m_{\mathrm{D}} \equiv y_\nu v/\sqrt{2}$, originating from electroweak symmetry breaking, is, in general, non-Hermitian and nonsymmetric, while the Majorana mass matrix M is symmetric, i.e. $M = M^{\mathrm{T}}$. We expect the eigenvalues of M to be of order M_{GUT} or more because the Majorana masses of ν^c are SU(3) $\times$ SU(2) $\times$ U(1) invariant, and hence unprotected and naturally of the order of the cutoff of the low-energy theory. Since all ν^c are very heavy, we can integrate them away. For this purpose, we write down the equations of motion for ν^c in the static limit, i.e. neglecting their kinetic terms:

$$-\frac{\partial \mathcal{L}}{\partial \nu^c} = y_\nu (Hl) - M \nu^c = 0 . \qquad (6.3)$$

From this, by solving for ν^c, we obtain

$$\nu^c = M^{-1} y_\nu (Hl) . \qquad (6.4)$$

We now substitute this expression for ν^c in the Lagrangian (6.2) and obtain the operator O_5 of (6.1), where

$$\frac{2\lambda}{\Lambda} = -y_\nu^{\mathrm{T}} M^{-1} y_\nu , \qquad (6.5)$$

and the resulting neutrino mass matrix reads

$$m_\nu = m_{\mathrm{D}}^{\mathrm{T}} M^{-1} m_{\mathrm{D}} \ . \tag{6.6}$$

This is the well-known result of the seesaw mechanism [3]: the light-neutrino masses are quadratic in the Dirac masses and inversely proportional to the large Majorana mass. If some ν^c were massless or light, they would not be integrated away but would simply be added to the light neutrinos. Notice that the above results hold true for any number n of heavy neutral fermions R coupled to the three known neutrinos. In this more general case, M is an $n \times n$ symmetric matrix and the coupling between heavy and light fields is described by the rectangular $n \times 3$ matrix m_{D}. Note that for $m_\nu \approx \sqrt{\Delta m_{\mathrm{atm}}^2} \approx 0.05 \, \mathrm{eV}$ and $m_\nu \approx m_{\mathrm{D}}^2/M$, where $m_{\mathrm{D}} \approx v \approx 200 \, \mathrm{GeV}$, we find $M \approx 10^{15} \, \mathrm{GeV}$, which indeed is an impressive indication of M_{GUT}.

If the additional nonrenormalizable contributions to O_5 (6.1) are comparatively nonnegligible, they should simply be added. After elimination of the heavy right-handed fields, the two types of terms are equivalent at the level of the effective low-energy theory. In particular, they have identical transformation properties under a chiral change of basis in flavor space. The difference is, however, that in the seesaw mechanism, the Dirac matrix m_{D} is presumably related to ordinary fermion masses because both types of terms are generated by the Higgs mechanism and both must obey GUT-induced constraints. Thus, if we assume the seesaw mechanism, more constraints are implied.

6.3 Baryogenesis via Leptogenesis from Heavy-ν^c Decay

In the universe, we observe an apparent excess of baryons over antibaryons. The idea of explaining the observed baryon asymmetry by dynamical evolution starting from an initial state of the universe with zero baryon number (baryogenesis) is appealing. For baryogenesis, one needs the three well-known Sakharov conditions: nonconservation of the baryon number B, CP violation, and no thermal equilibrium. These necessary requirements may occur at different epochs in the history of the universe. Note, however, that the asymmetry generated in one epoch could be erased in a following epoch if not protected for some dynamical reason. In principle, these conditions could be satisfied in the Standard Model (SM) at the electroweak phase transition. Conservation of B is violated by instantons when kT is of the order of the weak scale (but $B - L$ is conserved), CP is violated by the CKM phase, and conditions sufficiently far from equilibrium could occur during the electroweak phase transition. So the conditions for baryogenesis at the weak scale in the SM appear, superficially, to be present. However, a more quantitative analysis [7] shows that baryogenesis is not possible in the SM, because there is not enough CP violation and the phase transition is not sufficiently strongly first order, unless $m_{\mathrm{H}} < 80 \, \mathrm{GeV}$, which has now been completely excluded by LEP. In SUSY extensions of the SM, in particular in the MSSM (Minimal

Supersymmetric Standard Model), there are additional sources of CP violation and the bound on m_H is modified by a sufficient amount by the presence of scalars with large couplings to the Higgs sector, typically the s-top. What is required is that the mass of the lightest Higgs boson $m_h \sim 80-110\,\text{GeV}$, that the s-top is not heavier than the top quark, and, preferably, that $\tan\beta$ is small. This possibility has now become very marginal in view of the results of LEP2.

If baryogenesis at the weak scale is excluded by the data, baryogenesis could still occur at or just below the GUT scale, after inflation. But only that part with $|B - L| > 0$ would survive and not be erased at the weak scale by instanton effects. Thus baryogenesis at $kT \sim 10^{10}-10^{15}\,\text{GeV}$ needs $B - L$ nonconservation at some stage, like that required for m_v if neutrinos are Majorana particles. The two effects could be related if baryogenesis arises from leptogenesis, which is then converted into baryogenesis by instantons [4]. Recent results on neutrino masses (Chap. 3) are compatible with this elegant possibility. Thus the case for baryogenesis through leptogenesis has been boosted by recent results on neutrinos [8].

In leptogenesis, the departure from equilibrium is determined by the deviation from the average number density induced by the decay of the heavy neutrinos. The Yukawa interactions of the heavy Majorana neutrinos v^c lead to the decays $v^c \to lH$ (where l is a lepton) and $v^c \to \overline{lH}$, with CP violation. The violation of L conservation arises from the $\Delta L = 2$ terms that produce the Majorana mass terms. The rates of the various interaction processes involved are temperature-dependent with different powers of T, so that the equilibrium densities and the temperatures of decoupling from equilibrium during the expansion of the universe are different for different particles and interactions. The rates $\Gamma_{\Delta L}(T)$ of $\Delta L = 2$ processes depend also on the neutrino masses and mixings, so that the observed values of the baryon asymmetry are related to neutrino processes. Precisely, $\Gamma_{\Delta L}(T) \sim T^3/\Lambda^2$, where Λ is the large scale that appears in (6.1) and also in the expression for light-neutrino masses $m_v \sim v^2/\Lambda$. The out-of-equilibrium condition $\Gamma_{\Delta L}(T) < \Gamma_{\text{exp}}$, where $\Gamma_{\text{exp}} \sim T^2/M_{\text{Pl}}$ is the expansion rate of the universe, leads to $T \lesssim \Lambda^2/M_{\text{Pl}}$, which then implies the following relation (when the correct proportionality factors and the sum over flavors are included):

$$\sum_i m_{v_i}^2 \lesssim \left[0.2\,\text{eV} \left(\frac{10^{12}\,\text{GeV}}{T} \right)^{1/2} \right]^2 . \tag{6.7}$$

What exactly is the temperature T that is relevant to leptogenesis depends on the thermal history of the early universe and goes beyond the realm of neutrino physics. But if $T \lesssim \Lambda^2/M_{\text{Pl}}$ and $\Lambda \sim M_{\text{GUT}}$, then the upper limit is significant and is compatible with present neutrino data.

If the RH neutrinos are produced thermally, then the mass of the RH neutrino that drives L nonconservation is limited by the reheat temperature after inflation, which in turn is typically required not to exceed $10^8-10^{10}\,\text{GeV}$.

This limit can be avoided if the RH neutrinos are instead produced by large inflaton oscillations during the preheating stage [9].

6.4 Models with Four (or More) Neutrinos

The LSND signal [5] has not been confirmed by KARMEN [10]. It will soon be double-checked by MiniBooNE [11]. Perhaps it will fade away. But if an oscillation with $\Delta m^2 \approx 1\,\text{eV}^2$ is confirmed, the simplest possibility, in the presence of three distinct frequencies for the LSND, atmospheric [1], and solar [2] neutrino oscillations, is to introduce at least four light neutrinos. Since LEP has limited to three the number of "active" neutrinos (that is, neutrinos with weak interactions or, equivalently, with nonvanishing weak isospin, the only possible gauge charge of neutrinos) the additional light neutrino(s) ν_s must be "sterile", i.e. with vanishing weak isospin. Note that the ν^c that appears in the seesaw mechanism, if it exists, is indeed a sterile neutrino, but a heavy one.

A possible way to accommodate the atmospheric, solar, and LSND evidence for neutrino oscillations without introducing one or more sterile neutrinos is to invoke CPT violation [12]. The independent frequencies required are provided by different neutrino and antineutrino masses, and the fit to the present data has a good quality [13].

A typical pattern of masses that works for 4ν models consists of two pairs of neutrinos [14] with a mass separation between the two pairs of order 1 eV, corresponding to the LSND frequency. The upper doublet is almost degenerate, with a value of m^2 of order $1\,\text{eV}^2$, and is split only by the mass-squared difference corresponding to either the atmospheric or solar neutrino frequency, while the lower doublet is split by the other of the two frequencies. An alternative to this 2–2 spectrum is provided by a 3–1 pattern, where the "1" is a nearly pure sterile neutrino, separated by the LSND frequency from the "3". The 3–1 spectrum leads to an overall quality of fit comparable to that of the 2–2 pattern (i.e. poor). These mass configurations may be compatible with the existence of an important fraction of hot dark matter in the universe. A complication is that the data appear to be incompatible with pure 2ν ν_e–ν_s oscillations for solar neutrinos [15] and ν_μ–ν_s oscillations for atmospheric neutrinos [16]. There are, however, viable alternatives (although they have become only marginally viable after the SNO results). One possibility is obtained by using the large freedom allowed by the presence of six mixing angles in the most general 4ν mixing matrix. If at least four angles are significantly different from zero, one can go beyond pure 2ν oscillations and, for solar neutrino oscillations for example, ν_e can transform into a mixture of ν_a and ν_s, where ν_a is an active neutrino, itself a superposition of ν_μ and ν_τ (mainly ν_τ) [14]. A different alternative is to have many interfering sterile neutrinos: this is the case in the interesting class of models with large extra dimensions, where a whole tower of Kaluza–Klein (KK) neutrinos is introduced. This picture of sterile neutrinos from extra dimensions appears exciting, and we now discuss it in some detail [18].

The context is theories with large extra dimensions. Gravity propagates in all D dimensions (i.e. in the bulk), while SM particles live on a 4d brane. As is well known [19], this can make the fundamental scale of gravity M_D much smaller than the Planck mass M_{Pl}. In fact, for $D = \delta + 4$ we have a geometrical factor V_δ, the volume of the compact dimensions, that suppresses gravity, so that

$$(M_D)^\delta V_\delta = (M_{Pl}/M_D)^2 \, , \tag{6.8}$$

and, as a result, M_D can be as small as ~ 1 TeV. For neutrino phenomenology we need a really large extra dimension, with a radius R at least of the order of the scale set by the observed solar oscillation frequencies, so that $1/R \lesssim 0.01$ eV or $R \gtrsim 0.02$ mm. If we insist on having M_D around 1 TeV, we can assume, for instance, one compact dimension with radius R and $\delta - 1$ dimensions with a common radius R', such that the volume $V_\delta = (2\pi)^\delta R R'^{\delta-1}$ fits (6.8). In string theories of gravity, there are always scalar fields associated with gravity, together with their SUSY fermionic partners (dilatini and modulini) [20]. These are particles that propagate in the bulk, have no gauge interactions, and could well play the role of sterile neutrinos. The models based on this framework [21, 22] have some good features that make them very appealing at first sight. They provide a "physical" picture for v_s. In the simplest case, the theory includes a 5d fermion $\Psi(x, y)$, which decomposes into two 4d Weyl spinors $v_s(x, y)$ and $v'_s(x, y)$ and contains a KK tower of recurrences of sterile neutrinos:

$$v_s(x, y) = \frac{1}{\sqrt{2\pi R}} \sum_n v_s^{(n)}(x) e^{iny/R} \, . \tag{6.9}$$

The tower mixes with the ordinary light active neutrinos in the lepton doublet l:

$$S_{mix} = \int d^4x \frac{h}{\sqrt{M_5}} v_s(x, 0) H(x) l(x) \, . \tag{6.10}$$

The interaction is restricted to the 4d brane at $y = 0$ where the SM fields live. Since the 5d spinor $v_s(x, y)$ has mass dimension 2, we need the mass parameter $M_5 \equiv M_D^\delta (2\pi R')^{\delta-1}$ to keep the Yukawa coupling constant dimensionless. From (6.8)–(6.10), after electroweak symmetry breaking, we find

$$S_{mix} = \int d^4x \sum_n \frac{hv}{\sqrt{2}} \frac{M_D}{M_{Pl}} v_s^{(n)}(x) v_a(x) \, , \tag{6.11}$$

where $\langle H \rangle \equiv v/\sqrt{2}$ and v_a is the active neutrino embedded in l. Note that the geometrical factor M_D/M_{Pl}, which automatically suppresses the Yukawa coupling h, arises naturally from the fact that the sterile-neutrino tower lives in the bulk. An additional mass parameter μ, related to a possible bulk mass term for the 5d fermion $\Psi(x,y)$, is also allowed (in more realistic realizations, more 5d fields and L-nonconserving interactions can be present).

The pattern of oscillations results from the superposition of an infinite number of components with increasing frequencies $\sim n^2$ and decreasing amplitudes $\sim 1/n^2$. The leading oscillation frequency $\sqrt{\Delta m^2}$ and the dominant mixing angle are determined by μ and $m = hvM_D/M_{\mathrm{Pl}}$, whereas the number of KK excitations that effectively take part in the oscillation is controlled by $1/R$. Indeed, if $1/R \gg \sqrt{\Delta m^2}$, the KK modes, whose masses are approximately given by n/R, decouple, with the possible exception of the lightest mode. If, on the contrary, $1/R \lesssim \sqrt{\Delta m^2}$, then several KK levels participate in the oscillation, and the resulting energy dependence of the survival/conversion probability can differ appreciably from that of the two-level case. Indeed, the contribution from a few KK states makes the solar oscillation spectrum more compatible with the data. We note in passing that the ν_s mixings must be small, owing to existing limits from weak processes, supernovae, and nucleosynthesis [22], so that the preferred solution for this KK neutrino model is the MSW small-angle (SA) solution. Instead, the KK states should decouple in the case of atmospheric neutrino oscillations. These constraints fix the range of admissible values of R, as specified above.

In spite of its good properties, there are problems with this picture, in our opinion. The first property of models with large extra dimensions that we do not like is that the connection with GUTs is lost. In particular, the elegant explanation of the smallness of neutrino masses in terms of the large scale where the conservation of L is violated evaporates in general. Since $M_D \sim 1\,\mathrm{TeV}$ is small, what forbids an operator of the form $(Hl)_i^{\mathrm{T}} \lambda_{ij} (Hl)_j/M_D$ on the brane, which would lead to neutrino masses that were far too large? One must impose L conservation on the brane by hand and also require that it is broken only by some Majorana masses of sterile neutrinos in the bulk, which we find somewhat ad hoc. Another problem is that we would expect gravity to know nothing about flavor, but here we would need RH partners for ν_e, ν_μ, and ν_τ. Also, a single large extra dimension has problems because it implies [23] a linear evolution of the gauge couplings with energy from $0.01\,\mathrm{eV}$ to $M_D \sim 1\,\mathrm{TeV}$. But for more large extra dimensions the KK recurrences do not decouple fast enough. Perhaps a compromise at $d = 2$ (d is the number of large extra dimensions) is possible. In conclusion, the models with large extra dimensions are interesting because they are speculative and fascinating, but the more conventional framework still appears more plausible on closer inspection.

6.5 Three-Neutrino Models

We now assume that the LSND signal will not be confirmed, so that there are only two distinct neutrino oscillation frequencies, the atmospheric and the solar frequencies. These two frequencies can be reproduced with the known three light-neutrino species (for other reviews of three-neutrino models, see [24, 25]).

Neutrino oscillations are due to a misalignment between the flavor basis, $v' \equiv (v_e, v_\mu, v_\tau)$, where v_e is the partner of the mass and flavor eigenstate e^- in a left-handed (LH) weak isospin SU(2) doublet (and similarly for v_μ and v_τ), and the mass eigenstates $v \equiv (v_1, v_2, v_3)$ [26, 27], described by

$$v' = Uv \,, \tag{6.12}$$

where U is the unitary 3×3 mixing matrix. Given the definition of U and the transformation properties of the effective light-neutrino mass matrix m_v,

$$v'^{\mathrm{T}} m_v v' = v^{\mathrm{T}} U^{\mathrm{T}} m_v U v \,, \tag{6.13}$$

$$U^{\mathrm{T}} m_v U = \mathrm{Diag}\,(m_1, m_2, m_3) \equiv m_{\mathrm{diag}} \,,$$

we obtain the following general form of m_v (i.e. of the light-neutrino mass matrix in the basis where the charged-lepton mass is a diagonal matrix):

$$m_v = U m_{\mathrm{diag}} U^{\mathrm{T}} \,. \tag{6.14}$$

The matrix U can be parameterized in terms of three mixing angles θ_{12}, θ_{23}, and θ_{13} ($0 \le \theta_{ij} \le \pi/2$) and one phase φ ($0 \le \varphi \le 2\pi$) [28], exactly as for the quark mixing matrix V_{CKM}. The following definition of mixing angles can be adopted:

$$U = \begin{pmatrix} 1 & 0 & 0 \\ 0 & c_{23} & s_{23} \\ 0 & -s_{23} & c_{23} \end{pmatrix} \begin{pmatrix} c_{13} & 0 & s_{13}\,e^{i\varphi} \\ 0 & 1 & 0 \\ -s_{13}\,e^{-i\varphi} & 0 & c_{13} \end{pmatrix} \begin{pmatrix} c_{12} & s_{12} & 0 \\ -s_{12} & c_{12} & 0 \\ 0 & 0 & 1 \end{pmatrix} \,, \tag{6.15}$$

where $s_{ij} \equiv \sin\theta_{ij}$ and $c_{ij} \equiv \cos\theta_{ij}$. In addition, we have the relative phases among the Majorana masses m_1, m_2, and m_3. If we choose m_3 real and positive, these phases are carried by $m_{1,2} \equiv |m_{1,2}|e^{i\varphi_{1,2}}$.[1] Thus, in general, nine parameters are added to the SM when nonvanishing neutrino masses are included: three eigenvalues, three mixing angles, and three CP-violating phases.

In our notation, the two frequencies $\Delta m_I^2/4E$ ($I = \mathrm{sun}$, atm) are parameterized in terms of the neutrino mass eigenvalues by

$$\Delta m_{\mathrm{sun}}^2 \equiv |\Delta m_{12}^2| \,, \qquad \Delta m_{\mathrm{atm}}^2 \equiv |\Delta m_{23}^2| \,, \tag{6.16}$$

where $\Delta m_{12}^2 = |m_2|^2 - |m_1|^2$ and $\Delta m_{23}^2 = m_3^2 - |m_2|^2$. The numbering 1, 2, 3 corresponds to our definition of the frequencies and, in principle, need not coincide with the ordering from the lightest to the heaviest state.

From experiment (see Table 6.1), we know that $c_{23} \sim s_{23} \sim 1/\sqrt{2}$, corresponding to nearly maximal atmospheric neutrino mixing, and that s_{13}

---

[1] Mass matrices with a general dependence on φ and $\varphi_{1,2}$ have been analyzed in [29].

Table 6.1. Squared-mass differences and mixing angles [15–17]

	Lower limit (3σ)	Best value	Upper limit (3σ)
$(\Delta m^2_{\text{sun}})_{\text{LA}}$ (10^{-5} eV^2)	2.3	5	37
$(\Delta m^2_{\text{sun}})_{\text{LOW}}$ (10^{-8} eV^2)	3.5	8	12
Δm^2_{atm} (10^{-3} eV^2)	1	3	6
$(\tan^2 \theta_{12})_{\text{LA}}$	0.24	0.4	0.89
$(\tan^2 \theta_{12})_{\text{LOW}}$	0.43	0.6	0.86
$\tan^2 \theta_{23}$	0.33	0.8	3.3
$\tan^2 \theta_{13}$	0	0	0.07

is small (according to CHOOZ, $s_{13} < 0.2$ [30]). The solar angle θ_{12} is probably large (the MSW LMA, LOW, and VO solutions), or even maximal for the LOW and VO solutions, but could, alternatively, be very small ($s_{12}^2 \sim O(10^{-3})$ [15]), if the now disfavored MSW SA solution is also kept in our list. If we assume that s_{23} is maximal and keep only linear terms in $u = s_{13}\, e^{i\varphi}$, we find from experiment the following structure of the mixing matrix U_{fi} (f = e, μ, τ; i = 1, 2, 3), apart from redefinitions of sign conventions:

$$
U_{fi} = \begin{pmatrix}
c_{12} & s_{12} & u \\
-(s_{12} + c_{12}u^*)/\sqrt{2} & (c_{12} - s_{12}u^*)/\sqrt{2} & 1/\sqrt{2} \\
(s_{12} - c_{12}u^*)/\sqrt{2} & -(c_{12} + s_{12}u^*)/\sqrt{2} & 1/\sqrt{2}
\end{pmatrix}. \tag{6.17}
$$

Given the observed frequencies as defined in (6.16), there are three possible patterns of mass eigenvalues:

$$
\begin{aligned}
\text{degenerate:} \quad & |m_1| \sim |m_2| \sim |m_3| \gg |m_i - m_j|\,, \\
\text{inverted hierarchy:} \quad & |m_1| \sim |m_2| \gg |m_3|\,, \\
\text{normal hierarchy:} \quad & |m_3| \gg |m_{2,1}|\,.
\end{aligned} \tag{6.18}
$$

In the following, we shall discuss the phenomenology for these three different cases and the respective advantages and problems.

6.5.1 Degenerate Neutrinos

For degenerate neutrinos, the average m^2 is much larger than the splittings. At first sight, the degenerate case is the most appealing: the observation of nearly maximal atmospheric neutrino mixing and the experimental indication that the solar mixing is also large (at present, the MSW SA solution for the solar neutrino oscillations appears disfavored by the data [15]) suggest that all neutrino masses are nearly degenerate. Moreover, a common value of $|m_\nu|$

could be compatible with a large fraction of hot dark matter in the universe if $|m_\nu| \sim 1–2\,\mathrm{eV}$. In this case, however, the existing limits [31] on the absence of $0\nu\beta\beta$ decay ($|m_{ee}| < 0.2\,\mathrm{eV}$ or, to be more conservative, $|m_{ee}| < 0.3–0.5\,\mathrm{eV}$) imply [32] doubly maximal mixing (bimixing) for solar and atmospheric neutrinos. In fact, the quantity which is bounded by experiment is the 11 entry of the neutrino mass matrix, which, in general, from (6.13) and (6.15), is given by

$$|m_{ee}| = |(1 - s_{13}^2)(m_1 c_{12}^2 + m_2 s_{12}^2) + m_3 e^{2i\varphi} s_{13}^2| \,, \qquad (6.19)$$

which in this particular case (m_3 cannot compensate for the smallness of s_{13}^2) becomes approximately

$$|m_{ee}| \approx |m_1 c_{12}^2 + m_2 s_{12}^2| \lesssim 0.3–0.5\,\mathrm{eV} \,. \qquad (6.20)$$

To satisfy this constraint, one needs $m_1 \approx -m_2$ (recall that a relative phase $\varphi_2 - \varphi_1$ is allowed between m_1 and m_2) and $c_{12}^2 \approx s_{12}^2$ to a good accuracy (in fact, we need $\sin^2 2\theta_{12} > 0.96$ in order that $|\cos 2\theta_{12}| = |\cos^2\theta_{12} - \sin^2\theta_{12}| < 0.2$). This is exemplified by the following texture:

$$m_\nu = m \begin{pmatrix} 0 & -1/\sqrt{2} & 1/\sqrt{2} \\ -1/\sqrt{2} & (1+\eta)/2 & (1+\eta)/2 \\ 1/\sqrt{2} & (1+\eta)/2 & (1+\eta)/2 \end{pmatrix} \,, \qquad (6.21)$$

where $\eta \ll 1$, corresponding to an exact bimaximal mixing, $s_{13} = 0$, and the eigenvalues are $m_1 = m$, $m_2 = -m$, and $m_3 = (1+\eta)m$. This texture has been proposed in the context of a spontaneously broken SO(3) flavor symmetry and has been studied to analyze the stability of the degenerate spectrum against radiative corrections [33]. A more realistic mass matrix can be obtained by adding small perturbations to m_ν in (6.21):

$$m_\nu = m \begin{pmatrix} \delta & -1/\sqrt{2} & (1-\epsilon)/\sqrt{2} \\ -1/\sqrt{2} & (1+\eta)/2 & (1+\eta-\epsilon)/2 \\ (1-\epsilon)/\sqrt{2} & (1+\eta-\epsilon)/2 & (1+\eta-2\epsilon)/2 \end{pmatrix} \quad (\mathtt{D1}) \,, \qquad (6.22)$$

where ϵ parameterizes the leading flavor-dependent radiative corrections (mainly induced by the τ Yukawa coupling), and δ controls m_{ee}. (The symbol $\mathtt{D1}$ is explained in Sect. 6.5.4.) Consider first the case $\delta \ll \epsilon$. To first approximation, θ_{12} remains maximal. We obtain $\Delta m_{\mathrm{sun}}^2 \approx m^2\epsilon^2/\eta$ and

$$\theta_{13} \approx \left(\frac{\Delta m_{\mathrm{sun}}^2}{\Delta m_{\mathrm{atm}}^2} \right)^{1/2} \,, \qquad m_{ee} \ll m \left(\frac{\Delta m_{\mathrm{atm}}^2 \Delta m_{\mathrm{sun}}^2}{m^4} \right)^{1/2} \,. \qquad (6.23)$$

If instead we assume $\delta \gg \epsilon$, we find $\Delta m_{\mathrm{sun}}^2 \approx 2m^2\delta$, $\theta_{23} \approx \pi/4$, and $\sin^2 2\theta_{12} \approx 1 - \delta^2/4$. Also, in this case the solar mixing angle remains close to $\pi/4$. We obtain

$$\theta_{13} \approx 0 \,, \qquad m_{ee} \approx \frac{\Delta m_{\mathrm{sun}}^2}{2m} \,, \qquad (6.24)$$

which is too small for detection if the average neutrino mass m is around the eV scale. This example shows that there is no guarantee that $m_{\rm ee}$ will be close to the range of experimental interest, even with degenerate neutrinos where the masses involved are much larger than the oscillation frequencies. However, an almost maximal solar mixing angle, such as that implied by the previous analysis, is difficult to reconcile with the MSW LA solution. Of course, the strong constraint $s_{12}^2 = c_{12}^2$ can be relaxed if the common mass is below the hot-dark-matter maximum. It is true in any case that a signal of 0νββ decay near the present limit (similarly to a large relic density of hot dark matter) would be an indication of nearly degenerate neutrinos.

In general, for reasons of naturalness, the splittings cannot be too small with respect to the common mass, unless there is a protective symmetry [33, 34]. This is because the wide differences in fermion masses, in particular charged-lepton masses, would tend to create neutrino mass splittings via renormalization-group running effects, even starting from degenerate masses at a large scale. For example, for $m \approx 1$ eV, the VO solution for solar neutrino oscillations would imply $\Delta m/m \sim 10^{-9}-10^{-11}$, which is difficult to obtain. Even in the previous example, where, for $\delta \ll \epsilon$, the corrections to $\Delta m_{\rm sun}^2$ are quadratic in ϵ rather than linear, we would need $\epsilon < (10^{-3}/m({\rm eV}))^2$ in order to have $\Delta m_{\rm sun}^2 < 10^{-9}$ eV^2. In this respect, the MSW LA or LOW solution would be favored, but, if we insist that $|m_{\rm v}| \sim 1{-}2$ eV, it is not clear that the mixing angle preferred by the data is sufficiently maximal. To summarize, degenerate models with $|m| \sim 1{-}2$ eV, as required if neutrinos are a cosmologically important source of hot dark matter, have some problems related to 0νββ limits and to naturalness. In comparison, degenerate models with a sub-eV common mass appear simpler to realize.

It is clear that in the degenerate case, the most likely origin of neutrino masses is from some dimension-5 operators $(Hl)_i^{\rm T}\lambda_{ij}(Hl)_j/\Lambda$ not related to the seesaw result $m_{\rm v} = m_{\rm D}^{\rm T}M^{-1}m_{\rm D}$. In fact, we expect the Dirac mass $m_{\rm D}$ of the neutrinos not to be degenerate, as for all other fermions, and a conspiracy to reinstate a nearly perfect degeneracy between $m_{\rm D}$ and M, which would arise from completely different physics, looks very implausible (see, however, [35]). Thus, in degenerate models, in general, there is no direct relation to the Dirac masses of quarks and leptons, and the possibility of a simultaneous description of all fermion masses within a grand unified theory is more remote.[2]

The degeneracy of neutrinos should be guaranteed by some slightly broken symmetry. Models based on discrete or continuous symmetries have been proposed. For example, in the models of [37], the symmetry is SO(3). In the unbroken limit, neutrinos are degenerate and charged leptons are massless. When the symmetry is broken, the charged-lepton masses are much larger than the neutrino splittings because the former are first order while the latter are second order in the electroweak symmetry-breaking.

---

[2] Examples of degenerate models are described in [36].

A model which is simple to describe but difficult to derive in a natural way is one [38] where up quarks, down quarks, and charged leptons have "democratic" mass matrices, with all entries equal (to a first approximation):

$$
m_f = \hat{m}_f \begin{pmatrix} 1 & 1 & 1 \\ 1 & 1 & 1 \\ 1 & 1 & 1 \end{pmatrix} + \delta m_f \, ,
\tag{6.25}
$$

where $\hat{m}_f$ ($f = \mathrm{u, d, e}$) are three overall mass parameters and δm_f denotes small perturbations. If we neglect δm_f, the eigenvalues of m_f are given by $(0, 0, 3\,\hat{m}_f)$. The mass matrix m_f is diagonalized by a unitary matrix U_f which is in part determined by the small term δm_f. If $\delta m_\mathrm{u} \approx \delta m_\mathrm{d}$, the CKM matrix, given by $V_\mathrm{CKM} = U_\mathrm{u}^\dagger U_\mathrm{d}$, is nearly diagonal, owing to a compensation between the large mixings contained in U_u and U_d. When the small terms δm_f are diagonal and of the form $\delta m_f = \mathrm{Diag}(-\epsilon_f, \epsilon_f, \delta_f)$, the matrices U_f are approximately given by the following (note the analogy with the quark model eigenvalues π^0, η and η'):

$$
U_f^\dagger \approx \begin{pmatrix} 1/\sqrt{2} & -1/\sqrt{2} & 0 \\ 1/\sqrt{6} & 1/\sqrt{6} & -2/\sqrt{6} \\ 1/\sqrt{3} & 1/\sqrt{3} & 1/\sqrt{3} \end{pmatrix} .
\tag{6.26}
$$

At the same time, the lightest quarks and charged leptons acquire a nonvanishing mass. The leading part of the mass matrix in (6.25) is invariant under a discrete $S_\mathrm{3L} \times S_\mathrm{3R}$ permutation symmetry. The same requirement leads to the general neutrino mass matrix

$$
m_\mathrm{v} = m \left[\begin{pmatrix} 1 & 0 & 0 \\ 0 & 1 & 0 \\ 0 & 0 & 1 \end{pmatrix} + r \begin{pmatrix} 1 & 1 & 1 \\ 1 & 1 & 1 \\ 1 & 1 & 1 \end{pmatrix} \right] + \delta m_\mathrm{v} \quad (\mathrm{D2}) \, ,
\tag{6.27}
$$

where δm_v is a small symmetry-breaking term and the two independent invariants are allowed by the Majorana nature of the light neutrinos. If r vanishes, the neutrinos are almost degenerate. In the presence of δm_v, the permutation symmetry is broken and the degeneracy is removed. If, for example, we choose $\delta m_\mathrm{v} = \mathrm{Diag}(0, \epsilon, \eta)$, with $\epsilon < \eta \ll 1$ and $r \ll \epsilon$, the solar and atmospheric oscillation frequencies are determined by ϵ and η, respectively. The mixing angles are almost entirely due to the charged-lepton sector. A diagonal δm_e will lead to a neutrino mixing matrix $U \approx U_\mathrm{e}^\dagger$ characterized by an almost maximal θ_{12}, $\tan^2 \theta_{23} \approx 1/2$, and

$$
\theta_{13} \approx \sqrt{m_\mathrm{e}/m_\mu} \, .
\tag{6.28}
$$

By going to the basis where the charged leptons are diagonal, we can see that m_ee is close to m and independent of the parameters that characterize the oscillation phenomena.

The parameter r receives radiative corrections [39] that, at leading order, are logarithmic and proportional to the square of the τ lepton Yukawa coupling. It is important to guarantee that this correction does not spoil the relation $r \ll \epsilon$, whose violation would lead to a completely different mixing pattern. This raises a "naturalness" problem for the LOW and VO solutions. We conclude by stressing that a nonvanishing Δm^2_{atm}, maximal θ_{12}, large θ_{23}, and vanishing θ_{13} are not determined by the symmetric limit, but only by a specific choice of the parameter r and of the perturbations that cannot be easily justified on theoretical grounds. It would be desirable to provide a more sound basis for the choice of the small terms in this scenario, which is quite favorable to signals both in $0\nu2\beta$ decay and in subleading oscillations controlled by θ_{13}.

Anarchical models [40] can be considered as particular cases of degenerate models with $m^2 \sim \Delta m^2_{atm}$. In this class of models, one assumes that all mass matrices are structureless in the leptonic sector. At present, the data appear to indicate the MSW LA solution as the most likely one. For this solution, the ratio of the solar and atmospheric frequencies is not so small: typically $(\Delta m^2_{sun})_{LA}/\Delta m^2_{atm} \sim 0.01$–$0.2$ and two out of the three mixing angles are large. One important observation is that the seesaw mechanism tends to enhance the ratio of eigenvalues: this ratio is quadratic in m_D, so that a hierarchy factor f in m_D becomes f^2 in m_v, and the presence of the Majorana matrix M results in a further widening of the distribution. Another squaring takes place in going from the masses to the oscillation frequencies, which are quadratic. As a result, a random generation of the matrix elements of m_D and M leads to a distribution of $(\Delta m^2_{sun})_{LA}/\Delta m^2_{atm}$ that peaks around 0.1. At the same time, the distribution of $\sin^2 \theta_{ij}$ is peaked around 1 for all three mixing angles. Clearly, the smallness of θ_{13} is problematic. This can be turned into the prediction that in anarchical models θ_{13} must be near the present bound (after all, the value 0.2 for $\sin \theta_{13}$ is not very much smaller than the maximal value 0.701). In conclusion, there is a nonnegligible probability that if the MSW LA solution is correct and θ_{13} is near the present bound, then the neutrino masses and mixings, interpreted by means of the seesaw mechanism, just arise from structureless underlying Dirac and Majorana matrices.

6.5.2 Inverted Hierarchy

The inverted-hierarchy configuration $|m_1| \sim |m_2| \gg |m_3|$ consists of two levels m_1 and m_2, with a small splitting $\Delta m^2_{12} = \Delta m^2_{sun}$ and a common mass given by $|m^2_{1,2}| \sim |\Delta m^2_{atm}| \sim 3 \times 10^{-3}$ eV^2 (there is no large hot-dark-matter component in this case). One particularly interesting example of this sort [41], which leads to doubly maximal mixing, is obtained with the phase choice $m_1 = -m_2$ so that, approximately,

$$m_{diag} = \text{Diag}(\sqrt{2}m, -\sqrt{2}m, 0) . \tag{6.29}$$

The effective light-neutrino mass matrix

$$m_\nu = U m_{\text{diag}} U^\text{T} ,$$ (6.30)

which corresponds to the mixing matrix for doubly maximal mixing, obtained by setting $c_{12} = s_{12} = 1/\sqrt{2}$ and $s_{13} = u = 0$ in (6.17), i.e.

$$U_{fi} = \begin{pmatrix} 1/\sqrt{2} & 1/\sqrt{2} & 0 \\ -1/2 & 1/2 & 1/\sqrt{2} \\ 1/2 & -1/2 & 1/\sqrt{2} \end{pmatrix} ,$$ (6.31)

is given by

$$m_\nu = m \begin{pmatrix} 0 & -1 & 1 \\ -1 & 0 & 0 \\ 1 & 0 & 0 \end{pmatrix} .$$ (6.32)

The structure of m_ν can be reproduced by imposing a flavor symmetry $L_e - L_\mu - L_\tau$ starting either from $(Hl)_i^\text{T} \lambda_{ij} (Hl)_j / \Lambda$ or from RH neutrinos via the seesaw mechanism. The 1–2 degeneracy remains stable under radiative corrections. The preferred solar solutions are the VO and LOW solutions. The MSW LA solution could be also compatible if the mixing angle is large enough.

The leading texture given in (6.32) can be perturbed by adding small terms:

$$m_\nu = m \begin{pmatrix} \delta & -1 & 1 \\ -1 & \eta & \eta \\ 1 & \eta & \eta \end{pmatrix} \quad (\text{I}) ,$$ (6.33)

where δ and η are small ($\ll 1$), real parameters defined up to coefficients of order 1 that can differ in the various matrix elements. The perturbations leave Δm_{atm}^2 and θ_{23} unchanged, in the first approximation. We obtain $\tan^2 \theta_{12} \approx 1 + \delta + \eta$ and $\Delta m_{\text{sun}}^2 / \Delta m_{\text{atm}}^2 \approx \eta + \delta$, where coefficients of order one have been neglected. In addition, $\theta_{13} \approx \eta$. If $\eta \gg \delta$, we have

$$\theta_{13} \approx \frac{\Delta m_{\text{sun}}^2}{\Delta m_{\text{atm}}^2} , \qquad m_{ee} \ll \sqrt{\Delta m_{\text{sun}}^2} \left(\frac{\Delta m_{\text{sun}}^2}{\Delta m_{\text{atm}}^2} \right)^{1/2} .$$ (6.34)

In the other case, $\eta \ll \delta$, we obtain

$$\theta_{13} \ll \frac{\Delta m_{\text{sun}}^2}{\Delta m_{\text{atm}}^2} , \qquad m_{ee} \approx \frac{1}{2} \sqrt{\Delta m_{\text{sun}}^2} \left(\frac{\Delta m_{\text{sun}}^2}{\Delta m_{\text{atm}}^2} \right)^{1/2} .$$ (6.35)

There is a well-known difficulty in making this scenario fit the MSW LA solution [41, 42]. Indeed, barring cancellation between the perturbations, in order to obtain a value of Δm_{sun}^2 close to the best-fit MSW LA value, η and δ need to be smaller than about 0.1, and this keeps the value of $\sin^2 2\theta_{12}$ very close to 1, somewhat in disagreement with global fits of solar data [15]. Even by allowing a value of $\Delta m^2_{\rm sun}$ in the upper range of the MSW LA solution, or by some fine-tuning between η and δ, we would need large values of the perturbations to fit the MSW LA solution [43]. In contrast, the LOW solution can be accommodated, but in this case the values of θ_{13} and $m_{\rm ee}$ estimated from (6.34) and (6.35) are too small to be detected by planned experiments.

6.5.3 Normal Hierarchy

We now discuss the class of models which we consider to be of particular interest, which provide the most constrained framework that allows a comprehensive combined study of all fermion masses in GUTs. We assume three widely split neutrinos and the existence of an RH neutrino for each generation, as required to complete a 16-dimensional representation of SO(10) for each generation. We then assume the dominance of the seesaw mechanism result $m_\nu = m_{\rm D}^{\rm T} M^{-1} m_{\rm D}$. We know that the third-generation eigenvalue of the Dirac mass matrix of up and down quarks and of the Dirac mass matrix of charged leptons is, in all cases, systematically the largest one. It is natural to imagine that this property could also be true for the Dirac mass of neutrinos $m_{\rm D}^{\rm diag} \sim {\rm Diag}(0, 0, m_{\rm D3})$. After the operation of the seesaw mechanism, we expect m_ν to be even more hierarchical, being quadratic in $m_{\rm D}$ (barring fine-tuned compensations between $m_{\rm D}$ and M). The amount of hierarchy, $m_3^2/m_2^2 = \Delta m^2_{\rm atm}/\Delta m^2_{\rm sun}$, depends on which solar neutrino solution is adopted: the hierarchy is maximal for the VO and LOW solutions, is moderate for the MSW solution in general, and could become quite mild for the upper $\Delta m^2_{\rm sun}$ domain of the MSW LA solution. A possible difficulty is that one is used to expecting that large splittings correspond to small mixings, because normally only close-by states are strongly mixed. In the context of a 2×2 matrix, the requirement of a large splitting and large mixings leads to the condition of a vanishing determinant and large off-diagonal elements. For example, the matrix

$$\begin{pmatrix} x^2 & x \\ x & 1 \end{pmatrix} \tag{6.36}$$

has eigenvalues 0 and $1 + x^2$, and for x of order 1 the mixing is large. Thus, in the limit of neglecting small mass terms of order $m_{1,2}$, the demands of large atmospheric neutrino mixing and dominance of m_3 translate into the condition that the 2×2 subdeterminant 23 of the 3×3 mixing matrix approximately vanishes. The problem is to show that this vanishing can be arranged in a natural way without fine tuning. Once near-maximal atmospheric neutrino mixing is reproduced, the solar neutrino mixing can be arranged to be either small or large without difficulty, by imposing suitable relations among the small mass terms.

It is not difficult to imagine mechanisms that naturally lead to the approximate vanishing of the 23 subdeterminant. For example, [44] assumes that one $\nu^{\rm c}$ is particularly light and is coupled to μ and τ. In a 2×2 simplified context, if we have

$$M \propto \begin{pmatrix} \epsilon & 0 \\ 0 & 1 \end{pmatrix}, \qquad M^{-1} \approx \begin{pmatrix} 1/\epsilon & 0 \\ 0 & 0 \end{pmatrix}, \qquad m_D = \begin{pmatrix} a & b \\ c & d \end{pmatrix}, \qquad (6.37)$$

then for a generic m_D we find

$$m_\nu = m_D^T M^{-1} m_D \approx \frac{1}{\epsilon} \begin{pmatrix} a^2 & ab \\ ab & b^2 \end{pmatrix}. \qquad (6.38)$$

A different possibility that we find attractive is that, in the limit of neglecting terms of order $m_{1,2}$ and in the basis where charged leptons are diagonal, the Dirac matrix m_D, defined by $\nu^c m_D \nu$, takes the following approximate form, called "lopsided" [44–48]:

$$m_D \propto \begin{pmatrix} 0 & 0 & 0 \\ 0 & 0 & 0 \\ 0 & x & 1 \end{pmatrix}. \qquad (6.39)$$

This matrix has the property that, for a generic Majorana matrix M, one finds

$$m_\nu = m_D^T M^{-1} m_D \propto \begin{pmatrix} 0 & 0 & 0 \\ 0 & x^2 & x \\ 0 & x & 1 \end{pmatrix}. \qquad (6.40)$$

The only condition on M^{-1} is that the 33 entry is nonzero. However, when the approximately vanishing matrix elements are replaced by small terms, one must also assume that no new $O(1)$ terms are generated in m_ν by a compensation between small terms in m_D and large terms in M^{-1}. It is important for the following discussion to observe that the m_D given by (6.39) transforms under a change of basis as $m_D \to V^\dagger m_D U$, where V and U rotate the right and left fields, respectively. It is easy to check that we need large left mixings (i.e. large off-diagonal terms in the matrix that rotates LH fields) in order to make m_D diagonal. Thus, the question is how to reconcile large LH mixings in the leptonic sector with the observed near-diagonal form of V_{CKM}, the quark mixing matrix. Strictly speaking, since $V_{\mathrm{CKM}} = U_u^\dagger U_d$, the individual matrices U_u and U_d need not be near-diagonal, but V_{CKM} does need to be, while the analogue for leptons apparently cannot be near-diagonal. However, for quarks, nothing forbids the possibility that, in the basis where m_u is diagonal, the d quark matrix has large nondiagonal terms that can be rotated away by a pure RH rotation. We suggest that this is so and that, in some way, RH mixings for quarks correspond to LH mixings for leptons.

In the context of (SUSY) SU(5), there is a very attractive hint of how the present mechanism could be realized [49, 50]. In the $\bar{5}$ of SU(5), the d^c singlet appears together with the lepton doublet (ν, e). The (u, d) doublet and e^c belong to the 10 and ν^c to the 1, and similarly for the other families. As a consequence, in the simplest model with mass terms arising from only Higgs pentaplets, the Dirac matrix of down quarks is the transpose of the charged-lepton matrix: $m_d = (m_l)^T$. Thus, indeed, a large mixing for RH down quarks corresponds to a large LH mixing for charged leptons. At leading order, we may have a lopsided texture:

$$m_d = (m_l)^T = \begin{pmatrix} 0 & 0 & 0 \\ 0 & 0 & 1 \\ 0 & 0 & 1 \end{pmatrix} v_d . \tag{6.41}$$

In the same, simplest approximation with 5 or $\bar{5}$ Higgs, the up quark mass matrix is symmetric, so that the left and right mixing matrices are equal in this case. Small mixings for up quarks and small LH mixings for down quarks are then sufficient to guarantee small V_{CKM} mixing angles even for large d quark RH mixings. It is well known that a model where the down and charged-lepton matrices are exactly the transpose of one another cannot be exactly true, because of the e/d and μ/s mass ratios. It is also known that one remedy to this problem is to add some Higgs component in the 45 representation of SU(5) [51]. But this symmetry under transposition can still be a good guideline if we are interested only in the order of magnitude of the matrix entries and not in their exact values. Similarly, the Dirac neutrino mass matrix m_D is the same as the up quark mass matrix in the very crude model where the Higgs pentaplets come from a pure 10 representation of SO(10), i.e. $m_D = m_u$. For m_D, the dominance of the third-family eigenvalue and the near-diagonal form could be an order-of-magnitude remnant of this broken symmetry. Thus, if we neglect small terms, the neutrino Dirac matrix in the basis where charged leptons are diagonal can be directly obtained in the form of (6.39).

To obtain a realistic mass matrix, we allow deviations from the symmetric limit in (6.40), where we take $x = 1$. For instance, we can consider those models where the neutrino mass matrix elements are dominated, via the seesaw mechanism, by the exchange of two right-handed neutrinos [45]. Since the exchange of a single RH neutrino gives a good first approximation for the texture, we are encouraged to continue along this line. Thus, we add a subdominant contribution from a second RH neutrino, assuming that the third one gives a negligible contribution to the neutrino mass matrix, because it has much smaller Yukawa couplings or is much heavier than the first two. The Lagrangian that describes this plausible subset of seesaw models, written in the mass eigenstate basis of RH neutrinos and charged leptons, is

$$\mathcal{L} = y_i \nu^c H l_i + y'_i \nu^{c'} H l_i + \frac{M}{2} \nu^{c2} + \frac{M'}{2} \nu^{c'2} , \tag{6.42}$$

leading to

$$(m_\nu)_{ij} \propto \frac{y_i y_j}{M} + \frac{y'_i y'_j}{M'} , \tag{6.43}$$

where $i, j = \{e, \mu, \tau\}$. In particular, if $y_e \ll y_\mu \approx y_\tau$ and $y'_\mu \approx y'_\tau$, we obtain

$$m_v = m \begin{pmatrix} \delta & \epsilon & \epsilon \\ \epsilon & 1+\eta & 1+\eta \\ \epsilon & 1+\eta & 1+\eta \end{pmatrix} \quad \text{(N)} , \tag{6.44}$$

where coefficients of order one multiplying the small quantities δ, ϵ, and η have been omitted. The mass matrix in (6.44) does not describe the most general perturbation of the zeroth-order texture (6.40). We have implicitly assumed a symmetry between v_μ and v_τ which is preserved by the perturbations, at least at the level of orders of magnitude. The perturbed texture (6.44) can also arise when the zeros of the lopsided Dirac matrix in (6.39) are replaced by small quantities. It is possible to construct models along this line on the basis of a spontaneously broken $U(1)_F$ flavor symmetry, where δ, ϵ, and η are given by positive powers of one or more symmetry-breaking parameters. Moreover, by playing with the $U(1)_F$ charges, we can adjust, to certain extent, the relative hierarchy between η, ϵ, and δ [44, 45, 47–50], as we shall see in Sect. 6.8. The texture (6.44) can also be generated in SUSY models with R-parity violation [52].

After a first rotation by an angle θ_{23} close to $\pi/4$ and a second rotation by $\theta_{13} \approx \epsilon$, we obtain

$$m_v \approx m \begin{pmatrix} \delta + \epsilon^2 & \epsilon & 0 \\ \epsilon & \eta & 0 \\ 0 & 0 & 2 \end{pmatrix} , \tag{6.45}$$

up to coefficients of order one in the small entries. To obtain a large solar mixing angle, we need $|\eta - \delta| < \epsilon$. In realistic models, there is no reason for a cancellation between independent perturbations, and thus we assume both $\delta \leq \epsilon$ and $\eta \leq \epsilon$.

Consider first the case $\delta \approx \epsilon$ and $\eta < \epsilon$. The solar mixing angle θ_{12} is large but not maximal, as preferred by the MSW LA solution. We also have $\Delta m^2_{\text{atm}} \approx 4m^2$, $\Delta m^2_{\text{sun}} \approx \Delta m^2_{\text{atm}} \epsilon^2$, and

$$m_{\text{ee}} \approx \sqrt{\Delta m^2_{\text{sun}}} . \tag{6.46}$$

If $\eta \approx \epsilon$ and $\delta \ll \epsilon$, we still have a large solar mixing angle and $\Delta m^2_{\text{sun}} \approx \epsilon^2 \Delta m^2_{\text{atm}}$, as before. However, m_{ee} will be much smaller than the estimate in (6.46). Unfortunately, this is the case for the models based on the above-mentioned $U(1)_F$ flavor symmetry, which, at least in its simplest realization, tends to predict $\delta \approx \epsilon^2$. In this class of models, we find

$$m_{\text{ee}} \approx \sqrt{\Delta m^2_{\text{sun}}} \left(\frac{\Delta m^2_{\text{sun}}}{\Delta m^2_{\text{atm}}} \right)^{1/2} , \tag{6.47}$$

below the sensitivity of the next generation of planned experiments. It is worth mentioning that in both of the cases discussed above, we have

$$\theta_{13} \approx \left(\frac{\Delta m^2_{\text{sun}}}{\Delta m^2_{\text{atm}}} \right)^{1/2} , \tag{6.48}$$

which might be very close to the present experimental limit.

If both δ and η are much smaller than ϵ, the 12 block of m_ν has an approximately pseudo-Dirac structure and the angle θ_{12} becomes maximal. This situation is typical of some models where leptons have $U(1)_F$ charges of both signs but where the order parameters of $U(1)_F$ breaking all have charges of the same sign [49]. We have two eigenvalues given approximately by $\pm m\epsilon$. As an example, we consider the case where $\eta = 0$ and $\delta \approx \epsilon^2$. We find, that $\sin^2 2\theta_{12} \approx 1 - \epsilon^2/4$, $\Delta m^2_{\mathrm{sun}} \approx m^2\epsilon^3$, and

$$\theta_{13} \approx \left(\frac{\Delta m^2_{\mathrm{sun}}}{\Delta m^2_{\mathrm{atm}}}\right)^{1/3}, \qquad m_{ee} \approx \sqrt{\Delta m^2_{\mathrm{sun}}}\left(\frac{\Delta m^2_{\mathrm{sun}}}{\Delta m^2_{\mathrm{atm}}}\right)^{1/6}. \qquad (6.49)$$

In order to recover the MSW LA solution, we would need a relatively large value of ϵ. This is in general not acceptable, because, on the one hand the presence of a large perturbation raises doubts about the consistency of the whole approach and, on the other hand, in existing models where all fermion sectors are related to each other, ϵ is never larger than the Cabibbo angle. Therefore, the case $\delta, \eta \ll \epsilon$ can be more easily adapted to fit the LOW solution, where the solar frequency is small. As a consequence, m_{ee} is beyond the reach of the next generation of experiments, whereas θ_{13} might be tested at future facilities.

6.5.4 Summary

Given the present experimental knowledge, which favors $\Delta m^2_{\mathrm{atm}}$ as the leading oscillation frequency and two large mixing angles, θ_{23} and θ_{12}, it is natural to define a zeroth-order approximation to the theory, where $\Delta m^2_{\mathrm{sun}}$ and θ_{13} vanish (which allows us to neglect the CP-breaking parameter φ), whereas θ_{23} and θ_{12} are maximal. For each pattern of neutrino masses, we have considered the most interesting textures that arise in this limit.

This approximation is, of course, not realistic and should be regarded only as a limiting case, possibly arising from an underlying symmetry. Many effects can perturb this limit, such as small symmetry-breaking terms, radiative corrections, and effects from residual rotations needed to diagonalize the charged-lepton mass matrix or to render the leptonic kinetic terms canonical. Some of these will be discussed more in detail in the following sections. It turns out that in most of the existing models the leading textures are modified by small perturbations that have a simple structure, such as those labeled D1, D2, I, and N in (6.22), (6.27), (6.33), and (6.44).

We have analyzed these perturbations, in the hope that the results will be sufficiently representative of the many existing models. Of course, there is no guarantee that this discussion can cover all theoretical possibilities. Moreover, if $\Delta m^2_{\mathrm{sun}}$ and θ_{13} were as large as experimentally allowed, the perturbations would become large, the whole approach could become questionable, and the data would be more appropriately described by an anarchical framework.

A remarkable feature is that most models continue to predict an almost maximal solar angle, even after inclusion of the perturbations. This is often due to an approximately pseudo-Dirac structure in the 12 sector, which, at leading order, forces $\theta_{12} = \pi/4$. Exceptions to this trend exist in some of the possibilities offered by the normal hierarchy, for which θ_{12} is undetermined at leading order.

It is also apparent that, apart from the case of a degenerate spectrum with a texture similar to that of flavor democracy (D2), the possibility of measuring m_{ee} with the next generation of experiments seems to be significant only if the solar oscillation frequency is very close to the upper part of the range allowed by the MSW LA solution.

6.6 Importance of Neutrinoless Double Beta Decay

The discovery of $0\nu\beta\beta$ decay would be very important because it would establish lepton number nonconservation and the Majorana nature of neurinos. Indeed, oscillation experiments cannot distinguish between pure Dirac and Majorana neutrinos. Moreover, the search for $0\nu\beta\beta$ decay provides information about the absolute spectrum, while neutrino oscillations are sensitive only to mass differences. Complementary information about the sum of neutrino masses is also provided by the galaxy power spectrum combined with measurements of the cosmic microwave background anisotropy [53]. As already mentioned, the present limit from $0\nu\beta\beta$ decay is $|m_{ee}| < 0.2\,\mathrm{eV}$ or, to be more conservative, $|m_{ee}| < 0.3-0.5\,\mathrm{eV}$ [31]. In this respect, it is interesting to see what is the level at which a signal can be expected, or at least not excluded, in the various classes of models (see [54] for a recent analysis; see also [55]). For three-neutrino models with degenerate, inverse-hierarchy, or normal-hierarchy mass patterns, it is simple to derive the following bounds, starting from the general formula (6.19).

(a) *Degenerate case.* If $|m|$ is the common mass, apart from a phase, and if we take $s_{13} = 0$, which, as already observed, is a safe approximation in this case, we have $m_{ee} = |m|(c_{12}^2 \pm s_{12}^2)$. Here the phase ambiguity has been reduced to a sign ambiguity, which is sufficient for deriving bounds. So, depending on the sign, we have $m_{ee} = |m|$ or $m_{ee} = |m|\cos 2\theta_{12}$. We conclude that in this case m_{ee} could be very close to the present experimental limit because $|m|$ could be sizably larger than the $0\nu\beta\beta$ bound ($|m| < O(1\,\mathrm{eV})$), but should be at least of order $\sqrt{\Delta m_{\mathrm{atm}}^2} \sim 10^{-2}\,\mathrm{eV}$ unless the solar angle is practically maximal, in which case the minus-sign option can be as small as required.

(b) *Inverse-hierarchy case.* In this case, the same approximate formula $m_{ee} = |m|(c_{12}^2 \pm s_{12}^2)$ holds because m_3 is small and s_{13} can be neglected. The difference is that here we know that $|m| \approx \sqrt{\Delta m_{\mathrm{atm}}^2}$ so that $|m_{ee}| < \sqrt{\Delta m_{\mathrm{atm}}^2} \sim 0.05\,\mathrm{eV}$.

(c) *Normal-hierarchy case.* Here we cannot, in general, neglect the m_3 term. However, in this case $|m_{ee}| \sim \sqrt{\Delta m_{\mathrm{sun}}^2} s_{12}^2 + \sqrt{\Delta m_{\mathrm{atm}}^2} s_{13}^2$, and we have the bound $|m_{ee}| <$ a few times 10^{-3} eV.

Recently, evidence for $0\nu\beta\beta$ decay was claimed in [56] at the $2-3\sigma$ level, with $|m_{ee}| \sim 0.39$ eV. If confirmed, this would rule out cases (b) and (c), and point to case (a) or to models with more than three neutrinos.

6.7 Expectations for θ_{13}

The measurement of θ_{13} represents one of the main challenges for the next generations of experiments on neutrino oscillations, which, with the help of very intense neutrino beams [57], might reach a sensitivity of few percent for θ_{13}. A sizable θ_{13} would have an important impact on the observability of CP-violating effects in the leptonic sector. We collect together in Table 6.2 our estimates of θ_{13} for the various textures considered in Sect. 6.5 [58].

Figure 6.1 displays the expectations for θ_{13}. Such expectations are of course very rough and are not meant to be statistically meaningful. It is, however, interesting to note that the MSW LA solution favors values of θ_{13} in an experimentally accessible range for all textures but for the inverted hierarchy, where we find a strong suppression. The LOW solution prefers

Table 6.2. Order-of-magnitude estimates for θ_{13}

Texture		θ_{12}	θ_{13}	Perturbations
$m_\nu^{D1} = m \begin{pmatrix} \delta & -\frac{1}{\sqrt{2}} & \frac{(1-\epsilon)}{\sqrt{2}} \\ -\frac{1}{\sqrt{2}} & \frac{(1+\eta)}{2} & \frac{(1+\eta-\epsilon)}{2} \\ \frac{(1-\epsilon)}{\sqrt{2}} & \frac{(1+\eta-\epsilon)}{2} & \frac{(1+\eta-2\epsilon)}{2} \end{pmatrix}$		$\approx \pi/4$	$\approx \left(\frac{\Delta m_{\mathrm{sun}}^2}{\Delta m_{\mathrm{atm}}^2}\right)^{1/2}$	$\epsilon \gg \delta$
			≈ 0	$\epsilon \ll \delta$
$m_\nu^{D2} = m \left[\begin{pmatrix} 1 & 0 & 0 \\ 0 & 1 & 0 \\ 0 & 0 & 1 \end{pmatrix} + r \begin{pmatrix} 1 & 1 & 1 \\ 1 & 1 & 1 \\ 1 & 1 & 1 \end{pmatrix} \right] + \delta m_\nu$		$\approx \pi/4$	$\approx \left(\frac{m_e}{m_\mu}\right)^{1/2}$	
$m_\nu^{I} = m \begin{pmatrix} \delta & -1 & 1 \\ -1 & \eta & \eta \\ 1 & \eta & \eta \end{pmatrix}$		$\approx \pi/4$	$\approx \frac{\Delta m_{\mathrm{sun}}^2}{\Delta m_{\mathrm{atm}}^2}$	$\eta \gg \delta$
			$\ll \frac{\Delta m_{\mathrm{sun}}^2}{\Delta m_{\mathrm{atm}}^2}$	$\eta \ll \delta$
$m_\nu^{N} = m \begin{pmatrix} \delta & \epsilon & \epsilon \\ \epsilon & 1+\eta & 1+\eta \\ \epsilon & 1+\eta & 1+\eta \end{pmatrix}$		$O(1)$	$\approx \left(\frac{\Delta m_{\mathrm{sun}}^2}{\Delta m_{\mathrm{atm}}^2}\right)^{1/2}$	$\eta < \delta \approx \epsilon$ $\delta \approx \epsilon^2 \approx \eta^2$
		$\approx \pi/4$	$\approx \left(\frac{\Delta m_{\mathrm{sun}}^2}{\Delta m_{\mathrm{atm}}^2}\right)^{1/3}$	$\delta \approx \epsilon^2$, $\eta = 0$

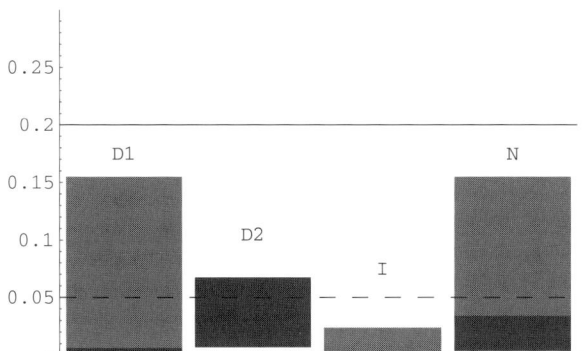

Fig. 6.1. Order-of-magnitude estimates of θ_{13}, for MSW LA solution (*gray*) and LOW solution (*black*). There is no difference between the MSW LA and LOW cases for the texture D2. The upper limits have been obtained by using the approximate expressions in Table 6.2 and the values $\Delta m^2_{\text{atm}} = 2.5 \times 10^{-3}\,\text{eV}^2$, $\Delta m^2_{\text{sun}} = 6.0 \times 10^{-5}\,\text{eV}^2$ for the MSW LA solution, and $\Delta m^2_{\text{sun}} = 10^{-7}\,\text{eV}^2$ for the LOW solution. The *continuous* and *dashed lines* show the present upper bound on θ_{13} and the possible reach of long-baseline experiments with high-intensity neutrino beams, respectively.

a smaller, unobservable θ_{13}, with the possible exception of the texture D2, corresponding to flavor democracy. We recall that if the neutrino mass matrix is structureless, as suggested by the anarchical framework, then θ_{13} is naturally expected to be very close to its present experimental bound.

6.8 Grand Unified Models of Fermion Masses

We have seen that the smallness of neutrino masses, interpreted via the seesaw mechanism, leads directly to a scale Λ for L nonconservation which is remarkably close to M_{GUT}. Thus neutrino masses and mixings should find a natural context in a GUT treatment of all fermion masses. The hierarchical pattern of quark and lepton masses, within each generation and across generations, requires some dynamical suppression mechanism that acts differently on the various particles. This hierarchy can be generated by a number of operators of different dimensions suppressed by inverse powers of the cutoff Λ_c of the theory. In some realizations, the different powers of $1/\Lambda_c$ correspond to different orders in some symmetry-breaking parameter v_f arising from the spontaneous breaking of a flavor symmetry. In the following subsections we describe some of the simplest models based on $\text{SU}(5) \times \text{U}(1)_F$ and on $\text{SO}(10)$ which illustrate these possibilities.[3]

---

[3] For models based on nonabelian flavor symmetries, see [59].

6.8.1 Models Based on Horizontal Abelian Charges

We discuss here some explicit examples of grand unified models in the framework of a unified SUSY SU(5) theory with an additional $U(1)_F$ flavor symmetry. The SU(5) generators act "vertically" inside one generation, while the $U(1)_F$ charges are different "horizontally" from one generation to another. If, for a given interaction vertex, the $U(1)_F$ charges do not add up to zero, the vertex is forbidden in the symmetric limit. But the symmetry is spontaneously broken by the vacuum expectation values (VEVs) v_f of a number of "flavon" fields with nonvanishing charge. A forbidden coupling is then rescued but is suppressed by powers of the small parameters v_f/Λ_c, with larger exponents for larger charge mismatch [60]. We expect $v_f \gtrsim M_{\mathrm{GUT}}$ and $\Lambda_c \lesssim M_{\mathrm{Pl}}$. Here, we discuss some aspects of the description of fermion masses in this framework.

In these models, the known generations of quarks and leptons are contained in triplets Ψ_i^{10} and $\Psi_i^{\bar 5}$ ($i = 1, 2, 3$) corresponding to the three generations, transforming as 10 and $\bar 5$ of SU(5), respectively. Three more SU(5) singlets Ψ_i^1 describe the RH neutrinos. In SUSY models we have two Higgs multiplets, which transform as 5 and $\bar 5$ in the minimal model. The two Higgs multiplets may have the same or different charges. We can arrange the unit of charge in such a way that the Cabibbo angle, which we consider as the typical hierarchy parameter of fermion masses and mixings, is obtained when the suppression exponent is unity. Remember that the Cabibbo angle is not too small, that $\lambda \sim 0.22$, and that in $U(1)_F$ models all mass matrix elements are of the form of a power of a suppression factor times a number of order unity, so that only their order of suppression is defined. As a consequence, in practice, we can limit ourselves to integral charges in our units (for example, $\sqrt{\lambda} \sim 1/2$ is already almost unsuppressed).

There are many variants of these models: fermion charges can be all nonnegative, with only negatively charged flavons, or there can be fermion charges of different signs with either flavons of both charges or flavons of only one charge. We can have the situation that only the top quark mass is allowed in the symmetric limit, or that other third-generation fermion masses are also allowed. The Higgs charges can be equal, in particular both of them can be vanishing, or they can be different. We can arrange that all the structure is in charged-fermion masses, while neutrinos are anarchical.

$F(\text{fermions}) \geq 0$. Consider, for example, a simple model where all charges of matter fields are nonnegative and which contains one single flavon $\bar\theta$ of charge $F = -1$. For maximum simplicity, we also assume that all the third-generation masses are directly allowed in the symmetric limit. This is realized by assuming vanishing charges for the Higgses and for the third-generation components Ψ_3^{10}, $\Psi_3^{\bar 5}$, and Ψ_3^1. For example, if we define

$F(\Psi_i^R) \equiv q_i^R$ $(R = 10, \bar{5}, 1; \; i = 1, 2, 3)$, we could take [47–49, 62] (see also [61])

$$(q_1^{10}, q_2^{10}, q_3^{10}) = (3, 2, 0) \,,$$
$$(q_1^{\bar{5}}, q_2^{\bar{5}}, q_3^{\bar{5}}) = (2, 0, 0) \,. \tag{6.50}$$

A generic mass matrix in this model has the form

$$m = \begin{pmatrix} y_{11}\lambda^{q_1+q_1'} & y_{12}\lambda^{q_1+q_2'} & y_{13}\lambda^{q_1+q_3'} \\ y_{21}\lambda^{q_2+q_1'} & y_{22}\lambda^{q_2+q_2'} & y_{23}\lambda^{q_2+q_3'} \\ y_{31}\lambda^{q_3+q_1'} & y_{32}\lambda^{q_3+q_2'} & y_{33}\lambda^{q_3+q_3'} \end{pmatrix} v \,, \tag{6.51}$$

where all the y_{ij} are of order 1, and (q_i, q_j') are the charges of (Ψ^{10}, Ψ^{10}) for $m_{\rm u}$, of $(\Psi^{\bar{5}}, \Psi^{10})$ for $m_{\rm d}$ or $m_l^{\rm T}$, of $(\Psi^1, \Psi^{\bar{5}})$ for $m_{\rm D}$ (the Dirac ν mass), and of (Ψ^1, Ψ^1) for M (the Majorana ν^c mass). We have $\lambda \equiv \langle \bar{\theta} \rangle / \Lambda_{\rm c}$, and the quantity v represents the appropriate VEV or mass parameter. It is important to observe that m can be written as

$$m = \lambda_R y \lambda_{R'} v \,, \tag{6.52}$$

where $\lambda_R = \mathrm{Diag}(\lambda^{q_1^R}, \lambda^{q_2^R}, \lambda^{q_3^R})$ and y is the matrix y_{ij}. The models with all charges nonnegative and one single flavon have particularly simple factorization properties. For example, if we start from the Dirac ν matrix $m_{\rm D} = \lambda_1 y_{\rm D} \lambda_{\bar{5}} v_{\rm u}$ and the ν^c Majorana matrix $M = \lambda_1 y_{\rm M} \lambda_1 \Lambda$ and write down the seesaw expression for $m_\nu = m_{\rm D}^{\rm T} M^{-1} m_{\rm D}$, we find that the dependence on the charges q^1 drops out and only the dependence arising from $q^{\bar{5}}$ remains. As a consequence, the effective light-neutrino Majorana mass matrix m_ν can be written in terms of $q^{\bar{5}}$ only: $m_\nu = \lambda_{\bar{5}}(y_{\rm D}^{\rm T} y_{\rm M}^{-1} y_{\rm D})\lambda_{\bar{5}} v_{\rm u}^2 / \Lambda$. In addition, for the neutrino mixing matrix U_{ij}, which is determined by m_ν in the basis where the charged leptons are diagonal, one can prove that $U_{ij} \approx \lambda^{|q_i^{\bar{5}} - q_j^{\bar{5}}|}$, which is in terms of the differences between the $\bar{5}$ charges, when terms that are small by powers of the small parameter λ are neglected. Similarly, the CKM matrix elements are approximately determined by only the 10 charges [60]: $V_{ij}^{\rm CKM} \approx \lambda^{|q_i^{10} - q_j^{10}|}$. With these results in mind, we can understand that the assignments of the charge q^{10} in (6.50) are determined by requiring $V_{\rm us} \sim \lambda$, $V_{\rm cb} \sim \lambda^2$ and $V_{\rm ub} \sim \lambda^3$. However, the same charges q^{10} also fix the ratio $m_{\rm u} : m_{\rm c} : m_{\rm t} \sim \lambda^6 : \lambda^4 : 1$. The experimental value of $m_{\rm u}$ (the relevant mass values are on GUT scale, $m = m(M_{\rm GUT})$ [63]) favours $q_1^{10} = 4$. Taking into account this indication and the presence of the unknown coefficients $y_{ij} \sim 1$, it is difficult to decide between $q_1^{10} = 3$ or 4, and both are acceptable.

Turning to the $\bar{5}$ charges, the entries $q_2^{\bar{5}} = q_3^{\bar{5}} = 0$ have been selected in (6.50), so that the 22, 23, 32, and 33 entries of the effective light-neutrino mass matrix m_ν are all $O(1)$ in order to accommodate the nearly maximal value of s_{23}. The small nondiagonal terms of the charged-lepton mass matrix cannot change this. In fact, for q_i^{10}, $q_i^{\bar{5}}$ chosen as in (6.50), we obtain

$$m_d = \begin{pmatrix} \lambda^5 & \lambda^4 & \lambda^2 \\ \lambda^3 & \lambda^2 & 1 \\ \lambda^3 & \lambda^2 & 1 \end{pmatrix} v_d = (m_l)^{\mathrm{T}}, \qquad m_\nu = \begin{pmatrix} \lambda^4 & \lambda^2 & \lambda^2 \\ \lambda^2 & 1 & 1 \\ \lambda^2 & 1 & 1 \end{pmatrix} \frac{v_u^2}{\Lambda}, \quad (6.53)$$

where $v_{u,d}$ are the VEVs of the Higgs doublets. Note that the patterns $m_d : m_s : m_b \sim m_e : m_\mu : m_\tau \sim \lambda^5 : \lambda^2 : 1$ are acceptable (but $q_1^{\bar{5}} = 3$ would also be possible). One difficulty is that for m_ν, the subdeterminant 23 is not suppressed in this case, so that the splitting between the 2 and 3 light-neutrino masses is in general small. In spite of the fact that m_D is, to a first approximation, of the form given in (6.39), the strong correlations between m_D and M implied by the simple charge structure of the model destroy the vanishing of the 23 subdeterminant that would be guaranteed for a generic M. Models of this sort have been proposed in the literature [47, 62]. The hierarchy between m_2 and m_3 is considered accidental and needs to be moderate. The preferred solar solution in this case is MSW SA, because if m_1 is suppressed (and some suppression is needed if we want s_{13} to be small) the solar mixing angle is typically small. However, if with a moderate fine tuning we stretch m_2 by hand to become sufficiently close to m_1, then the MSW LA solution can also be reproduced. From (6.53), taking $v_u \sim 250\,\mathrm{GeV}$, the mass scale Λ of the heavy Majorana neutrinos turns out to be close to the unification scale, i.e. $\Lambda \sim 10^{15}\,\mathrm{GeV}$.

A different interesting possibility [64] is to recover an anarchical picture of neutrinos by taking $(q_1^{10}, q_2^{10}, q_3^{10}) = (4, 2, 0)$ and $(q_1^{\bar{5},1}, q_2^{\bar{5},1}, q_3^{\bar{5},1}) = (0, 0, 0)$. The 10 charges lead to an acceptable pattern for the m_u matrix and V_{CKM}, as already discussed. For down quarks and charged leptons, we obtain a weakened hierarchy, essentially the square root of that of up quarks: $m_d : m_s : m_b \sim m_e : m_\mu : m_\tau \sim \lambda^4 : \lambda^2 : 1$. Finally, in the neutrino sector, the anarchical model is realized (both m_D and M are structureless).

Note that in all of the above cases we could add a constant to $q_i^{\bar{5}}$, for example by taking $(q_1^{\bar{5}}, q_2^{\bar{5}}, q_3^{\bar{5}}) = (4, 2, 2)$. This would have the consequence of leaving the top quark as the only unsuppressed mass and of decreasing the resulting value of $\tan \beta = v_u / v_d$ to $\lambda^2 m_t / m_b$. A constant shift of the charges q_i^1 might also provide a suppression of the leading ν^c mass eigenvalue, from Λ_c down to the appropriate scale Λ. One can also consider models where the 5 and $\bar{5}$ Higgs charges are different, as in the "realistic" SU(5) model of [65]. In these models also, the top mass could be the only one to be nonvanishing in the symmetric limit, and the value of $\tan \beta$ can be adjusted.

F(fermions) and F(flavons) of Both Signs. Models with naturally large 23 splittings are obtained if we allow negative charges and, at the same time, either introduce flavons of opposite charges or stipulate that matrix elements with overall negative charge are set to zero. For example, we can assign to the fermion fields the set of charges F given by

$$
\begin{aligned}
(q_1^{10}, q_2^{10}, q_3^{10}) &= (3, 2, 0) \,, \\
(q_1^{\bar{5}}, q_2^{\bar{5}}, q_3^{\bar{5}}) &= (b, 0, 0) \,, \qquad b \geq 2a > 0 \,, \\
(q_1^1, q_2^1, q_3^1) &= (a, -a, 0) \,.
\end{aligned}
\tag{6.54}
$$

We consider the Yukawa coupling allowed by $U(1)_F$-neutral Higgs multiplets in the 5 and $\bar{5}$ SU(5) representations and by a pair θ and $\bar{\theta}$ of SU(5) singlets with $F = 1$ and $F = -1$, respectively. If $b = 2$ or 3, the up, down, and charged-lepton sectors are not essentially different from what they were in the previous case. Also, in this case the $O(1)$ off-diagonal entry of m_l, typical of lopsided models, gives rise to a large LH mixing in the 23 block, which corresponds to a large RH mixing in the d mass matrix. In the neutrino sector, the Dirac and Majorana mass matrices are given by

$$
m_{\mathrm{D}} = \begin{pmatrix} \lambda^{a+b} & \lambda^a & \lambda^a \\ \lambda^{b-a} & \lambda'^a & \lambda'^a \\ \lambda^b & 1 & 1 \end{pmatrix} v_{\mathrm{u}} \,, \qquad M = \begin{pmatrix} \lambda^{2a} & 1 & \lambda^a \\ 1 & \lambda'^{2a} & \lambda'^a \\ \lambda^a & \lambda'^a & 1 \end{pmatrix} \Lambda \,, \tag{6.55}
$$

where λ' is given by $\langle \theta \rangle / \Lambda_{\mathrm{c}}$, and Λ, as before, denotes the large mass scale associated with the RH neutrinos, where $\Lambda \gg v_{\mathrm{u,d}}$. After diagonalization of the charged-lepton sector and integrating out the heavy RH neutrinos, we obtain the following neutrino mass matrix in the low-energy effective theory:

$$
m_{\mathrm{v}} = \begin{pmatrix} \lambda^{2b} & \lambda^b & \lambda^b \\ \lambda^b & 1 + \lambda^a \lambda'^a & 1 + \lambda^a \lambda'^a \\ \lambda^b & 1 + \lambda^a \lambda'^a & 1 + \lambda^a \lambda'^a \end{pmatrix} \frac{v_{\mathrm{u}}^2}{\Lambda} \,. \tag{6.56}
$$

The $O(1)$ elements in the 23 block are produced by combining the large LH mixing induced by the charged-lepton sector and the large LH mixing in m_{D}. A crucial property of m_{v} is that, as a result of the seesaw mechanism and of the specific $U(1)_F$ charge assignment, the determinant of the 23 block is automatically of $O(\lambda^a \lambda'^a)$ (the presence of negative charge values, leading to the presence of both λ and λ', is essential for this [48, 49]). The neutrino mass matrix of (6.56) is a particular case of the more general pattern presented in (6.44), from which it is obtained when $\delta \approx \lambda^{2b}$, $\epsilon \approx \lambda^b$, and $\eta \approx \lambda^a \lambda'^a$. If we take $\lambda \approx \lambda'$, it is easy to verify that the eigenvalues of m_{v} satisfy the relation

$$
m_1 : m_2 : m_3 = \lambda^{2(b-a)} : \lambda^{2a} : 1 \,. \tag{6.57}
$$

The atmospheric neutrino oscillations require $m_3^2 \sim 10^{-3}$ eV^2. The squared-mass difference between the lightest states is of $O(\lambda^{4a})m_3^2$, not far from the MSW solution to the solar neutrino problem if we choose $a = 1$. In general U_{e3} is nonvanishing, and of $O(\lambda^b)$. Finally, beyond the large mixing in the 23 sector, m_{v} provides a mixing angle $\theta_{12} \sim \lambda^{b-2a}$ in the 12 sector. For $b > 2a$, we recover a small solar mixing angle. For instance, taking $b = 3$ and $a = 1$,

θ_{12} becomes close to the range preferred by the MSW SA solution. When $b = 2a$, as for instance in the case $b = 2$ and $a = 1$, the MSW LA solution can be reproduced.

F(fermions) of Both Signs and F(flavons) < 0. A general problem common to all models dealing with flavor is that of recovering the correct vacuum structure by minimizing the effective potential of the theory. It may be noticed that the presence of two multiplets θ and $\bar{\theta}$ with opposite charges F could hardly be reconciled, without adding extra structure to the model, with a large common VEV for these fields, owing to possible analytic terms of the kind $(\theta\bar{\theta})^n$ in the superpotential. We find it instructive, therefore, to explore the consequences of allowing only the negatively charged $\bar{\theta}$ field in the theory.

It can be immediately recognized that, while the quark mass matrices previously discussed are unchanged, in the neutrino sector the Dirac and Majorana matrices are obtained from (6.55) by setting $\lambda' = 0$:

$$
m_{\rm D} = \begin{pmatrix} \lambda^{a+b} & \lambda^a & \lambda^a \\ \lambda^{b-a} & 0 & 0 \\ \lambda^b & 1 & 1 \end{pmatrix} v_{\rm u} , \qquad M = \begin{pmatrix} \lambda^{2a} & 1 & \lambda^a \\ 1 & 0 & 0 \\ \lambda^a & 0 & 1 \end{pmatrix} \Lambda . \tag{6.58}
$$

The zeros are due to an analytic property of the superpotential that makes it impossible to form the corresponding F invariant by using $\bar{\theta}$ alone. These zeros should not be taken literally, as they will eventually be filled by small terms that arise, for instance, from the diagonalization of the charged-lepton mass matrix and from the transformation that put the kinetic terms into canonical form. It is, however, interesting to work out, in a first approximation, the case of exactly zero entries in $m_{\rm D}$ and M, when forbidden by $U(1)_{\rm F}$. The neutrino mass matrix obtained from $m_{\rm D}$ and M via the seesaw mechanism has the same pattern as that displayed in (6.56). A closer inspection reveals that the determinant of the 23 block is identically zero, independent of λ. This leads to the following pattern of masses:

$$
m_1 : m_2 : m_3 = \lambda^b : \lambda^b : 1 , \qquad m_1^2 - m_2^2 = O(\lambda^{3b}) . \tag{6.59}
$$

Moreover, the mixing in the 12 sector is almost maximal:

$$
\theta_{12} = \frac{\pi}{4} + O(\lambda^b) . \tag{6.60}
$$

For $b = 3$ and $\lambda \sim 0.2$, both the squared-mass difference $(m_1^2 - m_2^2)/m_3^2$ and $\sin^2 2\theta_{12}$ are remarkably close to the values required by the VO solution to the solar neutrino problem. This property remains reasonably stable against the perturbations induced when small terms (of order λ^5) replace the zeros, where these small terms arise from the diagonalization of the charged-lepton sector and by the transformations that render the kinetic terms canonical. By choosing $b = 2$, we obtain the LOW solution. We find it quite interesting that the VO and LOW solutions, which require an intriguingly small mass difference and a bimaximal mixing, can also be described, at least on the level of orders of magnitude, in the context of a "minimal" model of flavor compatible with supersymmetric SU(5). In this case the role played by supersymmetry is essential; a nonsupersymmetric model with $\bar{\theta}$ alone is not distinguishable from the version with both θ and $\bar{\theta}$, as far as low-energy flavor properties are concerned.

In conclusion, models based on SU(5) $\times$ U(1)$_F$ are clearly toy models that can aim only at a semiquantitative description of fermion masses. In fact, only the order of magnitude of each matrix entry can be specified. However, it is rather impressive that a reasonable description of fermion masses, now also including neutrino masses and mixings, can be obtained in this simple context, which is suggestive of a deeper relation between gauge and flavor quantum numbers. There are 12 mass eigenvalues and six mixing angles that are specified, modulo coefficients of order 1, in terms of a set of integer numbers (from half a dozen to a dozen), the charges, and one or more scale parameters. In the neutrino sector, we have seen that the scheme is flexible enough to accommodate all the solutions that are still possible.

6.8.2 GUT Models Based on SO(10)

Models based on SO(10) times a flavor symmetry are more difficult to construct because a whole generation is contained in the representation **16**, so that, for U(1)$_F$ for example, one would have the same value of charge for all quarks and leptons of each generation, which is too rigid. But the mechanism discussed so far, based on asymmetric mass matrices, can be embedded in an SO(10) grand unified theory in a rather economical way [25, 46, 66, 67]. The 33 entries of the fermion mass matrices can be obtained through the coupling **16**$_3$**16**$_3$**10**$_H$ among the fermions in the third generation, **16**$_3$, and a Higgs tenplet **10**$_H$. The two independent VEVs of the tenplet v_u and v_d give mass to t/ν_τ and b/τ, respectively. The key point for obtaining an asymmetric texture is the introduction of an operator of the kind **16**$_2$**16**$_H$**16**$_3$**16**$'_H$. This operator is thought to arise from integrating out a heavy **10** that couples both to **16**$_2$**16**$_H$ and to **16**$_3$**16**$'_H$. If the **16**$_H$ develops a VEV that breaks SO(10) down to SU(5) at a large scale, then, in terms of SU(5) representations, we obtain an effective coupling of the kind $\bar{\mathbf{5}}_2\mathbf{10}_3\bar{\mathbf{5}}_H$, with a coefficient that can be of order one. This coupling contributes to the 23 entry of the down quark mass matrix and to the 32 entry of the charged-lepton mass matrix, which creates the desired asymmetry. To distinguish the lepton and quark sectors, one can introduce in addition an operator of the form **16**$_i$**16**$_j$**10**$_H$**45**$_H$ $(i, j = 2, 3)$, with the VEV of the **45**$_H$ pointing in the $B - L$ direction. Additional operators, still of the type **16**$_i$**16**$_j$**16**$_H$**16**$'_H$, can contribute to the matrix elements of the first generation. The mass matrices look like

$$m_u = \begin{pmatrix} 0 & 0 & 0 \\ 0 & 0 & \epsilon/3 \\ 0 & -\epsilon/3 & 1 \end{pmatrix} v_u \, , \qquad m_d = \begin{pmatrix} 0 & \delta & \delta' \\ \delta & 0 & \sigma + \epsilon/3 \\ \delta' & -\epsilon/3 & 1 \end{pmatrix} v_d \, , \quad (6.61)$$

$$m_D = \begin{pmatrix} 0 & 0 & 0 \\ 0 & 0 & -\epsilon \\ 0 & \epsilon & 1 \end{pmatrix} v_u , \qquad m_l = \begin{pmatrix} 0 & \delta & \delta' \\ \delta & 0 & -\epsilon \\ \delta' & \sigma + \epsilon & 1 \end{pmatrix} v_d . \qquad (6.62)$$

These matrices provide a good fit to the available data in the quark and charged-lepton sectors in terms of five parameters (one of which is complex). In the neutrino sector, one obtains a large mixing angle θ_{23}, $\sin^2 2\theta_{12} \sim 6.6 \times 10^{-3} \, \mathrm{eV}^2$, and θ_{13} of the same order as θ_{12}. Mass-squared differences are sensitive to the details of the Majorana mass matrix.

Looking at models with three light neutrinos only, i.e. no sterile neutrinos, from a more general point of view, we stress that in the above models the atmospheric neutrino mixing is considered large, in the sense of being of order one in some zeroth-order approximation. In other words, this mixing corresponds to off-diagonal matrix elements of the same order as the diagonal ones, although the mixing is not exactly maximal. The idea that all fermion mixings are small and induced by the observed smallness of the nondiagonal elements of the matrix V_{CKM} is then abandoned. An alternative is to argue that perhaps what appears to be large is not very large after all. The typical small parameter that appears in the mass matrices is $\lambda \sim \sqrt{m_d/m_s} \sim \sqrt{m_\mu/m_\tau} \sim 0.20$–$0.25$. This small parameter is not so small that it cannot become large owing to some peculiar accidental enhancement: this could arise from either a coefficient of order 3, an exponent of the mass ratio less than $1/2$ (due, for example, to a suitable charge assignment), or the addition in phase of an angle from the diagonalization of charged leptons and an angle from neutrino mixing. One might like this strategy of producing a large mixing by stretching small ones if, for example, one likes symmetric mass matrices, as in the case of left–right symmetry at the GUT scale. In left–right symmetric models, the smallness of left mixings implies that right-handed mixings are also small, so that all mixings tend to be small, unless nonrenormalizable mass operators with a suitable flavor pattern are introduced. Clearly, this set of models [67, 68] tends to favor moderate hierarchies and a single maximal mixing, so that the MSW SA solution of the solar neutrino problem is preferred.

6.9 Conclusion

There are now some rather convincing experimental indications of neutrino oscillations. The direct implication of these findings is that not all neutrino masses are vanishing. As a consequence, the phenomenology of neutrino masses and mixings has been brought to the forefront. This is a very interesting subject in many respects. It is a window on the physics of GUTs in that the extreme smallness of neutrino masses can only be explained in a natural way if lepton number conservation is violated. If so, neutrino masses are inversely proportional to the large scale where lepton number conservation is violated. Also, the pattern of neutrino masses and mixings, interpreted in a GUT framework, can provide new clues about the long-standing problem of understanding the origin of the hierarchical structure of quark and lepton mass matrices. Neutrino oscillation measurements determine only differences between m_i^2 values, and the actual scale of neutrino masses remains to be fixed experimentally. In particular, the scale of neutrino masses is important for cosmology as neutrinos are candidates for hot dark matter: nearly degenerate neutrinos with a common mass around 1–2 eV would significantly contribute to Ω_m, the matter density in the universe in units of the critical density. The detection of $0\nu\beta\beta$ decay would be extremely important for the determination of the overall scale of neutrino masses, the confirmation of their Majorana nature, and the experimental clarification of the ordering of levels in the associated spectrum. The recent indication of a signal of $0\nu\beta\beta$ decay with m_{ee} in a range around 0.4 eV, if confirmed, would point to a small but possibly nonnegligible contribution of neutrinos to Ω_m and, among models with three neutrinos, would favor those with a degenerate spectrum. The decay of heavy right-handed neutrinos with lepton number nonconservation can provide a viable and attractive model of baryogenesis through leptogenesis. The measured oscillation frequencies and mixings are remarkably consistent with this attractive possibility.

While the existence of oscillations appears to be on increasingly solid ground, many important experimental challenges remain. For atmospheric neutrino oscillations, the completion of the K2K experiment, now stopped by an accident that has seriously damaged the SuperKamiokande detector, is important for a terrestrial confirmation of the effect and for an independent measurement of the associated parameters. In the near future, the experimental study of atmospheric neutrinos will continue with long-baseline measurements by MINOS, OPERA, and ICARUS. For solar neutrinos, it is not yet clear which of the solutions, MSW SA, MSW LA, LOW, or VO, is correct, although a preference for the MSW LA solution appears to be indicated by the present data. This issue will presumably be clarified in the near future by the continuation of SNO and by the forthcoming data from KamLAND and Borexino. Finally, a clarification by MiniBooNE of the issue of the LSND alleged signal is necessary, in order to know whether three light neutrinos are sufficient or additional, sterile neutrinos must be introduced, in spite of the apparent lack of independent evidence in the data for such sterile neutrinos and the fact that attempts to construct plausible, natural theoretical models have not led so far to compelling results. Further in the future, there are projects for neutrino factories and/or superbeams, aimed at precision measurements of the oscillation parameters and possibly the detection of CP violation effects in the neutrino sector.

Pending the resolution of the existing experimental ambiguities, a large variety of theoretical models for the neutrino masses and mixings are still conceivable. For the three-neutrino models, we have described a variety of possibilities based on degenerate, inverted-hierarchy, and normal-hierarchy spectra. Most models prefer one or other of the possible experimental alternatives which are still open. It is interesting that the MSW LA solar-oscillation solution, which at present appears somewhat favored by the data, is perhaps the most constraining for theoretical models. For example, it is difficult to reproduce this solution in the inverted-hierarchy models. The MSW LA solution can be obtained in the degenerate case, including also the anarchical scenario, and in the normal-hierarchy case, but in those cases rather special conditions must be met. In many cases, the MSW LA solution corresponds to values of θ_{13} that are rather large and not far from the present bound. The values of m_{ee} (which determines the rate of $0\nu\beta\beta$ decay) that are found in the models leading to the MSW LA solution are typically of order $\sqrt{\Delta m_{sun}^2}$ in the normal-hierarchy case and can be even larger in the degenerate case, where the expected values are at least of order $\sqrt{\Delta m_{atm}^2}$ (assuming that for the MSW LA solution, s_{12} is indeed large but not close to maximal).

The fact that some neutrino mixing angles are large and even nearly maximal, while surprising at the start, was eventually found to be very compatible with a unified picture of quark and lepton masses within GUTs. The symmetry group at M_{GUT} could be either (SUSY) SU(5), SO(10), or a larger group. For example, we have presented a class of natural models where large right-handed mixings for quarks are transformed into large left-handed mixings for leptons by the transposition relation $m_d = m_e^T$, which is approximately obtained in SU(5) models. In particular, we have argued in favor of models with three widely split neutrinos. Reconciling large splittings with large mixing(s) requires some natural mechanism to impose a condition of a vanishing determinant. This can be obtained with the seesaw mechanism, for example, if one light right-handed neutrino is dominant, or a suitable texture of the Dirac matrix is imposed by an underlying symmetry. We have shown that these mechanisms can be implemented naturally by simple assignments of $U(1)_F$ horizontal charges that lead to a successful semiquantitative, unified description of all quark and lepton masses in SUSY $SU(5) \times U(1)_F$. Alternative realizations based on the SO(10) unification group have also been discussed. Models aiming at a realistic unification of electroweak and strong interactions should, of course, address other important questions, such as the doublet–triplet splitting and its stability against quantum corrections; a proton lifetime compatible with the existing limits, which now have become quite stringent for all minimal SUSY GUT models; a correct unification of gauge coupling constants; and consistency with present bounds on flavor nonconservation [69]. Encompassing all these features in a consistent and, if possible, simple model is a formidable task that might require us to go beyond the conventional formulation in terms of a four-dimensional quantum field theory [70].

In conclusion, the discovery of neutrino oscillations with frequencies that point to very small neutrino masses has opened a window on the physics beyond the Standard Model at very large energy scales. The study of the neutrino mass and mixing matrices, which is still in its early stages, may lead to particularly exciting insights into the theory at large energies, possibly as large as $M_{\rm GUT}$.

Acknowledgments

We thank J. Garcia-Bellido, A. Masiero, I. Masina, A. Riotto, A. Strumia, and F. Vissani for discussions. F.F. thanks the CERN Theoretical Division, where part of this work was done, for hospitality and financial support. F.F. is partially supported by the European Programs HPRN-CT-2000-00148 and HPRN-CT-2000-00149.

References

1. Super-Kamiokande Collaboration, Phys. Rev. Lett. **85**, 3999 (2000); T. Toshito (Super-Kamiokande Collaboration), arXiv:hep-ex/0105023; MACRO Collaboration, Phys. Lett. B **517**, 59 (2001); G. Giacomelli and M. Giorgini (MACRO Collaboration), arXiv:hep-ex/0110021.
2. B.T. Cleveland et al., Astrophys. J. **496**, 505 (1998); GALLEX Collaboration, Phys. Lett. B **447**, 127 (1999); SAGE Collaboration, Phys. Rev. C **60**, 055801 (1999); Super-Kamiokande Collaboration, arXiv:hep-ex/0103032; arXiv:hep-ex/0103033; GNO Collaboration, Phys. Lett. B **490**, 16 (2000); SNO Collaboration, Phys. Rev. Lett. **87**, 71301 (2001); Q.R. Ahmad et al. (SNO Collaboration), arXiv:nucl-ex/0204008; arXiv:nucl-ex/0204009.
3. T. Yanagida, in *Proceedings of the Workshop on Unified Theory and Baryon Number in the Universe*, KEK, March 1979; M. Gell-Mann, P. Ramond, and R. Slansky, in *Supergravity*, Stony Brook, Sep. 1979.
4. M. Fukugita and T. Yanagida, Phys. Lett. B **174**, 45 (1986).
5. LSND Collaboration, arXiv:hep-ex/0104049.
6. S. Weinberg, Phys. Rev. Lett. **43**, 1566 (1979).
7. For a review, see e.g. M. Trodden, Rev. Mod. Phys. **71**, 1463 (1999).
8. M.A. Luty, Phys. Rev. D **45**, 455 (1992); H. Murayama and T. Yanagida, Phys. Lett. B **322**, 349 (1994); M. Flanz, E.A. Paschos, and U. Sarkar, Phys. Lett. B **345**, 248 (1995); Phys. Lett. B **382**, 447 (1995) (erratum); M. Plumacher, Z. Phys. C **74**, 549 (1997); L. Covi, E. Roulet, and F. Vissani, Phys. Lett. B **384**, 169 (1996); E. Ma and U. Sarkar, Phys. Rev. Lett. **80**, 5716 (1998); A. Pilaftsis, Int. J. Mod. Phys. A **14**, 1811 (1999); A. Riotto and M. Trodden, Annu. Rev. Nucl. Part. Sci. **49**, 35 (1999); J.R. Ellis, S. Lola, and D.V. Nanopoulos, Phys. Lett. B **452**, 87 (1999); W. Buchmuller and M. Plumacher, Int. J. Mod. Phys. A **15**, 5047 (2000); D. Falcone and F. Tramontano, Phys. Lett. B **506**, 1 (2001); H.B. Nielsen and Y. Takanishi, Phys. Lett. B **507**, 241 (2001); A.S. Joshipura, E.A. Paschos, and W. Rodejohann, J. High Energy Phys. **0108**, 029 (2001); B. Brahmachari, E. Ma, and U. Sarkar, Phys. Lett. B **520**, 152 (2001); M. Hirsch and S.F. King, Phys. Rev. D **64**, 113005 (2001); F. Buccella, D. Falcone, and F. Tramontano, Phys. Lett. B **524**, 241 (2002);

W. Buchmuller and D. Wyler, Phys. Lett. B **521**, 291 (2001); M.S. Berger and K. Siyeon, Phys. Rev. D **65**, 053019 (2002); G.C. Branco, R. Gonzalez Felipe, F.R. Joaquim, and M.N. Rebelo, arXiv:hep-ph/0202030; M. Fujii, K. Hamaguchi, and T. Yanagida, arXiv:hep-ph/0202210; S. Davidson and A. Ibarra, arXiv:hep-ph/0202239; E.A. Paschos, arXiv:hep-ph/0204137; W. Buchmuller, arXiv:hep-ph/0204288.

9. G.F. Giudice, M. Peloso, A. Riotto, and I. Tkachev, J. High Energy Phys. **9908**, 014 (1999); J. Garcia-Bellido and E. Ruiz Morales, Phys. Lett. B **536**, 193 (2002).

10. KARMEN Collaboration, Nucl. Phys. Proc. Suppl. **91**, 191 (2000).

11. P. Spentzouris, Nucl. Phys. Proc. Suppl. **100**, 163 (2001).

12. H. Murayama and T. Yanagida, Phys. Lett. B **520**, 263 (2001); G. Barenboim, L. Borissov, J. Lykken, and A.Y. Smirnov, arXiv:hep-ph/0108199; A. de Gouvea, arXiv:hep-ph/0204077.

13. A. Strumia, arXiv:hep-ph/0201134.

14. N. Gaur, A. Ghosal, E. Ma, and P. Roy, Phys. Rev. D **58**, 071301 (1998); G.L. Fogli, E. Lisi, and A. Marrone, Phys. Rev. D **63**, 053008 (2001); O.L. Peres and A.Y. Smirnov, Nucl. Phys. B **599**, 3 (2001); C. Giunti, Nucl. Phys. Proc. Suppl. **100**, 244 (2001); W. Grimus and T. Schwetz, Eur. Phys. J. C **20**, 1 (2001); M.C. Gonzalez-Garcia, M. Maltoni, and C. Pena-Garay, arXiv:hep-ph/0108073; M. Maltoni, T. Schwetz, and J.W. Valle, arXiv:hep-ph/0112103; A. Strumia, arXiv:hep-ph/0201134.

15. G.L. Fogli, E. Lisi, D. Montanino, and A. Palazzo, Phys. Rev. D **64**, 093007 (2001); J.N. Bahcall, M.C. Gonzalez-Garcia, and C. Pena-Garay, J. High Energy Phys. **0108**, 014 (2001); arXiv:hep-ph/0111150; V. Barger, D. Marfatia, K. Whisnant, and B.P. Wood, arXiv:hep-ph/0204253; A. Bandyopadhyay, S. Choubey, S. Goswami, and D.P. Roy, arXiv:hep-ph/0204286; J.N. Bahcall, M.C. Gonzalez-Garcia, and C. Pena-Garay, arXiv:hep-ph/0204314; P. Aliani, V. Antonelli, R. Ferrari, M. Picariello, and E. Torrente-Lujan, arXiv:hep-ph/0205053.

16. G.L. Fogli, E. Lisi, and A. Marrone, Phys. Rev. D **64**, 093005 (2001); M.C. Gonzalez-Garcia, M. Maltoni, and C. Pena-Garay, arXiv:hep-ph/0108073. M.C. Gonzalez-Garcia and Y. Nir, arXiv:hep-ph/0202058.

17. G. Fogli and E. Lisi in "Neutrino Mass", chapter 6.

18. K.R. Dienes, E. Dudas, and T. Gherghetta, Nucl. Phys. B **557**, 25 (1999); N. Arkani-Hamed, S. Dimopoulos, G.R. Dvali, and J. March-Russell, Phys. Rev. D **65**, 024032 (2002).

19. N. Arkani-Hamed, S. Dimopoulos, and G.R. Dvali, Phys. Lett. B **429**, 263 (1998).

20. K. Benakli and A.Y. Smirnov, Phys. Rev. Lett. **79**, 4314 (1997).

21. A.E. Faraggi and M. Pospelov, Phys. Lett. B **458**, 237 (1999); G.R. Dvali and A.Y. Smirnov, Nucl. Phys. B **563**, 63 (1999); R.N. Mohapatra and A. Perez-Lorenzana, Nucl. Phys. B **576**, 466 (2000); Y. Grossman and M. Neubert, Phys. Lett. B **474**, 361 (2000); D.O. Caldwell, R.N. Mohapatra, and S.J. Yellin, Phys. Rev. D **64**, 073001 (2001); A.S. Dighe and A.S. Joshipura, Phys. Rev. D **64**, 073012 (2001); A. De Gouvea, G.F. Giudice, A. Strumia, and K. Tobe, Nucl. Phys. B **623**, 395 (2002); H. Davoudiasl, P. Langacker, and M. Perelstein, arXiv:hep-ph/0201128.

22. R. Barbieri, P. Creminelli, and A. Strumia, Nucl. Phys. B **585**, 28 (2000); A. Lukas, P. Ramond, A. Romanino, and G.G. Ross, Phys. Lett. B **495**, 136 (2000); J. High Energy Phys. **0104**, 010 (2001).

23. See e.g. I. Antoniadis and K. Benakli, Int. J. Mod. Phys. A **15**, 4237 (2000).

24. G. Altarelli and F. Feruglio, Phys. Rep. **320**, 295 (1999).

25. S.M. Barr and I. Dorsner, Nucl. Phys. B **585**, 79 (2000); I. Masina, Int. J. Mod. Phys. A **16**, 5101 (2001).

26. B. Pontecorvo, Sov. Phys. JETP **6**, 429 (1957) [Zh. Eksp. Teor. Fiz. **33**, 549 (1957)]; Z. Maki, M. Nakagawa, and S. Sakata, Prog. Theor. Phys. **28**, 870 (1962); B. Pontecorvo, Sov. Phys. JETP **26**, 984 (1968) [Zh. Eksp. Teor. Fiz. **53**, 1717 (1968)]; V.N. Gribov and B. Pontecorvo, Phys. Lett. B **28**, 493 (1969).

27. B.W. Lee, S. Pakvasa, R. Shrock, and H. Sugawara, Phys. Rev. Lett. **38**, 937 (1977); B.W. Lee and R. Shrock, Phys. Rev. D **16**, 1444 (1977).

28. N. Cabibbo, Phys. Lett. B **72**, 333 (1978).

29. M. Frigerio and A.Y. Smirnov, arXiv:hep-ph/0202247.

30. CHOOZ Collaboration, Phys. Lett. B **466**, 415 (1999); see also Palo Verde Collaboration, Phys. Rev. Lett. **84**, 3764 (2000).

31. H.V. Klapdor-Kleingrothaus et al. (Heidelberg–Moscow Collaboration), Eur. Phys. J. A **12**, 147 (2001); C.E. Aalseth et al. (16EX Collaboration), arXiv:hep-ex/0202026.

32. F. Vissani, arXiv:hep-ph/9708483; H. Georgi and S.L. Glashow, Phys. Rev. D **61**, 097301 (2000).

33. R. Barbieri, G.G. Ross, and A. Strumia, J. High Energy Phys. **9910**, 020 (1999).

34. J.R. Ellis and S. Lola, Phys. Lett. B **458**, 310 (1999); N. Haba, Y. Matsui, N. Okamura, and M. Sugiura, Eur. Phys. J. C **10**, 677 (1999); J.A. Casas, J.R. Espinosa, A. Ibarra, and I. Navarro, Nucl. Phys. B **569**, 82 (2000); J. High Energy Phys. **9909**, 015 (1999); Nucl. Phys. B **573**, 652 (2000); N. Haba and N. Okamura, Eur. Phys. J. C **14**, 347 (2000); P.H. Chankowski, W. Krolikowski, and S. Pokorski, Phys. Lett. B **473**, 109 (2000); T.K. Kuo, J. Pantaleone, and G.H. Wu, Phys. Lett. B **518**, 101 (2001); P.H. Chankowski and S. Pokorski, Int. J. Mod. Phys. A **17**, 575 (2002); S. Antusch, J. Kersten, M. Lindner, and M. Ratz, arXiv:arXiv:hep-ph/0203233.

35. M. Jezabek, arXiv:hep-ph/0205234.

36. A. Ioannisian and J.W.F. Valle, Phys. Lett. B **332**, 93 (1994); R.N. Mohapatra and S. Nussinov, Phys. Rev. D **60**, 013002 (1999); Phys. Lett. B **441**, 299 (1998); G.C. Branco, M.N. Rebelo, and J.I. Silva-Marcos, Phys. Rev. D **62**, 073004 (2000); C. Wetterich, Phys. Lett. B **451**, 397 (1999); G. Perez, J. High Energy Phys. **0012**, 027 (2000).

37. R. Barbieri, L.J. Hall, G.L. Kane, and G.G. Ross, arXiv:hep-ph/9901228.

38. H. Fritzsch and Z.Z. Xing, Phys. Lett. B **372**, 265 (1996); H. Fritzsch and Z.Z. Xing, Phys. Lett. B **440**, 313 (1998); H. Fritzsch and Z.Z. Xing, Prog. Part. Nucl. Phys. **45**, 1 (2000); M. Fukugita, M. Tanimoto, and T. Yanagida, Phys. Rev. D **57**, 4429 (1998); Phys. Rev. D **59**, 113016 (1999); M. Tanimoto, T. Watari, and T. Yanagida, Phys. Lett. B **461**, 345 (1999); S.K. Kang and C.S. Kim, Phys. Rev. D **59**, 091302 (1999); M. Tanimoto, Phys. Lett. B **483**, 417 (2000); Y. Koide and A. Ghosal, Phys. Lett. B **488**, 344 (2000); E.K. Akhmedov, G.C. Branco, F.R. Joaquim, and J.I. Silva-Marcos, Phys. Lett. B **498**, 237 (2001). For models based on discrete symmetries, see

also P.F. Harrison, D.H. Perkins, and W.G. Scott, Phys. Lett. B **530**, 167 (2002); P.F. Harrison and W.G. Scott, arXiv:hep-ph/0203209; E.K. Akhmedov, G.C. Branco, F.R. Joaquim, and J.I. Silva-Marcos, Phys. Lett. B **498**, 237 (2001); arXiv:arXiv:hep-ph/0008010.

39. N. Haba, Y. Matsui, N. Okamura, and T. Suzuki, Phys. Lett. B **489**, 184 (2000).

40. L.J. Hall, H. Murayama, and N. Weiner, Phys. Rev. Lett. **84**, 2572 (2000); N. Haba and H. Murayama, Phys. Rev. D **63**, 053010 (2001); M.S. Berger and K. Siyeon, Phys. Rev. D **63**, 057302 (2001); F. Vissani, Phys. Lett. B **508**, 79 (2001).

41. R. Barbieri, L.J. Hall, D.R. Smith, A. Strumia, and N. Weiner, J. High Energy Phys. **9812**, 017 (1998); R. Barbieri, L.J. Hall, and A. Strumia, Phys. Lett. B **445**, 407 (1999). See also Ya.B. Zeldovich, Dokl. Akad. Nauk. SSSR **86**, 505 (1952); E.J. Konopinsky and H. Mahmound, Phys. Rev. **92**, 1045 (1953); A. Zee, Phys. Lett. B **93**, 389 (1980); S. Petcov, Phys. Lett. B **100**, 245 (1982); A.S. Joshipura and S.D. Rindani, Eur. Phys. J. C **14**, 85 (2000); R.N. Mohapatra, A. Perez-Lorenzana, and C.A. de Sousa Pires, Phys. Lett. B **474**, 355 (2000); Q. Shafi and Z. Tavartkiladze, Phys. Lett. B **482**, 145 (2000); arXiv:hep-ph/0101350; A. Ghosal, Phys. Rev. D **62**, 092001 (2000); L. Lavoura, Phys. Rev. D **62**, 093011 (2000); L. Lavoura and W. Grimus, J. High Energy Phys. **0009**, 007 (2000); Phys. Rev. D **62**, 093012 (2000); S.F. King and N.N. Singh, Nucl. Phys. B **596**, 81 (2001); J.F. Oliver and A. Santamaria, Phys. Rev. D **65**, 033003 (2002); K.S. Babu and R.N. Mohapatra, arXiv:hep-ph/0201176.

42. I. Dorsner and S.M. Barr, Nucl. Phys. B **617**, 493 (2001).

43. H.J. He, D.A. Dicus, and J.N. Ng, Phys. Lett. B **536**, 83 (2002).

44. S.F. King, Phys. Lett. B **439**, 350 (1998); S. Davidson and S.F. King, Phys. Lett. B **445**, 191 (1998); Q. Shafi and Z. Tavartkiladze, Phys. Lett. B **451**, 129 (1999).

45. S.F. King, Nucl. Phys. B **562**, 57 (1999); Nucl. Phys. B **576**, 85 (2000); arXiv:hep-ph/0204360.

46. C.H. Albright and S.M. Barr, Phys. Rev. D **58**, 013002 (1998); C.H. Albright, K.S. Babu, and S.M. Barr, Phys. Rev. Lett. **81**, 1167 (1998); P.H. Frampton and A. Rasin, Phys. Lett. B **478**, 424 (2000).

47. N. Irges, S. Lavignac, and P. Ramond, Phys. Rev. D **58**, 035003 (1998).

48. G. Altarelli and F. Feruglio, Phys. Lett. B **439**, 112 (1998); G. Altarelli and F. Feruglio, J. High Energy Phys. **9811**, 021 (1998).

49. G. Altarelli and F. Feruglio, Phys. Lett. B **451**, 388 (1999).

50. J. Sato and T. Yanagida, Phys. Lett. B **430**, 127 (1998); J.K. Elwood, N. Irges, and P. Ramond, Phys. Rev. Lett. **81**, 5064 (1998); Z. Berezhiani and A. Rossi, J. High Energy Phys. **9903**, 002 (1999); K. Hagiwara and N. Okamura, Nucl. Phys. B **548**, 60 (1999); M. Bando and T. Kugo, Prog. Theor. Phys. **101**, 1313 (1999); W. Grimus and L. Lavoura, J. High Energy Phys. **0107**, 045 (2001); arXiv:hep-ph/0204070.

51. H. Georgi and C. Jarlskog, Phys. Lett. B **86**, 297 (1979).

52. M. Drees, S. Pakvasa, X. Tata, and T. ter Veldhuis, Phys. Rev. D **57**, 5335 (1998); E.J. Chun, S.K. Kang, C.W. Kim, and U.W. Lee, Nucl. Phys. B **544**, 89 (1999); B. Mukhopadhyaya, S. Roy, and F. Vissani, Phys. Lett. B **443**, 191 (1998); O.C. Kong, Mod. Phys. Lett. A **14**, 903 (1999); M. Bisset, O.C. Kong,

C. Macesanu, and L.H. Orr, Phys. Rev. D **62**, 035001 (2000); A.S. Joshipura and S.K. Vempati, Phys. Rev. D **60**, 111303 (1999); A. Datta, B. Mukhopadhyaya, and S. Roy, Phys. Rev. D **61**, 055006 (2000); B. Mukhopadhyaya, Pramana **54**, 147 (2000); G. Bhattacharyya, H.V. Klapdor-Kleingrothaus, and H. Pas, Phys. Lett. B **463**, 77 (1999); J.C. Romao, M.A. Diaz, M. Hirsch, W. Porod, and J.W. Valle, Phys. Rev. D **61**, 071703 (2000); A. Abada and M. Losada, Nucl. Phys. B **585**, 45 (2000); S. Davidson and M. Losada, J. High Energy Phys. **0005**, 021 (2000); A. Abada and G. Bhattacharyya, Phys. Rev. D **63**, 017701 (2001); V.D. Barger, T. Han, S. Hesselbach, and D. Marfatia, arXiv:hep-ph/0108261; Y. Koide and A. Ghosal, arXiv:hep-ph/0203113.

53. W. Hu, D.J. Eisenstein, and M. Tegmark, Phys. Rev. Lett. **80**, 5255 (1998); O. Elgaroy et al., arXiv:astro-ph/0204152.

54. F. Feruglio, A. Strumia, and F. Vissani, arXiv:hep-ph/0201291.

55. S.T. Petcov and A.Y. Smirnov, Phys. Lett. B **322**, 109 (1994); S.M. Bilenky, A. Bottino, C. Giunti, and C.W. Kim, Phys. Rev. D **54**, 1881 (1996); S.M. Bilenky, C. Giunti, C.W. Kim, and S.T. Petcov, Phys. Rev. D **54**, 4432 (1996); F. Vissani, J. High Energy Phys. **9906**, 22 (1999); T. Fukuyama, K. Matsuda, and H. Nishiura, Phys. Rev. D **57**, 5884 (1998); V.D. Barger and K. Whisnant, Phys. Lett. B **456**, 194 (1999); F. Vissani, arXiv:hep-ph/9904349; C. Giunti, Phys. Rev. D **61**, 036002 (2000); M. Czakon, J. Gluza, and M. Zralek, Phys. Lett. B **465**, 211 (1999); S.M. Bilenky, C. Giunti, W. Grimus, B. Kayser, and S.T. Petcov, Phys. Lett. B **465**, 193 (1999); H.V. Klapdor-Kleingrothaus, H. Pas, and A.Y. Smirnov, Phys. Rev. D **63**, 73005 (2001); W. Rodejohann, Nucl. Phys. B **597**, 110 (2001); K. Matsuda, N. Takeda, T. Fukuyama, and H. Nishiura, Phys. Rev. D **64**, 13001 (2001); D. Falcone and F. Tramontano, Phys. Rev. D **64**, 077302 (2001); S.M. Bilenky, S. Pascoli, and S.T. Petcov, Phys. Rev. D **64**, 53010 (2001); S.M. Bilenky, S. Pascoli, and S.T. Petcov, Phys. Rev. D **64**, 113003 (2001); P. Osland and G. Vigdel, Phys. Lett. B **520**, 143 (2001); S. Pascoli, S.T. Petcov, and L. Wolfenstein, Phys. Lett. B **524**, 319 (2002); S. Pascoli and S.T. Petcov, arXiv:hep-ph/0205022.

56. H.V. Klapdor-Kleingrothaus et al., Mod. Phys. Lett. A **37**, 2409 (2001). See also C.E. Aalseth et al., arXiv:hep-ex/0202018; F. Feruglio, A. Strumia, and F. Vissani, arXiv:hep-ph/0201291 (appendix).

57. J.J. Gomez-Cadenas et al. (CERN Working Group on Super Beams Collaboration), arXiv:hep-ph/0105297; Y. Itow et al., arXiv:hep-ex/0106019.

58. See also E.K. Akhmedov, G.C. Branco, and M.N. Rebelo, Phys. Rev. Lett. **84**, 3535 (2000). S.M. Barr and I. Dorsner, Nucl. Phys. B **585**, 79 (2000);

59. D.B. Kaplan and M. Schmaltz, Phys. Rev. D **49**, 3741 (1994); M. Schmaltz, Phys. Rev. D **52**, 1643 (1995); P.H. Frampton and T.W. Kephart, Int. J. Mod. Phys. A **10**, 4689 (1995); K.C. Chou and Y.L. Wu, Phys. Rev. D **53**, 3492 (1996); P.H. Frampton and O.C. Kong, Phys. Rev. Lett. **77**, 1699 (1996); K.S. Babu and S.M. Barr, Phys. Lett. B **387**, 87 (1996); R. Barbieri, L.J. Hall, S. Raby, and A. Romanino, Nucl. Phys. B **493**, 3 (1997); R. Barbieri, L.J. Hall, and A. Romanino, Phys. Lett. B **401**, 47 (1997); T. Blazek, S. Raby, and K. Tobe, Phys. Rev. D **60**, 113001 (1999); C.D. Carone and R.F. Lebed, Phys. Rev. D **60**, 096002 (1999); P.H. Frampton and A. Rasin, Phys. Lett. B **478**, 424 (2000); R. Dermisek and S. Raby, Phys. Rev. D **62**, 015007 (2000); A. Aranda, C.D. Carone, and R.F. Lebed, Phys. Rev. D **62**, 016009 (2000); Phys. Lett.

B **474**, 170 (2000); D. Falcone, Phys. Rev. D **64**, 117302 (2001); A. Aranda, C.D. Carone, and P. Meade, Phys. Rev. D **65**, 013011 (2002); Phys. Rev. D **64**, 117302 (2001); Y. Koide and A. Ghosal, Phys. Rev. D **63**, 037301 (2001).

60. C.D. Froggatt and H.B. Nielsen, Nucl. Phys. B **147**, 277 (1979).
61. J. Bijnens and C. Wetterich, Nucl. Phys. B **147**, 292 (1987); M. Leurer, Y. Nir, and N. Seiberg, Nucl. Phys. B **398**, 319 (1993); Nucl. Phys. B **420**, 468 (1994); L.E. Ibanez and G.G. Ross, Phys. Lett. B **332**, 100 (1994); P. Binetruy and P. Ramond, Phys. Lett. B **350**, 49 (1995); E. Dudas, S. Pokorski, and C.A. Savoy, Phys. Lett. B **356**, 45 (1995); G.K. Leontaris and J. Rizos, Nucl. Phys. B **567**, 32 (2000); J.M. Mira, E. Nardi, and D.A. Restrepo, Phys. Rev. D **62**, 016002 (2000).
62. P. Binetruy, S. Lavignac, and P. Ramond, Nucl. Phys. B **477**, 353 (1996); P. Binetruy, S. Lavignac, S.T. Petcov, and P. Ramond, Nucl. Phys. B **496**, 3 (1997); Z. Berezhiani and Z. Tavartkiladze, Phys. Lett. B **396**, 150 (1997); M. Jezabek and Y. Sumino, Phys. Lett. B **440**, 327 (1998); Y. Grossman, Y. Nir, and Y. Shadmi, J. High Energy Phys. **9810**, 007 (1998); W. Buchmuller and T. Yanagida, Phys. Lett. B **445**, 399 (1999); C.D. Froggatt, M. Gibson, and H.B. Nielsen, Phys. Lett. B **446**, 256 (1999); K. Choi, K. Hwang, and E.J. Chun, Phys. Rev. D **60**, 031301 (1999); S. Lola and G.G. Ross, Nucl. Phys. B **553**, 81 (1999); Y. Nir and Y. Shadmi, J. High Energy Phys. **9905**, 023 (1999); G. Altarelli, F. Feruglio, and I. Masina, Phys. Lett. B **472**, 382 (2000); B. Stech, Phys. Rev. D **62**, 093019 (2000); S.F. King and M. Oliveira, Phys. Rev. D **63**, 095004 (2001); M. Tanimoto, Phys. Lett. B **501**, 231 (2001); M. Bando and N. Maekawa, Prog. Theor. Phys. **106**, 1255 (2001); H.B. Nielsen and Y. Takanishi, Phys. Scr. T **93**, 44 (2001).
63. H. Fusaoka and Y. Koide, Phys. Rev. D **57**, 3986 (1998).
64. H. Murayama, arXiv:hep-ph/0201022.
65. G. Altarelli, F. Feruglio, and I. Masina, J. High Energy Phys. **0011**, 040 (2000).
66. J.A. Harvey, D.B. Reiss, and P. Ramond, Nucl. Phys. B **199**, 223 (1982); Z. Berezhiani and Z. Tavartkiladze, Phys. Lett. B **409**, 220 (1997); B. Brahmachari and R.N. Mohapatra, Phys. Rev. D **58**, 015001 (1998); Y. Nomura and T. Yanagida, Phys. Rev. D **59**, 017303 (1999); K.Y. Oda, E. Takasugi, M. Tanaka, and M. Yoshimura, Phys. Rev. D **59**, 055001 (1999); N.N. Singh, arXiv:hep-ph/0009087; N. Maekawa, Prog. Theor. Phys. **106**, 401 (2001); K.S. Babu and S.M. Barr, Phys. Lett. B **525**, 289 (2002).
67. K.S. Babu, J.C. Pati, and F. Wilczek, Nucl. Phys. B **566**, 33 (2000).
68. See e.g. S. Lola and G.G. Ross, Nucl. Phys. B **553**, 81 (1999).
69. W. Buchmuller, D. Delepine, and F. Vissani, Phys. Lett. B **459**, 171 (1999); J.L. Feng, Y. Nir, and Y. Shadmi, Phys. Rev. D **61**, 113005 (2000); J.R. Ellis, M.E. Gomez, G.K. Leontaris, S. Lola, and D.V. Nanopoulos, Eur. Phys. J. C **14**, 319 (2000); W. Buchmuller, D. Delepine, and L.T. Handoko, Nucl. Phys. B **576**, 445 (2000); G. Barenboim, K. Huitu, and M. Raidal, Phys. Rev. D **63**, 055006 (2001); A. de Gouvea, S. Lola, and K. Tobe, Phys. Rev. D **63**, 035004 (2001); J. Sato and K. Tobe, Phys. Rev. D **63**, 116010 (2001); D.F. Carvalho, M.E. Gomez, and S. Khalil, J. High Energy Phys. **0107**, 001 (2001); J.A. Casas and A. Ibarra, Nucl. Phys. B **618**, 171 (2001); S.W. Baek, T. Goto, Y. Okada, and K.I. Okumura, Phys. Rev. D **64**, 095001 (2001); S. Lavignac, I. Masina, and C.A. Savoy, Phys. Lett. B **520**, 269 (2001); A. Kageyama, S. Kaneko, N. Shimoyama, and M. Tanimoto, Phys. Lett. B **527**, 206 (2002);

D.F. Carvalho, M.E. Gomez, and J.C. Romao, arXiv:hep-ph/0202054; S. Lavignac, I. Masina, and C.A. Savoy, arXiv:hep-ph/0202086.

70. Y. Kawamura, Prog. Theor. Phys. **105**, 999 (2001); G. Altarelli and F. Feruglio, Phys. Lett. B **511**, 257 (2001); L.J. Hall and Y. Nomura, Phys. Rev. D **64**, 055003 (2001); A. Hebecker and J. March-Russell, Nucl. Phys. B **613**, 3 (2001); R. Barbieri, L.J. Hall, and Y. Nomura, arXiv:hep-ph/0106190; A. Hebecker and J. March-Russell, Nucl. Phys. B **625**, 128 (2002); T.-j. Li, Phys. Lett. B **520**, 377 (2001); L.J. Hall, H. Murayama, and Y. Nomura, hep-th/0107245; C. Csaki, G.D. Kribs, and J. Terning, Phys. Rev. D **65**, 015004 (2002); H.C. Cheng, K.T. Matchev, and J. Wang, Phys. Lett. B **521**, 308 (2001); N. Haba, T. Kondo, Y. Shimizu, T. Suzuki, and K. Ukai, Prog. Theor. Phys. **106**, 1247 (2001); T. Asaka, W. Buchmuller, and L. Covi, Phys. Lett. B **523**, 199 (2001); L.J. Hall, Y. Nomura, T. Okui, and D.R. Smith, Phys. Rev. D **65**, 035008 (2002); T.-j. Li, Nucl. Phys. B **619**, 75 (2001); R. Dermisek and A. Mafi, Phys. Rev. D **65**, 055002 (2002); T. Watari and T. Yanagida, Phys. Lett. B **519**, 164 (2001); Y. Nomura, Phys. Rev. D **65**, 085036 (2002); G. Bhattacharyya and K. Sridhar, arXiv:hep-ph/0111345; C.S. Huang, J. Jiang, T.-j. Li, and W. Liao, Phys. Lett. B **530**, 218 (2002); A. Hebecker and J. March-Russell, arXiv:hep-ph/0204037; T. Asaka, W. Buchmuller, and L. Covi, arXiv:hep-ph/0204358. S.M. Barr and I. Dorsner, arXiv:hep-ph/0205088; C. Biggio, arXiv:hep-ph/0205142; C. Biggio and F. Feruglio, arXiv:hep-th/0207014.

7 The Physics Potential of Future Long-Baseline Neutrino Oscillation Experiments

Manfred Lindner

We discuss in detail several proposed long-baseline neutrino oscillation setups and show their remarkable potential for very precise measurements of mass splittings and mixing angles. Furthermore, it will be possible to make precise tests of coherent forward scattering and MSW effects, which will allow to determine the sign of Δm^2. Finally, strong limits or measurements of leptonic CP violation will be possible, which is very interesting, since this is most likely connected to the baryon asymmetry of the universe.

7.1 Introduction

Since the existing evidence for atmospheric neutrino oscillations includes some sensitivity to the characteristic L/E dependence of the oscillations [1], there is little doubt that the observed flavor transitions are due to neutrino oscillations. Recently, it has been established reliably that solar neutrinos undergo flavor transitions [2,3], most likely due to oscillations. This solves the long-standing solar neutrino problem, even though the characteristic L/E dependence of oscillations has not yet been established in this case. However, oscillation is by far the most plausible explanation when all alternatives are considered, and global oscillation fits to all available data clearly favor the LMA solution for the mass splittings and mixings [4–7]. The CHOOZ reactor experiment [8] currently provides in addition the most stringent upper bound on the subleading element U_{e3} of the neutrino mixing matrix. The global pattern of neutrino oscillation parameters seems therefore quite well known, and one may ask how precisely future experiments will ultimately be able to measure mass splittings and mixings and what can be learned from such precise measurements.

The characteristic length scale L of oscillations is given by $\Delta m^2 L/E_\nu = \pi/2$, and the known atmospheric value of Δm^2_{31} thus leads to an oscillation length scale L_{atm} as a function of energy. For $\Delta m^2_{31} \simeq 3 \times 10^{-3}$ eV and neutrino energies $E_\nu \simeq 10$ GeV, one finds $L_{\text{atm}} \simeq \mathcal{O}(2000)$ km, i.e. distances and energies which can be realized by sending neutrino beams from one point on the earth to another. Such long-baseline (LBL) experiments have the advantage that the source can, in principle, be controlled and understood very precisely. In contrast, natural neutrino sources such as the sun or the atmosphere cannot be controlled directly, and they involve assumptions and indirect measurements. The precision of future solar and atmospheric oscillation experiments is thus limited systematically by the source at some level, which is, in principle, not the case for LBL experiments. The solar value of Δm_{21}^2 is, for the favored LMA solution, about two orders of magnitude smaller than the atmospheric value of Δm_{31}^2, resulting, for the same energies, in oscillations at scales $L_{sol} \simeq (10-1000)L_{atm}$. The solar oscillations will thus not develop fully in such LBL experiments on earth, but subleading effects will nevertheless play an important role in precision experiments. Another modification arises from the fact that the neutrino beams of LBL experiments traverse the matter of the earth. Coherent forward scattering in matter must therefore be taken into account in precision experiments. This makes the analysis more involved, but, as we shall see, it also offers unique opportunities.

The existing K2K experiment [9], and also MINOS [10] and CNGS [11], which are both under construction, are a promising first generation of LBL experiments which will lead to improved oscillation parameters. We shall discuss the remarkable potential of future LBL experiments in some detail in this chapter, and further details can be found in [12]. One important point is that the increased precision will allow us to test in detail the three-flavoredness of oscillations. We shall also see that it will be possible to limit or measure θ_{13} very much better than today, that it will be possible to study MSW matter effects in detail [13–16], and that it will be possible to extract sign (Δm_{31}^2), i.e. the mass ordering of the neutrino states. For the now favored LMA solution, it will also be possible to measure leptonic CP violation. The precise neutrino masses, mixings, and CP phases which will be obtained in this way will provide extremely valuable information about flavor physics, since unlike the case for quarks, these parameters are not obscured by hadronic uncertainties. These parameters can then be evolved with the renormalization group to the GUT scale, for example,[1] where they can be compared with mass models based on flavor symmetries or other models of neutrino and charged-lepton masses. Moreover, leptonic CP violation is related to leptogenesis [19–21], the currently most plausible mechanism for the generation of the baryon asymmetry of the universe. LBL experiments therefore offer, in a unique way, precise knowledge about extremely interesting and valuable physics parameters.

7.2 Beams and Detectors

Precise experiments in combination with an exact theoretical description are extremely valuable, since they allow precision determinations of the underlying quantities. Long-baseline neutrino oscillation experiments are, in prin-

---

[1] See [17, 18] for examples.

ciple, of this type, since unlike the case for quarks there are no hadronic uncertainties on theoretical side. Experimentally, LBL experiments have the advantage that both the source and the detector can be kept under precise conditions. This includes, among other things, for the source, a precise knowledge of the mean neutrino energy E_ν and of the neutrino flux and spectrum, as well as the flavor composition and contamination of the beam. Another important aspect is the question if neutrino and antineutrino beam data can be obtained symmetrically, such that systematic uncertainties cancel at least partly in an analysis. Precise measurements require also a sufficient luminosity of the beam and a large enough detector, such that sufficient statistics can be obtained. On the detector side, there are a number of issues which have to be understood or determined very precisely, such as the detection threshold function, the energy calibration, the resolution, and the particle identification capabilities (flavor, charge, event reconstruction, and understanding backgrounds). Another source of uncertainty in the detection process lies in the knowledge of neutrino cross sections, especially at low energies [22]. The potential source and detector combinations of future LBL experiments are, furthermore, constrained by the available technology. We shall restrict ourselves in the following to some specific types of neutrino sources and detectors for LBL experiments. However, it is important to keep potential new source and detector developments in mind. An example is provided by liquid-argon detectors such as ICARUS [23]. Detectors of this type are not included in this study, but they may become extremely valuable detectors or detector components in this context.

The first type of source considered is a conventional neutrino or antineutrino beam. An intense proton beam is typically directed onto a massive target, producing mostly pions and some K mesons, which are captured by an optical system of magnets in order to obtain a beam. The pions (and K mesons) decay in a decay pipe, yielding essentially a muon neutrino beam, which can undergo oscillations as shown in Fig. 7.1. The $\nu_\mu \rightarrow \nu_\mu$ disappearance channel and the $\nu_\mu \rightarrow \nu_e$ appearance channels are the most interesting. The neutrino beam is, however, contaminated by approximately 0.5% of electron neutrinos, which also produce electron reactions in the disappearance channel, thus limiting the precision in the extraction of $\nu_\mu \rightarrow \nu_e$ oscillation parameters. The energy spectrum of the muon beam can be controlled over a wide range: it depends on the incident proton energy, the optical system, and the precise direction of the beam axis compared with the direction of the detector. It is possible to produce broadband high-energy beams, such as the CNGS beam [24, 25], or narrowband lower-energy beams, such as those obtained in some configurations of the NuMI beam [26]. Reversing the electrical current in the lens system results in an antineutrino beam. The neutrino and antineutrino beams have significant differences, such that errors do not cancel systematically in ratios or differences. The neutrino and antineutrino

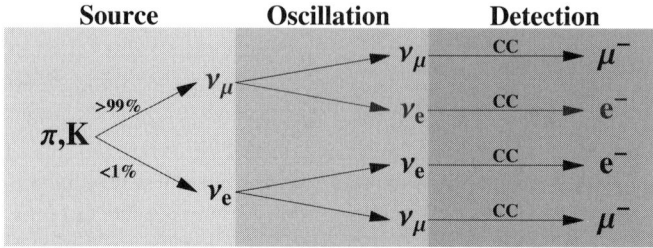

Fig. 7.1. Schematic overview of neutrino production, oscillation, and detection via charged-current interactions for conventional beams and superbeams. The interesting channels are the $\nu_\mu \to \nu_\mu$ and $\nu_\mu \to \nu_e$ disappearance and appearance channels. The ν_e beam contamination at the level of $< 1\%$ limits the ability to identify the $\nu_\mu \to \nu_e$ appearance oscillation, since this contamination produces electrons without oscillation. The ν_τ oscillation channel is not shown here, but it would become very important if tau lepton detection were feasible. This requires energies sufficient for tau production and detectors with suitable tau lepton detection capabilities

beams must therefore be considered more or less as independent sources with different systematic errors.

The so-called "superbeams" are based on the same beam dump techniques for producing neutrino beams, but at much larger luminosities [24–27]. Superbeams are thus a technological extrapolation of conventional beams, but use a proton beam intensity close to the mechanical stability limit of the target at a typical thermal power of 0.7 MW to 4 MW. The much higher neutrino luminosity allows the use of the decay kinematics of pions to produce "off-axis beams", where the detector is located a few degrees off the main beam axis. This reduces the neutrino flux and the average neutrino energy, but leads to a more monoenergetic beam and a significant suppression of the electron neutrino contamination. Several off-axis superbeams with energies of about 1 GeV to 2 GeV have been proposed in Japan [28, 29], America [30], and Europe [31, 32].

The most sensitive neutrino oscillation channel for subleading oscillation parameters is the $\nu_\mu \to \nu_e$ appearance transition. Therefore, the detector should have excellent electron and muon charged-current identification capabilities. In addition, an efficient rejection of neutral-current events is required, because the neutral-current interaction mode is flavor-blind. With low statistics, the magnitude of the contamination itself limits the sensitivity to the $\nu_\mu \to \nu_e$ transition severely, while the insufficient knowledge of its magnitude constrains the sensitivity for high statistics. A near detector allows a substantial reduction of the background uncertainties [28, 33] and plays a crucial role in controlling other systematic errors, such as the flux normalization, the spectral shape of the beam, and the neutrino cross section at low energies. At energies of about 1 GeV, the dominant charged-current interaction mode is quasi-elastic scattering, which suggests that water Cherenkov detectors are the optimal type of detector. At these energies, a baseline of about 300 km would be optimal for measuring at the first maximum of the oscillation. At about 2 GeV, there is already a considerable contribution from inelastic scattering to the charged-current interactions, which means that it would be useful to measure the energy of the hadronic part of the cross section. This favors low-Z hadron calorimeters, which also have a neutral-current rejection capability a factor of ten better than water Cherenkov detectors [30]. In this case, the optimum baseline is around 600 km. The matter effects are expected to be small in these experiments for two reasons. First of all, the energy of about 1 GeV to 2 GeV is small compared with the MSW resonance energy of approximately 13 GeV in the upper mantle of the earth. The second reason is that the baseline is too short to produce significant matter effects.

The second type of source considered is the so-called neutrino factory, where muons are stored in the long straight sections of a storage ring. The decaying muons produce muon and electron antineutrinos in equal numbers [34]. The muons are produced by pion decays, where the pions are produced by the same technique as for superbeams. After being collected, they have to be cooled and reaccelerated very quickly. This has not yet been demonstrated, and it is the major technological challenge for neutrino factories [35]. The spectrum and flavor content of the beam are completely characterized by the muon decay and are therefore very precisely known [36]. The only adjustable parameter is the muon energy E_μ; the values usually considered are in the range from 20 to 50 GeV. In a neutrino factory, it is also possible to produce and store antimuons in order to obtain a CP-conjugated beam. The symmetric use of both beams leads to a cancellation or significant reduction of errors and systematic uncertainties. We shall discuss in the following the neutrino beam, which always implies implicitly the CP-conjugate channel as well, unless otherwise stated.

The decay of the muons and the relevant oscillation channels are shown in Fig. 7.2. Amongst all flavors and interaction types, muon charged-current events are the easiest to detect. The appearance channel with the best sensitivity is thus the $\bar{\nu}_e \to \bar{\nu}_\mu$ transition, which produces so-called "wrong-sign muons". Therefore, a detector must be able to identify very reliably the charge of a muon in order to distinguish wrong-sign muons in the appearance channel from the higher rate of right-sign muons in the disappearance channels. The dominant charged-current interaction in the multi-GeV range is deep inelastic scattering, making a good energy resolution for the hadronic energy deposition necessary. Magnetized iron calorimeters are thus the favored choice for neutrino factory detectors. In order to achieve the required muon charge separation, it is necessary to impose a minimum muon energy cut at approximately 4 GeV [37, 38]. This leads to a significant loss of neutrino events in the range from about 4 GeV to 20 GeV, which means that a high muon energy $E_\mu = 50$ GeV is desirable. The first oscillation maximum then lies at approximately 3000 km. Matter effects are sizable at this baseline and

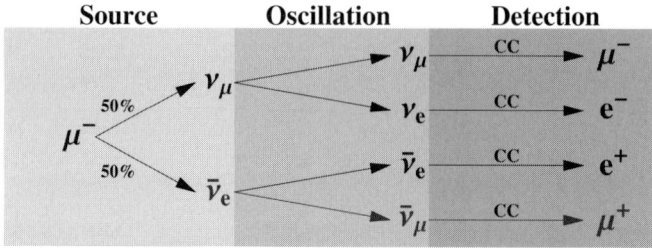

Fig. 7.2. Neutrino production, oscillation, and detection via charged-current inter-actions for a neutrino factory, for one polarity. $\bar{\nu}_e$ and ν_μ are produced in equal num-bers from μ decays and can undergo various oscillations. The $\nu_\mu \to \nu_\mu$ and $\bar{\nu}_e \to \bar{\nu}_\mu$ channels are the most interesting for detectors with μ identification. Note, however, that excellent charge identification capabilities are required to separate "wrong-sign muons" and "right-sign muons". The ν_τ oscillation channel is not shown here, but it would become important if detectors with tau identification capabilities could be built

energy, and the limited knowledge about the density profile of the matter in the earth becomes an additional source of errors.

Finally, the so-called β-beams are an interesting type of beam. The idea is to store radioactive isotopes in a storage ring, similarly to the muons in a neutrino factory, such that the β-decays of the radioactive elements lead to pure ν_e or $\bar{\nu}_e$ beams with $\gamma \simeq 100$ [39]. The energy spectrum of the neutrinos in the beam is determined by the neutrino energy of the decay at rest, boosted by the γ factor, resulting typically in beam energies of a few hundred MeV at acceleration energies of about $100\,\mathrm{GeV}$ per nucleon. There are technological and environmental challenges, and it is unclear whether β-beams can become an affordable and competitive neutrino source. We shall not include β-beams in our quantitative discussion, but we shall see that superbeams and neutrino factories already have an impressive potential, which could only be improved if β-beams were realized.

7.3 The Oscillation Framework

Most existing results on neutrino oscillations can, so far, be understood in an effective two-neutrino framework. Here the well-known oscillation probability for a baseline L and neutrino energy E_ν is

$$P(\nu_{f_1} \to \nu_{f_2}) = |\langle \nu_{f_1}(t)|\nu_{f_2}(t=0)\rangle|^2 = \sin^2 2\theta \sin^2\left(\frac{\Delta m^2 L}{4E_\nu}\right), \quad (7.1)$$

where θ is the mixing angle between the two flavor eigenstates f_1 and f_2 and $\Delta m^2 = m_2^2 - m_1^2$ is the difference between the mass eigenvalues. Precision measurements at future LBL experiments will involve a very precise knowl-edge of the source, the detector, and the oscillation framework in matter. An effective two-neutrino description is therefore definitively not adequate, and matter effects must be included in the three-neutrino oscillation framework.

The generalization of the formula for oscillation in vacuum to the case of N neutrinos leads to probabilities for flavor transitions $\nu_{f_l} \to \nu_{f_m}$ given by

$$P\left(\nu_{f_l} \to \nu_{f_m}\right)$$
$$= \underbrace{\delta_{lm} - 4\sum_{i>j} \mathrm{Re}\left(J_{ij}^{f_l f_m}\right)\sin^2 \Delta_{ij}}_{P_{\mathrm{CP}}} \underbrace{- 2\sum_{i>j} \mathrm{Im}\left(J_{ij}^{f_l f_m}\right)\sin 2\Delta_{ij}}_{P_{\mathrm{CP}}} , \quad (7.2)$$

where the shorthands $J_{ij}^{f_l f_m} := U_{li} U_{lj}^* U_{mi}^* U_{mj}$ and $\Delta_{ij} := \Delta m_{ij}^2 L/4E$ have been used. These generalized vacuum transition probabilities depend on all combinations of the quadratic mass differences $\Delta m_{ij}^2 = m_i^2 - m_j^2$ and on various products of elements of the leptonic mixing matrix U.

For the rest of this chapter, we shall assume a three-neutrino framework, which can easily be generalized to more neutrinos if necessary. We thus have $1 \le i, j \le 3$, and U simplifies to the 3×3 mixing matrix

$$U = \begin{pmatrix} c_{12}c_{13} & s_{12}c_{13} & s_{13}\,\mathrm{e}^{-i\delta} \\ -s_{12}c_{23} - c_{12}s_{23}s_{13}\,\mathrm{e}^{i\delta} & c_{12}c_{23} - s_{12}s_{23}s_{13}\,\mathrm{e}^{i\delta} & s_{23}c_{13} \\ s_{12}s_{23} - c_{12}c_{23}s_{13}\,\mathrm{e}^{i\delta} & -c_{12}s_{23} - s_{12}c_{23}s_{13}\,\mathrm{e}^{i\delta} & c_{23}c_{13} \end{pmatrix}, \quad (7.3)$$

which contains three leptonic mixing angles and one Dirac-like leptonic CP phase δ. Note that the most general mixing matrix for three Majorana neutrinos contains two further Majorana-like CP phases. However, it can easily be seen that these two extra diagonal Majorana phases do not enter into the above oscillation formulae, and therefore we can omit them safely. Thus, three-neutrino oscillations depend, in general, on only the three mixing angles and one CP phase. The disappearance probabilities, i.e. the probabilities for transitions $\nu_{f_l} \to \nu_{f_l}$, do not depend even on this CP phase, since $J_{ij}^{f_l f_l}$ is a function only of the moduli of elements of U. Appearance probabilities, such as that for $\nu_e \to \nu_\mu$, are therefore the place where leptonic CP violation can be studied.

From (7.2), the oscillation probabilities for neutrinos are $P(\nu_{f_l} \to \nu_{f_m}) = P_{\mathrm{CP}} + P_{\mathrm{CP}}$, and the corresponding probabilities for antineutrinos are $P(\overline{\nu}_{f_l} \to \overline{\nu}_{f_m}) = P_{\mathrm{CP}} - P_{\mathrm{CP}}$. Equation (7.2) thus has a CP-conserving part P_{CP} and a CP-violating part P_{CP}, and both terms depend on the CP phase δ. An extraction strategy for CP violation thus seems to be provided by looking at the CP asymmetries [40]

$$a^{\mathrm{CP}} := \frac{P(\nu_{f_l} \to \nu_{f_m}) - P(\overline{\nu}_{f_l} \to \overline{\nu}_{f_m})}{P(\nu_{f_l} \to \nu_{f_m}) + P(\overline{\nu}_{f_l} \to \overline{\nu}_{f_m})} = \frac{P_{\mathrm{CP}}}{P_{\mathrm{CP}}} . \quad (7.4)$$

Note, however, that the beams of an LBL experiment traverse the earth, and the presence of matter by itself violates CP, which modifies (7.2) and which makes a measurement of leptonic CP violation more involved. The above general formulae for oscillation in vacuum, (7.2), lead to well-known, but rather lengthy trigonometric expressions for the oscillation probabilities in vacuum. These expressions become even longer and do not exist in closed form when arbitrary matter corrections are taken into account. However, the problem simplifies somewhat under the assumption of a spherically symmetric distribution of the matter in the earth [41]; such a distribution is shown in Fig. 7.3 as function of the radius. There is a one-to-one correspondence between the baseline L and the angle at which the beam must enter the earth at the source. Some examples are given in Table 7.1, and obviously large values of L correspond to steep angles, resulting in technological and environmental challenges. Matter effects depend, in a first approximation, only on the average matter density along the path between source and detector, and fluctuations in the density profile partially average out, but density errors must nevertheless be taken into account [42]. Matter effects in the oscillation formulae generally grow with the distance L, while uncertainties in the average density

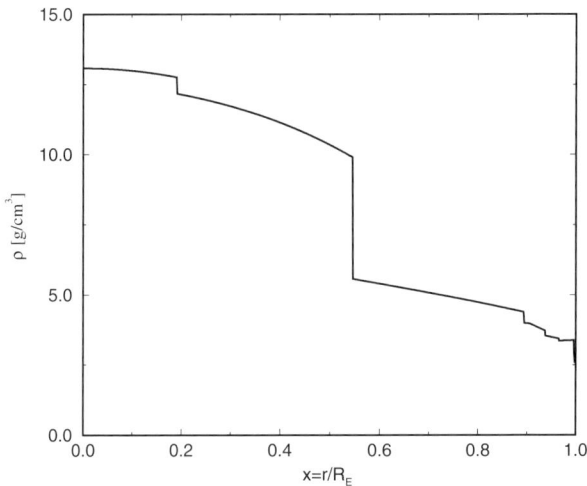

Fig. 7.3. The matter density profile of the earth as a function of the radius r [41]. Such matter density profiles are obtained by combining data from geology, materials science, seismology, and astronomy under the assumption of spherical symmetry

Table 7.1. Examples of oscillation lengths L of long-baseline experiments and the corresponding angle at which the beam enters and leaves the earth near the source and near the detector

Baseline L (km)	2800	7300	12750
Angle (degrees)	13	35	90

profile tend to decrease with L, leading approximately to a constant error, which we assume to be about 5%. Nonconstant matter profiles can in principle lead to very interesting oscillation effects [43–48]. Observing such effects would be very interesting, but they do not affect the experiments studied here and would just complicate the analysis of the oscillation parameter space. Figure 7.3 shows the pronounced density jump at $r/R_E \simeq 0.5$ between the mantle and the iron core, and the average-density approximation becomes much worse for beams which pass through the core. Avoiding the iron core and the associated density jump corresponds to a baseline of $L \leq \mathcal{O}(10\,000\,\mathrm{km})$. Matter effects lead to an MSW resonance at a characteristic energy. It is interesting to note that this resonance energy lies at approximately 10–15 GeV for the atmospheric value of Δm^2_{31} and an assumed density of matter in the crust of about $3.8\,\mathrm{g/cm^3}$.

The addition of arbitrary matter effects would make the oscillation formulae rather complicated in general, but the problem becomes much simpler in the case of three neutrinos and an approximately constant average matter density. The Hamiltonian describing three-neutrino oscillations in matter can be written in the flavor basis as

$$H = \frac{1}{2E_\nu} U \begin{pmatrix} m_1^2 & 0 & 0 \\ 0 & m_2^2 & 0 \\ 0 & 0 & m_3^2 \end{pmatrix} U^\mathrm{T} + \frac{1}{2E_\nu} \begin{pmatrix} A+A' & 0 & 0 \\ 0 & A' & 0 \\ 0 & 0 & A' \end{pmatrix} . \tag{7.5}$$

The first term describes oscillations in vacuum in the flavor basis. The quantities A and A' in the second term are given by the charged-current and neutral-current contributions to coherent forward scattering in matter. The charged-current contribution is given by

$$A = \pm \frac{2\sqrt{2}G_\mathrm{F} Y \rho E_\nu}{m_\mathrm{n}} , \tag{7.6}$$

where G_F is the Fermi constant, Y is the number of electrons per nucleon, m_n is the nucleon mass, and ρ is the matter density. A is positive for neutrinos in matter and antineutrinos in antimatter, while it is negative for antineutrinos in matter and neutrinos in antimatter. The flavor-universal neutral current contributions A' lead to an overall phase, which does not enter into the transition probabilities. The overall neutrino mass scale m_1^2 can be written as a term proportional to the unit matrix and can be removed in a similar way, such that only Δm^2_{21} and Δm^2_{31} remain in the first term of (7.5). Since $\Delta m^2_{21} \ll \Delta m^2_{31}$, we may make the further approximation $\Delta m^2_{21} \simeq 0$, and we thus obtain the approximately equivalent Hamiltonian

$$H' = \frac{1}{2E_\nu} U \begin{pmatrix} 0 & 0 & 0 \\ 0 & 0 & 0 \\ 0 & 0 & \Delta m^2_{31} \end{pmatrix} U^\mathrm{T} + \frac{1}{2E_\nu} \begin{pmatrix} A & 0 & 0 \\ 0 & 0 & 0 \\ 0 & 0 & 0 \end{pmatrix} . \tag{7.7}$$

The mixing matrix U can, furthermore, be written as a sequence of rotations R_{ij} in the two dimensional subspaces ij, namely

$$U = R_{23} R_{13} R_{12} \,, \tag{7.8}$$

where the 12 and 23 rotations are real, i.e. $R_{12}^{\mathrm{T}} = R_{12}^{-1}$ and $R_{23}^{\mathrm{T}} = R_{23}^{-1}$. This can be used to simplify the problem further, since R_{12} obviously commutes with $\mathrm{diag}(0, 0, \Delta m_{31}^2)$ and the θ_{12} dependence disappears from (7.7). Next, we observe that $\mathrm{diag}(A, 0, 0)$ commutes with R_{23}, such that R_{23} can be factored out from the complete Hamiltonian in (7.7), which can therefore be rewritten as

$$H' = R_{23} \left[\frac{1}{2E_\nu} R_{13} \begin{pmatrix} 0 & 0 & 0 \\ 0 & 0 & 0 \\ 0 & 0 & \Delta m_{31}^2 \end{pmatrix} R_{13}^{\mathrm{T}} + \frac{1}{2E_\nu} \begin{pmatrix} A & 0 & 0 \\ 0 & 0 & 0 \\ 0 & 0 & 0 \end{pmatrix} \right] R_{23}^{\mathrm{T}} \,. \tag{7.9}$$

Inside the square brackets of (7.9), the original mass matrix is rotated by R_{13}. The matter-dependent part is then added inside the square bracket, while R_{23} is factored out. H' is thus diagonalized by $R_{23} R_{13}'$, and it becomes clear that we are dealing with a modification of the diagonalization of the 1–3 subspace. Matter effects lead, therefore, to an A-dependent parameter mapping in the 1–3 subspace, which can be written as

$$\sin^2 2\theta_{13,\mathrm{m}} = \frac{\sin^2 2\theta_{13}}{C_\pm^2} \,, \tag{7.10}$$

$$\Delta m_{31,\mathrm{m}}^2 = \Delta m_{31}^2 \, C_\pm \,, \tag{7.11}$$

$$\Delta m_{32,\mathrm{m}}^2 = \frac{\Delta m_{31}^2 \, (C_\pm + 1) + A}{2} \,, \tag{7.12}$$

$$\Delta m_{21,\mathrm{m}}^2 = \frac{\Delta m_{31}^2 \, (C_\pm - 1) - A}{2} \,, \tag{7.13}$$

where the subscript "m" denotes effective quantities in matter, and

$$C_\pm^2 = \left(\frac{A}{\Delta m_{31}^2} - \cos 2\theta \right)^2 + \sin^2 2\theta \,. \tag{7.14}$$

Note that A in $C_\pm$ can change its sign, and therefore the mappings for neutrinos and antineutrinos are different, resulting in different effective mixings and masses. This is an important effect, which will allow detailed tests of coherent forward scattering of neutrinos in matter in future LBL experiments. Note that oscillations in matter, unlike vacuum oscillations, depend on the sign of Δm_{31}^2 via $C_\pm$. This is very interesting, since it opens up the possibility to extract $\mathrm{sign}(\Delta m_{31}^2)$ via matter effects.

Another very interesting question is whether it will be possible to establish leptonic CP violation in future LBL experiments. Therefore, we note that in the oscillation probabilities of (7.2), all CP-violating effects are proportional to the following two quantities:

$$D = \sin \Delta_{21} \sin \Delta_{32} \sin \Delta_{31} \ , \tag{7.15}$$

$$8J_{\mathrm{CP}} = \cos \theta_{13} \sin 2\theta_{13} \sin 2\theta_{12} \sin 2\theta_{23} \sin \delta \ . \tag{7.16}$$

We can immediately see from (7.15) that CP violation is possible only if all three masses are different, i.e. if none of the Δm_{ij}^2 vanishes. For the LBL experiments considered here, we have $\Delta_{32} \simeq \Delta_{31} \simeq 1$, while $\sin \Delta_{21} \approx \Delta_{21} \ll 1$, and there is thus a suppression due to the small solar mass splitting. Furthermore, we can see from (7.16) that J_{CP} is suppressed by the small value of θ_{13} and, for the small-mixing-angle solution, by θ_{12} in addition. The largest CP-violating effects thus occur when θ_{12} is large and for the largest possible solar value of Δm_{21}^2, i.e. the LMA solution, which, interestingly, happens to be the solution which is clearly favored by data.

Putting everything together still leads to quite lengthy expressions for the oscillation probabilities in matter, where it is not easy to obtain an overview of all effects. It is therefore instructive to simplify the problem to a point where an analytic understanding of all effects is possible, although quantitative statements should be obtained with the help of numerical evaluations using the full expressions. The key to further simplification is to expand the oscillation probabilities in small quantities. These expansion parameters are $\alpha = \Delta m_{21}^2/\Delta m_{31}^2 \simeq 10^{-2}$ and $\sin^2 2\theta_{13} \leq 0.1$. The matter effects can be parameterized by the dimensionless quantity $\hat{A} = A/\Delta m_{31}^2 = 2VE/\Delta m_{31}^2$, where $V = \sqrt{2}G_{\mathrm{F}}n_e$. If we write $\Delta \equiv \Delta_{31}$, the leading terms in the resulting expansion for $P(\nu_\mu \to \nu_\mu)$ and $P(\nu_e \to \nu_\mu)$, for example, are [49–51]

$$
\begin{aligned}
P(\ & \nu_\mu \to \nu_\mu) \\
& \approx 1 - \cos^2 \theta_{13} \sin^2 2\theta_{23} \sin^2 \Delta \\
& \quad + 2\alpha \cos^2 \theta_{13} \cos^2 \theta_{12} \sin^2 2\theta_{23} \Delta \cos \Delta \ ,
\end{aligned}
\tag{7.17}
$$

$$
\begin{aligned}
P(\ \nu_e \to \nu_\mu) &\approx \sin^2 2\theta_{13} \sin^2 \theta_{23} \frac{\sin^2\left[(1-\hat{A})\Delta\right]}{(1-\hat{A})^2} \\
&\pm \sin \delta \sin 2\theta_{13} \, \alpha \sin 2\theta_{12} \cos \theta_{13} \sin 2\theta_{23} \sin \Delta \frac{\sin \hat{A}\Delta \sin\left[(1-\hat{A})\Delta\right]}{\hat{A}(1-\hat{A})} \\
&+ \cos \delta \sin 2\theta_{13} \, \alpha \sin 2\theta_{12} \cos \theta_{13} \sin 2\theta_{23} \cos \Delta \frac{\sin \hat{A}\Delta \sin\left[(1-\hat{A})\Delta\right]}{\hat{A}(1-\hat{A})} \\
&+ \alpha^2 \sin^2 2\theta_{12} \cos^2 \theta_{23} \frac{\sin^2 \hat{A}\Delta}{\hat{A}^2} \ ,
\end{aligned}
\tag{7.18}
$$

where the plus sign in (7.18) corresponds to neutrinos and the minus sign to antineutrinos. The most important feature of (7.18) is that all interesting effects in the $\nu_e \to \nu_\mu$ transition depend crucially on θ_{13}. The size of $\sin^2 2\theta_{13}$ thus determines whether the total transition rate, matter effects, effects due to the sign of Δm_{31}^2, and CP-violating effects are measurable. One of the most important questions for future LBL experiments is therefore how far experiments can push the limit on θ_{13} below the current CHOOZ bound of approximately $\sin^2 2\theta_{13} < 0.1$.

Before we discuss further important features of (7.17) and (7.18) in more detail, we would like to comment once more on the underlying assumptions and on the reliability of these equations. First, (7.17) and (7.18) are expansions in terms of the small quantities α and $\sin 2\theta_{13}$. Higher-order terms are suppressed by at least another power of one of these small parameters, and these corrections are thus typically at the percent level. Second, the matter corrections in (7.17) and (7.18) have been derived for a constant average matter density. Numerical tests have shown that this approximation works quite well as long as the matter profile is reasonably smooth. Third, a number of very interesting effects that exist in general nonconstant matter distributions therefore cause only small theoretical uncertainties. An example is provided by asymmetric matter profiles, which lead to interesting T-violating effects [52], but these effects do not play a role here, since the earth is sufficiently symmetric.

Note that all of the results that are shown later in this chapter are based on numerical simulations of the full problem in matter. These results, therefore, do not depend on any approximation. Equations (7.17) and (7.18) will only be used to understand the problem analytically, which is extremely helpful for obtaining an overview of the parameter space, which has six or more dimensions. However, the full numerical analysis and (7.17) and (7.18) depend on the assumption of a standard three-neutrino scenario. It is thus assumed that the LSND signal [53] will not be confirmed by the MiniBooNE experiment [54]. Neutrinos could in principle decay. However, it is assumed in this chapter that neutrinos are stable; a combined treatment of oscillation and decay [55] would be much more involved. Further, neutrinos might have unusual properties and might, for example, violate CPT. In that case neutrinos and antineutrinos could have different properties, and LBL experiments can give very interesting limits on this possibility [56], but we shall assume in this study that CPT is preserved.

7.4 Correlations and Degeneracies

Equations (7.17) and (7.18) exhibit certain parameter correlations and degeneracies, which play an important role in the analysis of LBL experiments and would be hard to understand in a purely numerical analysis of the high-dimensional parameter space. The most important properties are:

- First, we observe that (7.17) and (7.18) depend only on the product $\alpha \sin 2\theta_{12}$ or, equivalently, $\Delta m_{21}^2 \sin 2\theta_{12}$. These are the parameters related to solar oscillations, which will be taken as an external input. The fact that only the product enters implies that it may be better determined than the product of the individual measurements of Δm_{21}^2 and $\sin 2\theta_{12}$.

- Next, we observe in (7.18) that the second and third terms both contain a factor $\sin \hat{A}\Delta$, while the last term contains a factor $\sin^2 \hat{A}\Delta$. Since $\hat{A}\Delta = 2VL$, we find that these factors depend only on L, resulting in a "magic baseline" when $2VL_{\text{magic}} = \pi/4V$, at which $\sin \hat{A}\Delta$ vanishes. At this magic baseline, only the first term in (7.18) survives and $P(\nu_e \to \nu_\mu)$ no longer depends on δ, α, and $\sin 2\theta_{12}$. This is in principle very important, since it implies that $\sin^2 2\theta_{13}$ can be determined at the magic baseline from the first term of (7.18) whatever the values and errors of δ, α, and $\sin 2\theta_{12}$ are. For the matter density of the earth, we find

$$L_{\text{magic}} = \pi/4V \simeq 8100 \,\text{km}, \tag{7.19}$$

which is an amazing number, since the value of V could have been such that L_{magic} was very different from the scales under discussion.

- Next, we observe that only the second and third terms of (7.18) depend on the CP phase δ, and both terms contain a factor $\sin 2\theta_{13} \times \alpha$, while the first and fourth terms of (7.18) do not depend on the CP phase δ, and contain factors of $\sin^2 2\theta_{13}$ and α^2, respectively. The extraction of CP violation is thus always suppressed by the product $\sin 2\theta_{13} \times \alpha$ and, furthermore, the CP-violating terms are obscured by large CP-independent terms if either $\sin^2 2\theta_{13} \ll \alpha^2$ or $\sin^2 2\theta_{13} \gg \alpha^2$. The determination of the CP phase δ is thus most easily possible if $\sin^2 2\theta_{13} \simeq 4\theta_{13}^2 \simeq \alpha^2$.

- Another observation is that the last term in (7.18), which is proportional to $\alpha^2 = (\Delta m_{21}^2)^2/(\Delta m_{31}^2)^2$, dominates in the limit of very small $\sin^2 2\theta_{13}$. The error in Δm_{21}^2 therefore limits the parameter extraction for small $\sin^2 2\theta_{13}$.

- Equations (7.17) and (7.18) have a structure which suggests that transformations exist which leave these equations invariant. We therefore expect degeneracies, i.e. parameter sets with identical oscillation probabilities for a given L/E. An example of such an invariance is provided by a simultaneous replacement of neutrinos by antineutrinos and Δm_{31}^2 by $-\Delta m_{31}^2$. This is equivalent to changing the sign of the second term of (7.18) and replacing α by $-\alpha$ and Δ by $-\Delta$, while $\hat{A}$ remains $\hat{A}$. It is easy to see that (7.17) and (7.18) are unchanged, but this constitutes no degeneracy, since we can distinguish neutrinos and antineutrinos experimentally.

- The first real degeneracy [57, 58] can be seen in the disappearance probability (7.17), which is invariant under the replacement $\theta_{23} \to \pi/2 - \theta_{23}$. Note that the second and third terms in (7.18) are not invariant under this transformation, but this change in the subleading appearance probability can be compensated approximately by small parameter shifts. However, the degeneracy can in principle be lifted by precision measurements in the disappearance channels.

– The second degeneracy can be found in the appearance probability (7.18) in the (δ–θ_{13}) plane [59]. In terms of θ_{13} (which is small) and δ, the four terms of (7.18) have the structure

$$P(\nu_e \to \nu_\mu) \approx \theta_{13}^2 F_1 + \theta_{13}(\pm \sin \delta F_2 + \cos \delta F_3) + F_4 , \qquad (7.20)$$

where the quantities F_i, $i = 1, \ldots, 4$, contain all the other parameters. The requirement $P(\nu_e \to \nu_\mu) = $ const. leads, for both neutrinos and anti-neutrinos, to parameter manifolds of degenerate or correlated solutions. If we have both neutrino and antineutrino beams, the two channels can be used independently, which is equivalent to considering (7.20) for $F_2 \equiv 0$ and $F_3 \equiv 0$ simultaneously. The requirement that these probabilities are now independently constant, i.e. $P(\nu_e \to \nu_\mu) = $ const. for $F_2 \equiv 0$ and $F_3 \equiv 0$, leads to more constraint manifolds in the (δ–θ_{13}) plane, but some degeneracies still survive.

– The third degeneracy [60] arises from the fact that a change in sign of Δm_{31}^2 can, essentially, be compensated by an offset in δ. Therefore, we note again that the transformation $\Delta m_{31}^2 \to -\Delta m_{31}^2$ leads to $\alpha \to -\alpha$, $\Delta \to -\Delta$, and $\hat{A} \to -\hat{A}$. All terms of the disappearance probability in (7.17) are invariant under this transformation. The first and fourth terms in the appearance probability in (7.17), which do not depend on the CP phase δ, are also invariant. The second and third terms of (7.17) depend on the CP phase and are changed by the transformation $\Delta m_{31}^2 \to -\Delta m_{31}^2$. The fact that these changes can be compensated by an offset in the CP phase δ is the third degeneracy.

– Altogether there exists, thus, an eightfold degeneracy [58], as long as only the $\nu_\mu \to \nu_\mu$, $\bar{\nu}_\mu \to \bar{\nu}_\mu$, $\nu_e \to \nu_\mu$, and $\bar{\nu}_e \to \bar{\nu}_\mu$ channels and one fixed value of L/E are considered. However, the structure of (7.17) and (7.18) makes it clear that the degeneracies can be broken by using information for different L/E values in a suitable way. This can be achieved in the total event rates by changing L or E [61, 62], but it can in principle also be done by using information in the event rate spectrum for a single baseline L, although this requires detectors with very good energy resolution [51]. Another strategy for breaking the degeneracies is to include further oscillation channels in the analysis ("silver channels") [61, 63].

The discussion in this section shows the strength of the analytic approximations, which allow us to understand the complicated interdependence of the parameters. This discussion should also help one to plan experimental setups optimally and to find strategies to resolve the degeneracies.

7.5 Event Rates

The experimentally detected event rates must be compared with theoretical expressions, which depend only indirectly on the above oscillation probabilities. Every event can be classified by the flavor of the detected neutrino and the type of interaction. The particles detected in an experiment are produced by neutral-current (NC), inelastic charged-current (CC), or quasi-elastic charged-current (QE) interactions. The contribution to each mode depends on a number of factors, such as the detector type, the neutrino energy, and the neutrino flavor. In order to calculate realistic event rates, we must first compute the number of events for each type of interaction in the fiducial mass of an ideal detector, for each neutrino flavor and each energy bin. Next, the deficiencies of a real detector, such as limited event reconstruction capabilities, must be included. The combined description leads to the differential event rate spectrum for each flavor and interaction mode, as it would be seen by a detector which was able to separate all those channels. Finally, different channels must be combined, since they cannot be observed separately. This can be due to physics, e.g. due to the flavor-blindness of NC interactions, or it can be a consequence of detector properties, e.g. due to charge misidentification. The differential event rates can thus be written, for each channel of some interaction type IT, as

$$
\frac{\mathrm{d}n_f^{\mathrm{IT}}}{\mathrm{d}E'} = N \sum_i \int\int \mathrm{d}E\,\mathrm{d}\hat{E}\ \underbrace{\Phi_i(E)}_{\text{Production}}
$$

$$
\times \underbrace{\frac{1}{L^2} P_{(i\to f)}(E, L, \rho; \theta_{23}, \theta_{12}, \theta_{13}, \Delta m_{31}^2, \Delta m_{21}^2, \delta)}_{\text{Propagation}}
$$

$$
\times \underbrace{\sigma_f^{\mathrm{IT}}(E) k_f^{\mathrm{IT}}(E - \hat{E})}_{\text{Interaction}} \times \underbrace{T_f(\hat{E}) V_f(\hat{E} - E')}_{\text{Detection}}, \qquad (7.21)
$$

where f and i stand for the final and initial neutrino flavors, respectively. E is the incident neutrino energy, $\Phi_i(E)$ is the flux from the source for the initial flavor i, L is the baseline, N is a normalization factor, and ρ is the density of matter in the earth. The interaction term is composed of two factors, which are the total cross section $\sigma_f^{\mathrm{IT}}(E)$ for the flavor f and interaction type IT, and the energy distribution of the secondary particle $k_f^{\mathrm{IT}}(E - \hat{E})$, where $\hat{E}$ is the energy of the secondary particle. The detector threshold is parameterized by the function $T_f(\hat{E})$, describing a limited resolution or cuts in the analysis. The energy resolution of the detector is parameterized by the function $V_f(\hat{E} - E')$ for the secondary particle, where E' is the reconstructed neutrino energy.

The numerical calculation of the double integral in (7.21) for all possible parameter combinations requires enormous computing power. Therefore, we use here an approximation in which we evaluate the integral over $\hat{E}$, where the only terms containing $\hat{E}$ are $k_f^{\mathrm{IT}}(E - \hat{E})$, $T_f(\hat{E})$, and $V_f(\hat{E} - E')$. We define

$$
R_f^{\mathrm{IT}}(E, E')\,\epsilon_f^{\mathrm{IT}}(E') \equiv \int \mathrm{d}\hat{E}\,T_f(\hat{E})\,k_f^{\mathrm{IT}}(E - \hat{E})\,V_f(\hat{E} - E')\,, \qquad (7.22)
$$

and approximate R_f^{IT} by the analytical expression

$$R_f^{\mathrm{IT}}(E, E') = \frac{1}{\sigma\sqrt{2\pi}} \exp\frac{(E - E')^2}{2\sigma^2} . \qquad (7.23)$$

For the QE interactions in the case of the beam from the JHF experiment, we shall use [28] $\sigma = 85\,\mathrm{MeV}$, and for the beams from a neutrino factory and from NuMI, we shall use ([64] and A. Para, private communication 2002) $\sigma = 0.15E$. Values of the effective relative energy resolution δE and the effective efficiency ϵ_f^{IT} can be found in the literature; they are given for the neutrino factory in [37, 38, 49, 65, 66] and for the superbeam setups in [28, 67–70]. The threshold for muon detection is an important parameter for neutrino factories, and we shall use essentially an interpolation between a more optimistic and a conservative estimate [38, 49]. For further details, see [12].

In order to include backgrounds, we group the channels into pairs of signal and background in an experiment-specific way. The backgrounds considered are NC events which are misidentified as CC events, and CC events identified with the wrong flavor or charge. For superbeams, we include also the background of CC events arising from the intrinsic contamination of the beam.

Finally, in the analysis we combine all available signal channels and perform a global fit to extract the physics parameters in an optimal way. For a neutrino factory, the relevant channels are the event rate spectra for the ν_μ–CC channel (disappearance) and the $\bar{\nu}_\mu$–CC channel (appearance), for each polarity of the beam. The backgrounds for these signals are NC events for all flavors, and misidentified ν_μ–CC events. For superbeam experiments, the signal is provided by the ν_μ–QE channel (disappearance) and the ν_e–CC channel (appearance), again for each polarity of the beam. The backgrounds here are NC events for all flavors, misidentified ν_μ–CC events, and, for the ν_e–CC channel, the ν_e–CC beam contamination.

7.6 The LBL Setups Considered

The sources and detectors discussed here allow several different LBL experiments, and it is interesting to compare their physics potential on an equal footing as realistically as possible. Studies at the level of probabilities are not sufficient, and the true potential must be evaluated at the level of event rates as described in Sect. 7.5, with realistic assumptions about the beams, detectors, and backgrounds. We present now the results of such an analysis, which is essentially based on [12], where we have calculated the oscillation probabilities numerically from the exact three-neutrino formulae for oscillation in matter, i.e. we use the approximations to the probabilities given in (7.17) and (7.18) only to obtain a qualitative understanding. The results shown below are therefore not affected by the approximations which were made in the derivation of the approximate analytic oscillation formulae (7.17) and (7.18). Sensitivities, etc., are defined by the ability to reextract the physics parameters from a simulation of event rates. Therefore, event rate distributions were generated for all possible parameter sets. Subsequently, a combined fit to these event rate distributions was performed, for the appearance and disappearance channels and both polarities simultaneously. This procedure uses all the available information in an optimal way. It includes spectral distributions when they are present, and it reduces to a fit of total rates when the total event rates are small. Suitable statistical methods, as described in [12], were used to deal with event distributions which have small event rates per bin in some regions. Systematic uncertainties were parameterized, and external input from geophysics was used in the form of the detailed matter profile and its errors, which were included in the analysis [12]. The ability to reextract the input parameters that were used to generate the event rate distributions of a simulation of the full experiment was used to define the sensitivities and precision.

An important aspect of such an analysis is the inclusion of external information. The LBL experiments discussed here could, in principle, measure the solar parameter Δm_{21}^2 and the mixing angle θ_{12}. However, the precision which can be obtained cannot compete with that expected from KamLAND [71]. We have therefore included as an external input the assumption that Kam-LAND will measure the solar parameters to be in the center of the LMA region with typical errors. Otherwise, the unknown parameters (such as the CP phase) have not been constrained and have therefore been left free, with all parameter degeneracies and error correlations being taken into account. All nuisance parameters were integrated out, and a projection onto the parameter of interest was performed. Altogether, we are dealing with six free parameters.

The beam characteristics of the three sources considered are shown in Figs. 7.4 (JHF), 7.5 (NuMI off-axis), and 7.6 (neutrino factory). We have also included uncertainties in these beam parameters, i.e. uncertainies in the ν_e background for the first two, conventional beams, flux uncertainties for all beams [28, 30, 34]. As detectors, we have considered water Cherenkov detectors, low-Z calorimeters, and magnetized iron detectors, with parameters as given in Table 7.2. For magnetized iron calorimeters, it is important to include realistic threshold effects. We have used a linear rise of the efficiency between 4 GeV and 20 GeV and have studied the sensitivity to the position of the threshold. We have not included liquid-argon time projection chambers in our analysis, but they would certainly be an important detector if this technology were to work. The beams and detectors that we have included allow several different interesting combinations, which are listed in Table 7.3. JHF-SK is the planned combination of the existing Super-Kamiokande detector and the JHF beam, while JHF-HK is the combination of an upgraded JHF beam with the proposed Hyper-Kamiokande detector. With typical parameters, JHF-HK has about 95 times more integrated luminosity than JHF-SK, and we have assumed that it operates partly with an antineutrino beam. Water Cherenkov detectors are ideal for the JHF beam, since charged-current

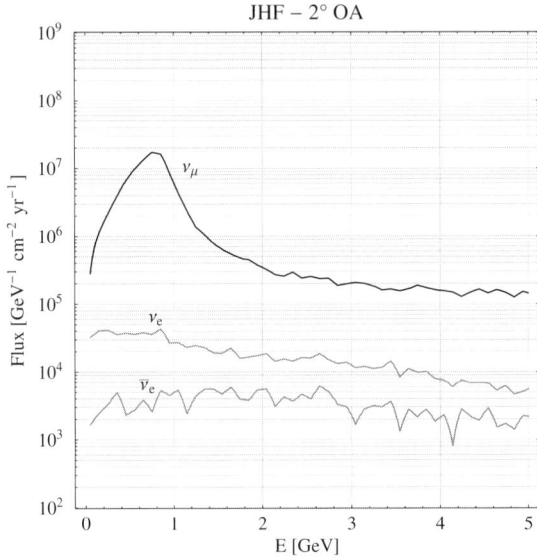

Fig. 7.4. The flux of the 2° off-axis JHF beam as a function of energy. The mean energy is 0.51 GeV, and the peak intensity is 1.7×10^7 GeV^{-1}cm^{-2}yr^{-1} at 0.78 GeV. The ν_e/ν_μ ratio is 0.2% at the peak

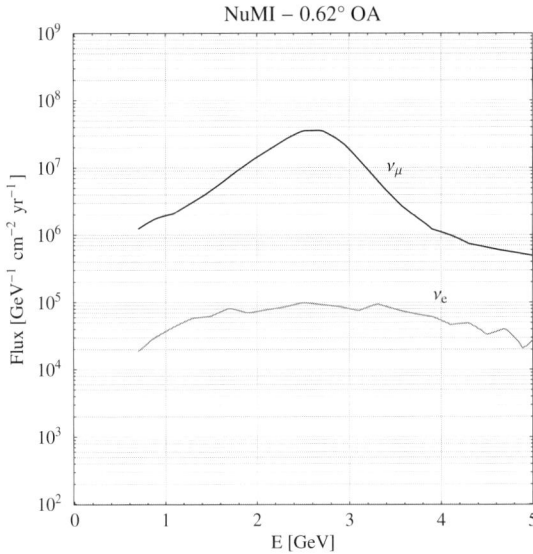

Fig. 7.5. The flux of the proposed NuMI off-axis beam, with a mean energy of 2.78 GeV, and a peak intensity of 3.6×10^7 GeV^{-1}cm^{-2}yr^{-1} at 2.18 GeV. The ν_e/ν_μ ratio is 0.2% at the peak

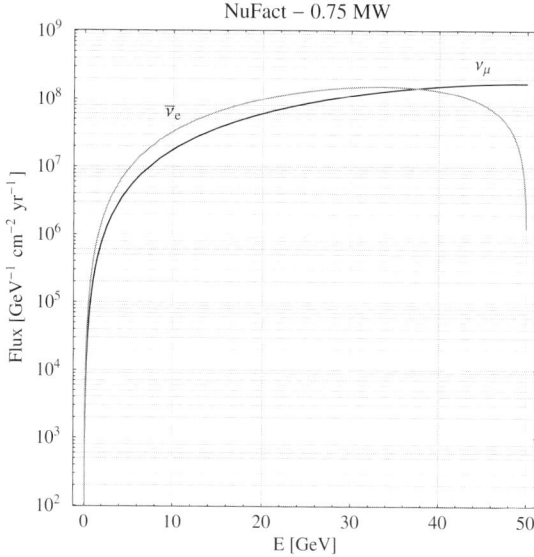

Fig. 7.6. The flux of a neutrino factory with $E_\mu = 50\,\text{GeV}$. The mean neutrino energy is about $30\,\text{GeV}$, and the peak intensity is $1.5 \times 10^8\,\text{GeV}^{-1}\text{cm}^{-2}\text{yr}^{-1}$ at $33.33\,\text{GeV}$. The ν_μ/ν_e ratio is 83% at the peak

Table 7.2. The detector types considered and their most important parameters

	Water Cherenkov, SK (HK)	Low-Z calorimeter	Magnetized iron calorimeter
Fiducial mass	22.5 kt (1000 kt)	20 kt	10 kt (50 kt)
Energy range	0.4–1.2 GeV	1–5 GeV	4–50 GeV
Energy resolution	85 MeV	$0.15E$	$0.15E$
Signal efficiency	0.5	0.5	0.45
NC rejection	0.01	0.001	$< 10^{-5}$
Charge identification	–	–	$< 10^{-5}$
Background uncertainty	5%	5%	5%

quasi-elastic scattering dominates. NuMI is the proposed combination of the NuMI off-axis beam with a low-Z calorimeter, which is better here, since the energy is higher and there is a considerable contribution of inelastic charged-current interactions. NuFact-I will be an initial neutrino factory, while NuFact-II will be a fully developed machine, with 42 times the luminosity of NuFact-I [28, 30, 37]. Deep inelastic scattering dominates for the even higher energies of these neutrino factories, and magnetized iron detectors are therefore considered in this case.

Table 7.3. The combinations of beams and detectors considered and their acronyms

Acronym	Detector	Baseline	Matter density	L/E_{peak}
JHF-SK	Water Cherenkov	295 km	$2.8\,\text{g cm}^{-3}$	$378\,\text{km GeV}^{-1}$
NuMI	Low-Z	735 km	$2.8\,\text{g cm}^{-3}$	$337\,\text{km GeV}^{-1}$
NuFact-I	10 kt magnetized iron	3000 km	$3.5\,\text{g cm}^{-3}$	$90\,\text{km GeV}^{-1}$
JHF-HK	Water Cherenkov	735 km	$2.8\,\text{g cm}^{-3}$	$295\,\text{km GeV}^{-1}$
NuFact-II	40 kt magnetized iron	3000 km	$3.5\,\text{g cm}^{-3}$	$90\,\text{km GeV}^{-1}$

Table 7.4. The expected signal and background event rates in the appearance channels for the scenarios considered

	JHF-SK	NuMI	JHF-HK	NuFact-I	NuFact-II
Signal	139.0	387.5	13 180.0	1522.8	64 932.6
Background	23.3	53.3	2204.6	4.2	180.3
S/N	6	6	6	360	360

7.7 The Qualitative Picture

A number of studies have analyzed various aspects of the above scenarios or some combined variants of them [12, 38, 49, 51, 58, 60, 72–89], and the results can all be understood in the same qualitative picture. The proposed setups lead, in general, to remarkably large event rates in the disappearance channel. This leads to many events per energy bin, and the spectral information allows very precise fits of the leading oscillation parameters Δm_{31}^2 and θ_{23}. A typical example is shown in Fig. 7.7 for a neutrino factory, for both polarities [77]. The situation is somewhat different in the appearance channels, where the event rates are small. The results for the appearance channels thus depend dominantly on total rates, with some spectral information. Note, however, that the available spectral information is very important, since it allows one to distinguish solutions which are degenerate on the basis of total event rates. Figure 7.8 shows, for a neutrino factory with typical parameters, how the total event rates depend on the presence of matter and on the CP phase. It can clearly be seen in Fig. 7.8 that matter effects grow with distance and become dominant at large baselines $L \gtrsim 3000$ km. In contrast, the CP phase δ affects the rates at shorter distances, while at medium distances the effects of matter and CP violation are comparable. The simplest strategy for separating matter effects and CP violation is thus to have one short baseline for CP-violating effects and another, large baseline for matter effects [49, 62]. Alternatively, one might use a single medium baseline, where the effects can be separated if the event rates are large enough that spectral information can be extracted [51]. Another alternative is, as discussed before, to use one baseline and further oscillation channels [63].

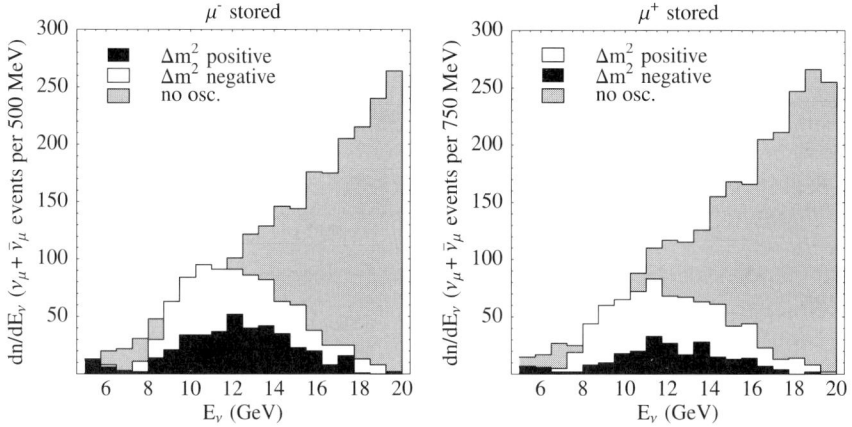

Fig. 7.7. ν_μ disappearance rates for a typical neutrino factory setup. The *gray* area marks the expected distribution in the absence of oscillations. The *black* and *white* *histograms* show the expected distributions, with oscillations, for both $\Delta m_{31}^2 > 0$ and $\Delta m_{31}^2 < 0$. Here, a relatively large value of $\sin^2 2\theta_{13} = 0.01$ has been chosen, where matter effects allow us to extract $\mathrm{sign}(\Delta m_{31}^2)$ even from the disappearance channel, which is not possible for smaller $\sin^2 2\theta_{13}$ [77]

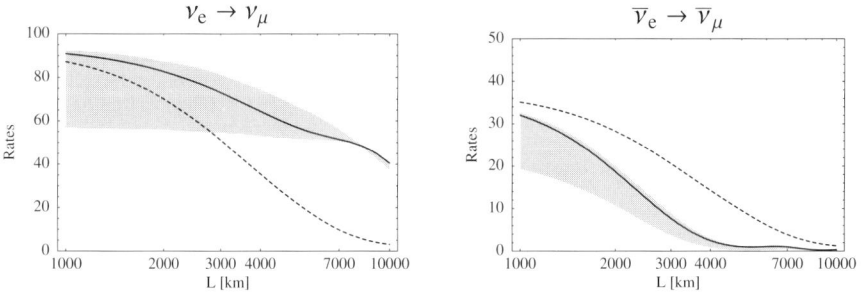

Fig. 7.8. The total event rates for the two polarities of a neutrino factory with $E_\mu = 50\,\mathrm{GeV}$, assuming $\sin^2 2\theta_{13} = 0.01$ and the LMA solution. The *solid lines* are for $\delta = 0$ with the matter corrections included. The *dashed lines* show, for comparison, the results for $\delta = 0$ in vacuum. The *gray band* shows the range within which the *solid line* moves when the CP phase is allowed to take all possible values, with matter included

It is interesting to understand the interplay of matter effects and CP-violating effects by use of our analytic formulae. First, we observe from (7.17) and (7.18), as well as Fig. 7.8, that matter effects grow with distance, and CP-violating effects are thus not affected by matter effects at the shortest distances. By inspecting (7.18) carefully, we found the existence of a magic baseline at larger distances in Sect. 7.4. The point was that $\hat{A} = 2VE/\Delta m_{31}^2$ and $\Delta = \Delta m_{31}^2 L/E$, so that $\hat{A}\Delta = 2VL$ depends only on the matter potential V and the baseline L. All terms in (7.18) except the first therefore vanish at

L_{magic} as given in (7.19), since they contain factors of $\sin \hat{A}\Delta = \sin 2VL$. All terms which contain the CP phase δ and the mass-splitting hierarchy parameter α thus vanish as a consequence of matter effects at $L_{\mathrm{magic}} \simeq 8100\,\mathrm{km}$, where a typical density in the earth's crust has been used to calculate this value. Matter effects and CP-violating effects can thus in principle be completely separated by performing measurements at a small baseline and at the magic baseline [90].

7.8 The Analysis

The sensitivity and precision of the measured quantities are defined, as explained in Sect. 7.5, via the ability to reextract the physics parameters from previously generated event rate distributions. In order to determine the sensitivity and precision in such an analysis, all input parameters are scanned, while systematic, statistical, and background errors are included. The best possible results are obtained by using optimally the information contained in the rates and energy spectra of all available channels in a combined analysis. At the same time, the parameter correlations and degeneracies mentioned above must be included, in addition to systematic errors such as normalization and calibration. However, the precise relative information contained in a spectrum allows very precise measurements even when an overall normalization error of 5% and an energy calibration error of 5% are included. The inclusion of the backgrounds limits the sensitivity. However, information in the energy spectrum helps to reduce the impact of the background, which has typically a different energy dependence.

There are many events per bin in the disappearance channels, which leads via (7.17) to a very precise determination of the leading oscillation parameters Δm_{31}^2 and $\sin^2 2\theta_{23}$. The combination of the available appearance channels allows one to determine or restrict θ_{13}, and, via the matter effects, $\mathrm{sign}(\Delta m_{31}^2)$. This is always possible, even when α is very small or negligible, for the disfavored SMA, LOW, and VAC solutions of the solar neutrino problem, since the first term in (7.18) does not depend on α. For the LMA solution (i.e. $\alpha \simeq 10^{-2}$), it is also possible to determine θ_{13} and $\mathrm{sign}(\Delta m_{31}^2)$, and it is in principle even possible to extract the solar parameters θ_{12} and Δm_{21}^2 from a combined fit of the appearance and disappearance channels without using external input. The precision which can be obtained for θ_{12} and Δm_{21}^2, however, cannot compete with the expected measurements from KamLAND. Therefore, we have assumed in our analysis that KamLAND will measure the solar parameters to be at the current best fit of the LMA solution, with typical errors, and have taken these values as an external input in our analysis. This allows us, for the favored LMA solution, to extract information about the CP phase δ, as expected from the second and third terms of (7.18).

It should be kept in mind that (7.17) and (7.18) exhibit parameter correlations and degeneracies, as discussed above, which must be taken into account in the analysis [12, 49, 51, 57–60]. The analytic formulae (7.17) and (7.18) are here very useful, since they allow us to understand qualitatively the highly nonlinear behavior and complex topology of the parameter manifolds. The dependence of the probabilities on the parameters leads, as discussed above, to three degeneracies in the following parameter planes:

– δ–θ_{13},
– δ–sign(Δm_{31}^2),
– θ_{23}–($\pi/2 - \theta_{23}$).

These three different degeneracies can lead to equivalent solutions that are rather close together in parameter space. In this case they effectively enlarge the allowed range of the combined solutions, as shown by an example in the δ–θ_{13} plane in the left plot of Fig. 7.9. In other cases, the degenerate solutions are well separated, as shown in the right plot of Fig. 7.9. In such cases it is more sensible to quote two values and their respective errors. In the case of a detailed single-baseline analysis, one can see, as discussed analytically in Sect. 7.4, that the δ–θ_{13} degeneracy typically becomes a parameter correlation when only a neutrino beam is used, while degenerate islands show up when both polarities are combined. Similarly, one can see that with one baseline, the δ–sign(Δm_{31}^2) degeneracy can never be removed. The θ_{23}–($\pi/2 - \theta_{23}$) degeneracy also cannot be removed if the analysis is dominated by the disappearance rates only. This degeneracy can be lifted for sufficiently high statistics in the appearance channel, where the degeneracy is in principle broken, as seen in the analytic discussion in Sect. 7.4.

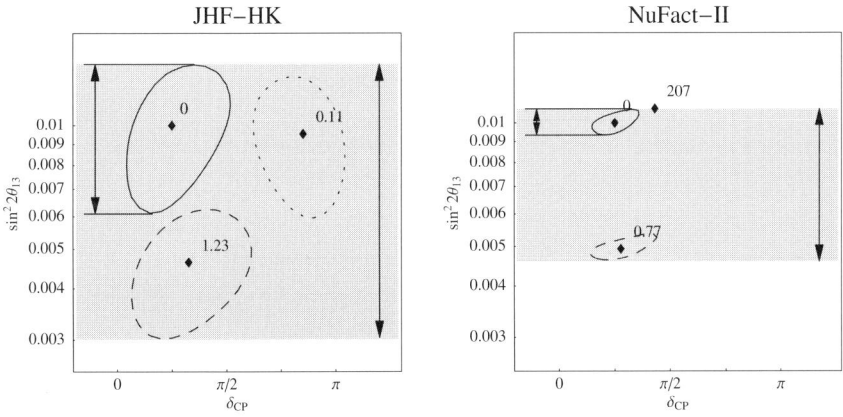

Fig. 7.9. Examples of degenerate solutions [12]. The *left plot* shows a case where the degenerate solutions are very close. Consequently, the combination of these adjacent solutions leads to an enlarged error because of the degeneracies. The *right plot* shows a case where two degenerate solutions are well separated. In this case, two solutions and their respective errors should be quoted

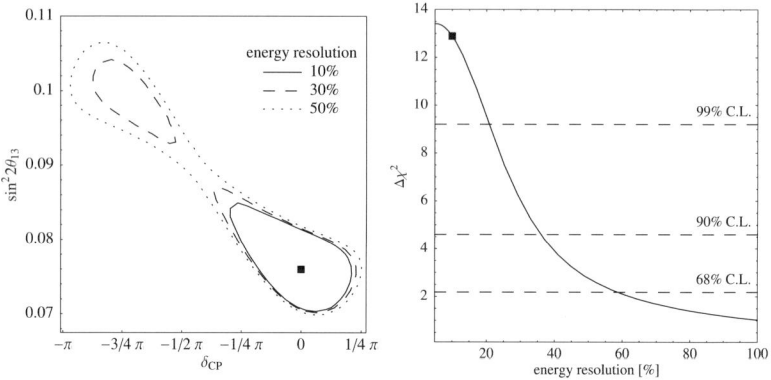

Fig. 7.10. The correlation or degeneracy of δ with θ_{13} as a function of the detector energy resolution for a neutrino factory with $E = 50\,\text{GeV}$, $L = 3000\,\text{km}$, $\Delta m_{31}^2 = 3.5 \times 10^{-3}\,\text{eV}^2$, $\Delta m_{21}^2 = 10^{-4}\,\text{eV}^2$, and $\theta_{23} = \theta_{12} = \pi/4$ [51]. For poor resolution, there is a correlation between δ and θ_{13}, which transforms into two degenerate solutions for medium resolution, and the degeneracy is finally lifted for an energy resolution better than 25%

The separation of matter and CP-violating effects is, as discussed in Sect. 7.4, not possible at the level of total event rates for one value of L/E, which translates into one baseline for fixed E. However, as mentioned before, there exist ways to break this degeneracy. One such strategy is to use a short baseline for CP-violating effects and a long baseline for matter effects. Another strategy is to use one medium baseline and the information contained in the beam spectrum. The point is that different parameter sets are degenerate at the level of total event rates for one value of L/E, but the event rate distributions differ significantly. The degeneracy may thus be lifted for one medium baseline in combination with good resolution and sufficient statistics. This is shown in Fig. 7.10 for the correlation of δ with θ_{13} [51]. This example shows nicely how degeneracies and correlations at the level of event rates can change in a real experiment when the spectral information is used.

7.9 Results

A realistic analysis of future LBL experiments requires a number of different aspects to be taken into account. It should be clear from the discussion above that it is not sufficient to quote limits which are based on oscillation probabilities or merely on the statistics of a single channel without backgrounds or systematics. Depending on the position in the space of physics parameters, the limiting factor may be the degeneracies or correlations, the backgrounds, the systematics, or the statistics. A reliable comparative study

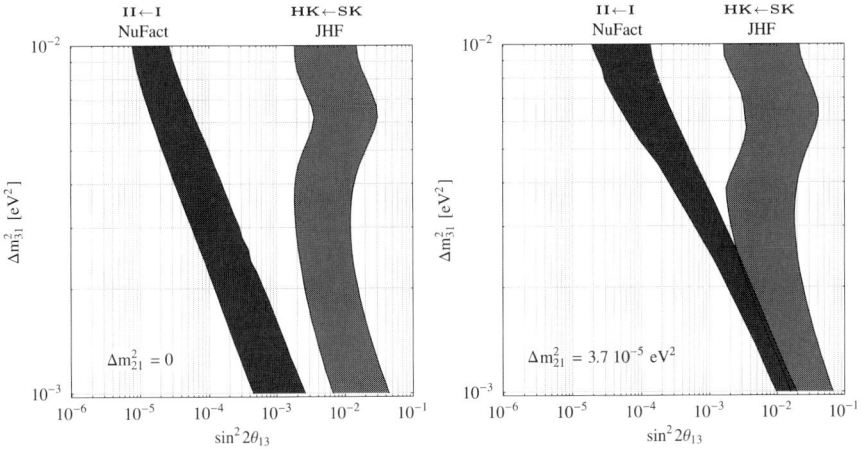

Fig. 7.11. The sensitivity to $\sin^2 2\theta_{13}$ as a function of Δm_{31}^2 for $\Delta m_{21}^2 = 0$ (*left plot*) and $\Delta m_{21}^2 = 3.7 \times 10^{-5} \, \mathrm{eV}^2$ (*right plot*) for the JHF and NuFact setups considered here. The *right* edges of the NuFact and JHF bands correspond to the less advanced options, NuFact I and JHF-SK, while NuFact II and JHF-HK are equivalent to the *left* edges of the bands [12]

of the LBL setups discussed here requires, therefore, a detailed analysis of the six-dimensional parameter space which includes all these effects on the same footing [12].

There is excellent precision for the leading oscillation parameters Δm_{31}^2 and $\sin^2 2\theta_{23}$, which will not be discussed further here. The more interesting sensitivity to the subleading parameter $\sin^2 2\theta_{13}$ is shown in Fig. 7.11, where we can see that the $\sin^2 2\theta_{13}$ sensitivity limit depends considerably on the value of Δm_{31}^2. From a comparison of the cases $\Delta m_{21}^2 = 0$ (left plot) and $\Delta m_{21}^2 = 3.7 \times 10^{-5} \, \mathrm{eV}^2$ (right plot), we find, moreover, a significant Δm_{21}^2 dependence of the sensitivity limits.[2] The $\sin^2 2\theta_{13}$ sensitivity limit can thus vary considerably, depending on what is found for Δm_{31}^2 and Δm_{21}^2, and these dependencies are strongest for short baselines.

Assuming that the leading parameters are measured to be $\Delta m_{31}^2 = 3 \times 10^{-3} \, \mathrm{eV}^2$ and $\sin^2 2\theta_{23} = 0.8$ and that KamLAND measures the solar parameters to be at the current best-fit point of the LMA region, i.e. $\Delta m_{21}^2 = 6 \times 10^{-5} \, \mathrm{eV}^2$ and $\sin 2\theta_{12} = 0.91$, we can translate this into a comparison of the $\sin^2 2\theta_{13}$ sensitivity limit for the different setups. The result is shown in Fig. 7.12. The individual contributions of the different sources of uncertainties are shown for every experiment; the left edge of every band in Fig. 7.12 corresponds to the sensitivity limit which would be obtained purely

---

[2] This translates into a corresponding Δm_{31}^2 and Δm_{21}^2 dependence of the obtainable precision in the case where $\sin^2 2\theta_{13}$ is large enough to be measured. The corresponding plots can be found in [12].

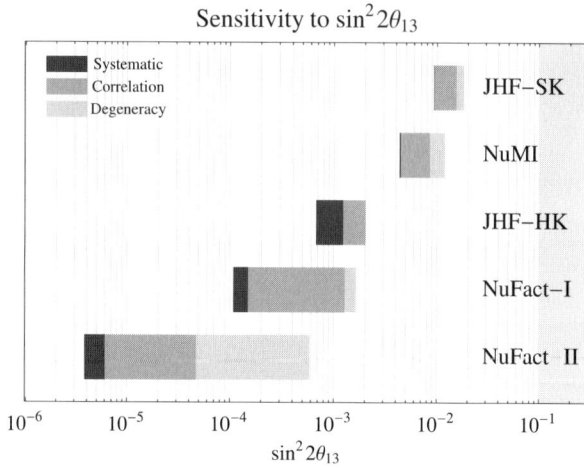

Fig. 7.12. The $\sin^2 2\theta_{13}$ sensitivity for all setups described in Sect. 7.6 at the 90% confidence level for $\Delta m_{31}^2 = 3 \times 10^{-3}$ eV^2 and $\sin^2 2\theta_{23} = 0.8$. The plot shows the deterioration of the sensitivity limits as the different error sources are successively switched on. The *left* edge of each bar is the sensitivity statistical limit. This limit becomes poorer as systematic, correlational, and degeneracy errors are switched on. The *right* edge is the final sensitivity limit [12]

on statistical grounds. This limit is successively made poorer by adding the systematic uncertainties of each experiment, the correlational errors, and finally the degeneracy errors. The right edge of each band constitutes the final error for the experiment under consideration. It is interesting to see how the errors of the different setups are made up. There are different sensitivity reductions due to systematic errors, correlations, and degeneracies. The largest sensitivity loss due to correlations and degeneracies occurs for NuFact-II, which is mostly a consequence of the uncertainty in Δm_{21}^2, which translates into an uncertainty in α, and which dominates the appearance probability (7.18) for small $\sin^2 2\theta_{13}$. Note that it is, in principle, possible to combine different experiments. If done correctly, this allows one to eliminate part or all of the correlational and degeneracy errors [61].

It is interesting to recall the existence of the magic baseline defined in (7.19), where the α and δ dependence drops out completely owing to matter effects. Correlational and degeneracy errors are then drastically reduced and a measurement of $\sin^2 2\theta_{13}$ becomes more precise, even though the event rates are smaller at this larger baseline. This improvement in sensitivity is shown for NuFact-II in Fig. 7.13, where a baseline of 3000 km is compared with the magic baseline of 8100 km.

Another challenge of future LBL experiments is to measure $\text{sign}(\Delta m_{31}^2)$ via matter effects, and the sensitivity which can be obtained for the setups under discussion is shown in Fig. 7.14. Taking all correlational and degeneracy

Sensitivity to $\sin^2 2\theta_{13}$ for NuFact$-$II

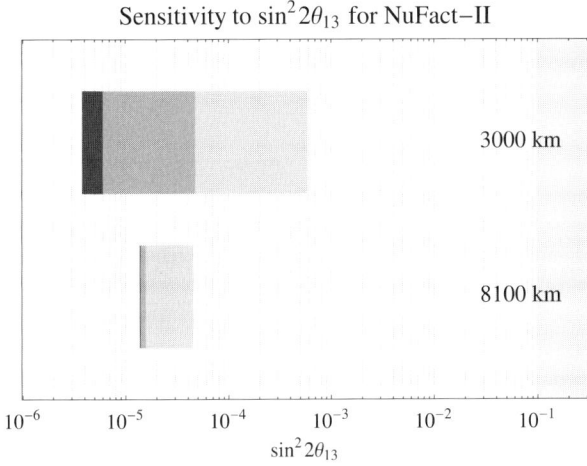

Fig. 7.13. Comparison of the NuFact-II setup for a baseline of 3000 km and the magic baseline of 8100 km. It can be seen that the statistical sensitivity (*left* edge of the bars) is reduced owing to smaller event rates, but the total sensitivity is increased since the correlational and degeneracy errors disappear almost completely at the magic baseline

Sensitivity to the sign of Δm_{31}^2

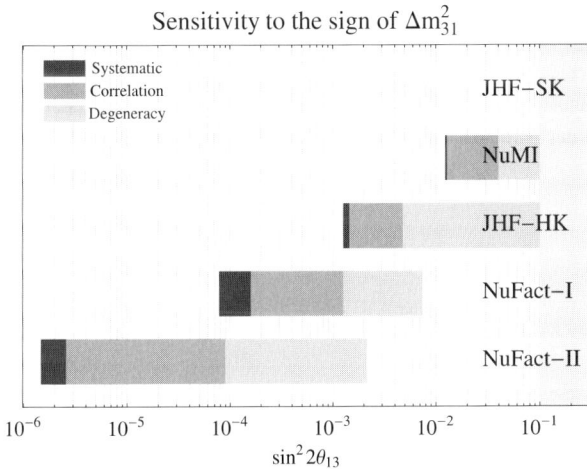

Fig. 7.14. The sensitivity of $\sin^2 2\theta_{13}$ to $\text{sign}(\Delta m_{31}^2)$ for the setups defined in Sect. 7.6. The *left* edge of each bar is the statistical sensitivity limit, which is successively impaired by systematic, correlational, and degeneracy errors. The *right* edge of each bar is the final limit

errors into account, we can see that it is very hard to determine $\text{sign}(\Delta m_{31}^2)$ with the superbeam setups considered. The main problem is the degeneracy with δ, which always allows the reverse $\text{sign}(\Delta m_{31}^2)$ for another CP phase.

Sensitivity to CP–Violation at $\delta_{CP} = +\pi/2$

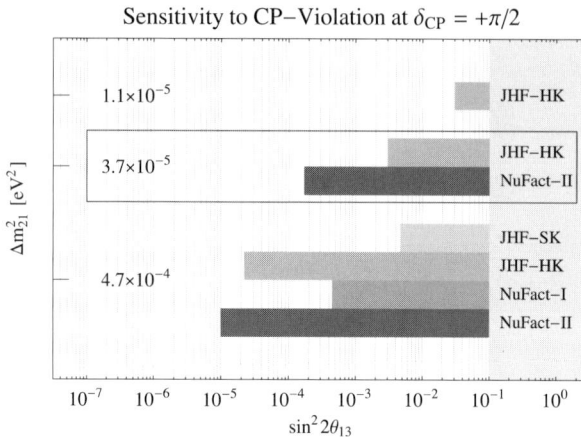

Fig. 7.15. The $\sin^2 2\theta_{13}$ sensitivity range for CP violation in the setups considered at 90% confidence level for different Δm_{21}^2 values. The *top row* corresponds to the lower bound $\Delta m_{21}^2 = 1.1 \times 10^{-5}$ eV^2, the *bottom row* to the upper bound $\Delta m_{21}^2 = 4.7 \times 10^{-4}$ eV^2, and the *middle row* to the best LMA fit, $\Delta m_{21}^2 = 3.7 \times 10^{-5}$ eV^2. Cases which do not have CP sensitivity are omitted from this plot. The chosen parameters are $\delta = +\pi/2$, $\Delta m_{31}^2 = 3 \times 10^{-3}$ eV^2, $\sin^2 2\theta_{23} = 0.8$, and a solar mixing angle corresponding to the current best fit in the LMA regime [12]

Note, however, that the situation could in principle be improved if different superbeam experiments were combined in such a way that this degeneracy error could be removed. Neutrino factories perform considerably better on $\text{sign}(\Delta m_{31}^2)$, particularly for larger baselines. Combination strategies would again lead to further improvements.

Coherent forward scattering of neutrinos and the corresponding MSW matter effects are, so far, experimentally untested. It is therefore very important to realize that matter effects will not only be useful for extracting $\text{sign}(\Delta m_{31}^2)$ but also will allow detailed tests of the theory of coherent forward scattering of neutrinos. This possibility has been studied in detail in [51, 74, 77, 82].

The Holy Grail of LBL experiments is the measurement of leptonic CP violation. The $\sin^2 2\theta_{13}$ sensitivity range for measurable CP violation is shown in Fig. 7.15 for $\delta = \pi/2$ for the various setups considered and for various values of Δm_{21}^2. It can be seen that measurements of CP violation are feasible in principle both with high-luminosity superbeams and with advanced neutrino factories. However, the sensitivity depends in a crucial way on Δm_{21}^2. For a low value $\Delta m_{21}^2 = 1.1 \times 10^{-5}$ eV^2, the sensitivity is almost completely lost, while the situation would be best for the largest value considered, $\Delta m_{21}^2 = 4.7 \times 10^{-4}$ eV^2. KamLAND has now confirmed the LMA region with a best fit value of $\Delta m_{21}^2 = 7 \times 10^{-5}$ [91], which implies that the prospects to measure leptonic CP violation are very promising. The sensitivities shown in Fig. 7.15 depend on the choice of δ. The value which has been used here is $\delta = \pi/2$, and the sensitivities become become worse for small CP phases close to zero or π.

7.10 Conclusions

We have discussed the potential of some proposed long-baseline neutrino oscillation experiments, where it will be possible to perform precision neutrino physics. The basic fact which makes this possible is that the atmospheric mass splitting $\Delta m_{31}^2 \simeq \Delta m_{\mathrm{atm}}^2$ leads, for typical neutrino energies $E_{\mathrm{v}} \simeq 1-100\,\mathrm{GeV}$, to oscillation baselines in the range of 100 km to 10 000 km. Moreover, beam sources have the advantage that, unlike the sun or the atmosphere, they can be controlled very precisely, such that unknowns of the neutrino source do not limit the precision. Equally precise detectors and an adequately precise theoretical framework for describing the oscillations (including three neutrinos and matter effects) must be used in order to exploit this precision. There exist other interesting sources for long-baseline oscillation experiments, such as reactors and β-beams, but we have restricted the discussion here to superbeams and neutrino factories. We have presented the issues which enter into realistic assessments of the potential of such experiments. The experiments discussed turn out to be very promising and they are expected to lead to very precise measurements of the leading oscillation parameters Δm_{31}^2 and $\sin^2 2\theta_{23}$. We have discussed in detail how the various setups may lead to very interesting measurements of or limits on θ_{13} and δ. It will also be possible to perform impressive tests of effects of the matter in the earth, allowing us to extract $\mathrm{sign}(\Delta m_{31}^2)$. The setups discussed in this chapter have an increasing potential and increasing technological challenges, but it seems possible to build them in stages. The results presented here are valid for each individual setup, and future results should of course be included in the analysis. This would be especially important if more LBL experiments were built, since this leads to optimization strategies depending on previous results. Further results from KamLAND are extremely important and will have considerable impact. KamLAND established that Δm_{21}^2 lies in the LMA regime, which is very important since CP violation can only be measured if that is the case. Within the LMA solution, it is now very important to find out whether Δm_{21}^2 lies close to the current best fit, on the high side, or on the low side. A value of Δm_{21}^2 on the high side (HLMA) would be ideal, since it would guarantee an extremely promising LBL program, with a chance to see leptonic CP violation with the JHF beam in the next decade.

Finally, we would like to stress that the LBL experiments will have a unique impact on physics. These experiments will lead to very precise values of neutrino mass splittings and very precise leptonic mixings. These measurements will yield directly the physics parameters of interest, which are (unlike those of quarks) not masked by any hadronic uncertainties. It should thus be possible to obtain very valuable lepton flavor information, which could be compared directly with models of masses and mixings and with renormalization group effects. Such precise leptonic flavor information might also prove more valuable than the corresponding information in the quark sector, since, in general, neutrino masses receive contributions from both Dirac and Majorana mass terms, and more might be learned in this way. Leptonic CP violation is also an extremely interesting issue, since it is related to leptogenesis, which is currently the most plausible mechanism to explain the baryon asymmetry of the universe. The experiments presented here also allow a number of other studies which have not been discussed here. Some examples are limits on nonstandard interactions, flavor changing neutral currents (FCNC), the possible existence of more than three neutrinos, and CPT violation. The results presented here should, however, make it clear that oscillation physics with long-baseline experiments by itself is very interesting, powerful, and important. The realization of the setups discussed here will not be easy, but it appears possible to do this in stages, guaranteeing a very promising future for neutrino physics.

Acknowledgments

I would especially like to thank M. Freund, P. Huber, and W. Winter for their collaboration in the studies on which this article is based.

References

1. T. Toshito (Super-Kamiokande Collaboration), arXiv:hep-ex/0105023.
2. Q.R. Ahmad et al. (SNO Collaboration), Phys. Rev. Lett. **89**, 011301 (2002); arXiv:nucl-ex/0204008.
3. Q.R. Ahmad et al. (SNO Collaboration), Phys. Rev. Lett. **89**, 011302 (2002); arXiv:nucl-ex/0204009.
4. V. Barger, D. Marfatia, K. Whisnant, and B.P. Wood, Phys. Lett. B **537**, 179 (2002); arXiv:hep-ph/0204253.
5. A. Bandyopadhyay, S. Choubey, S. Goswami, and D.P. Roy, arXiv:hep-ph/0204286.
6. J.N. Bahcall, M.C. Gonzalez-Garcia, and C. Peña-Garay, arXiv:hep-ph/0204314.
7. P.C. de Holanda and A.Yu. Smirnov, arXiv:hep-ph/0205241.
8. M. Apollonio et al. (CHOOZ Collaboration), Phys. Lett. B **466**, 415 (1999); arXiv:hep-ex/9907037.
9. K. Nakamura (K2K Collaboration), Nucl. Phys. Proc. Suppl. **91**, 203 (2001).
10. V. Paolone, Nucl. Phys. Proc. Suppl. **100**, 197 (2001).
11. A. Ereditato, Nucl. Phys. Proc. Suppl. **100**, 200 (2001).
12. P. Huber, M. Lindner, and W. Winter, arXiv:hep-ph/0204352.
13. L. Wolfenstein, Phys. Rev. D **17**, 2369 (1978).

14. L. Wolfenstein, Phys. Rev. D **20**, 2634 (1979).
15. S.P. Mikheev and A.Y. Smirnov, Sov. J. Nucl. Phys. **42**, 913 (1985).
16. S.P. Mikheev and A.Y. Smirnov, Nuovo Cim. C **9**, 17 (1986).
17. S. Antusch, J. Kersten, M. Lindner, and M. Ratz, arXiv:hep-ph/0206078.
18. S. Antusch, J. Kersten, M. Lindner, and M. Ratz, Phys. Lett. B **538**, 87 (2002); arXiv:hep-ph/0203233.
19. M. Fukugita and T. Yanagida, Phys. Lett. B **174**, 45 (1986).
20. W. Buchmuller, arXiv:hep-ph/0107153, and references therein.
21. See e.g. G.C. Branco, R. Gonzalez Felipe, F.R. Joaquim, and M.N. Rebelo, arXiv:hep-ph/0202030.
22. E.A. Paschos, arXiv:hep-ph/0204138.
23. F. Arneodo et al. (ICARUS Collaboration), Nucl. Instrum. Meth. A **471**, 272 (2000).
24. G. Acquistapace et al. (CNGS Collaboration), CERN-98-02 (1998).
25. R. Baldy et al. (CNGS Collaboration), CERN-SL-99-034-DI (1999).
26. J. Hylen et al. (NuMI Collaboration), FERMILAB-TM-2018 (1997).
27. K. Nakamura (K2K Collaboration), Nucl. Phys. A **663**, 795 (2000).
28. Y. Itow et al., arXiv:hep-ex/0106019.
29. M. Aoki, arXiv:hep-ph/0204008.
30. A. Para and M. Szleper, arXiv:hep-ex/0110032.
31. See e.g. J.J. Gomez-Cadenas et al. (CERN Super Beam Working Group), arXiv:hep-ph/0105297.
32. F. Dydak, Technical Report, CERN (2002), http://home.cern.ch/dydak/oscexp.ps.
33. M. Szleper and A. Para, arXiv:hep-ex/0110001.
34. S. Geer, Phys. Rev. D **57**, 6989 (1998); arXiv:hep-ph/9712290.
35. N. Holtkamp and D. Finley, Technical Report, Fermi National Accelerator Laboratory (2002), http://www.fnal.gov/projects/muon_collider/nu-factory/nu-factory.html.
36. D.E. Groom et al. (Particle Data Group), Eur. Phys. J. C **15**, 1 (2000); http://pdg.lbl.gov/.
37. A. Blondel et al., Nucl. Instrum. Meth. A **451**, 102 (2000).
38. C. Albright et al., arXiv:hep-ex/0008064, and references therein.
39. P. Zucchelli, arXiv:hep-ex/0107006.
40. K. Dick, M. Freund, M. Lindner, and A. Romanino, Nucl. Phys. B **562**, 29 (1999); arXiv:hep-ph/9903308.
41. F.D. Stacy, *Physics of the Earth*, 2nd edn. (Wiley, New York, 1977).
42. R.J. Geller and T. Hara, arXiv:hep-ph/0111342.
43. E.K. Akhmedov, Sov. J. Nucl. Phys. **47**, 301 (1988).
44. S.T. Petcov, Phys. Lett. B **434**, 321 (1998); arXiv:hep-ph/9805262.
45. E.K. Akhmedov, Nucl. Phys. B **538**, 25 (1999); arXiv:hep-ph/9805272.
46. M.V. Chizhov and S.T. Petcov, Phys. Rev. Lett. **83**, 1096 (1999); arXiv:hep-ph/9903399.
47. M.V. Chizhov and S.T. Petcov, Phys. Rev. D **63**, 073003 (2001); arXiv:hep-ph/9903424.
48. E.K. Akhmedov, Phys. Atom. Nucl. **64**, 787 (2001) [Yad. Fiz. **64**, 851 (2001)]; arXiv:hep-ph/0008134.
49. A. Cervera et al., Nucl. Phys. B **579**, 17 (2000); Nucl. Phys. B **593**, 731 (2001) (erratum); arXiv:hep-ph/0002108.

50. M. Freund, Phys. Rev. D **64**, 053003 (2001), arXiv:hep-ph/0103300.
51. M. Freund, P. Huber, and M. Lindner, Nucl. Phys. B **615**, 331 (2001); arXiv:hep-ph/0105071.
52. E. Akhmedov, P. Huber, M. Lindner, and T. Ohlsson, Nucl. Phys. B **608**, 394 (2001).
53. E.D. Church, K. Eitel, G.B. Mills, and M. Steidl, Phys. Rev. D **66**, 013001 (2002); arXiv:hep-ex/0203023.
54. E.A. Hawker, Int. J. Mod. Phys. A **16**(S1B), 755 (2001).
55. M. Lindner, T. Ohlsson, and W. Winter, Nucl. Phys. B **607**, 326 (2001); arXiv:hep-ph/0103170.
56. S.M. Bilenky, M. Freund, M. Lindner, T. Ohlsson, and W. Winter, Phys. Rev. D **65**, 073024 (2002).
57. G.L. Fogli and E. Lisi, Phys. Rev. D **54**, 3667 (1996).
58. V. Barger, D. Marfatia, and K. Whisnant, Phys. Rev. D **65**, 073023 (2002); arXiv:hep-ph/0112119.
59. J. Burguet-Castell, M.B. Gavela, J.J. Gomez-Cadenas, P. Hernandez, and O. Mena, Nucl. Phys. B **608**, 301 (2001); arXiv:hep-ph/0103258.
60. H. Minakata and H. Nunokawa, J. High Energy Phys. **10**, 001 (2001); arXiv:hep-ph/0108085.
61. J. Burguet-Castell, M.B. Gavela, J.J. Gomez-Cadenas, P. Hernandez, and O. Mena, arXiv:hep-ph/0207080.
62. V. Barger, D. Marfatia, and K. Whisnant, arXiv:hep-ph/0206038.
63. A. Donini, D. Meloni, and P. Migliozzi, arXiv:hep-ph/0206034.
64. E. Ables et al. (MINOS Collaboration), FERMILAB-PROPOSAL-P-875 (1995).
65. N.Y. Agafonova et al. (MONOLITH Collaboration), LNGS-P26-2000 (2000).
66. A. Cervera, F. Dydak, and J.J. Gomez-Cadenas, Nucl. Instrum. Meth. A **451**, 123 (2000).
67. M.D. Messier, Evidence for neutrino mass from observations of atmospheric neutrinos with Super-Kamiokande, Ph.D. thesis, Boston University (1999).
68. K. Ishihara, Study of $\nu_\mu \to \nu_\tau$ and $\nu_\mu \to \nu_{\text{sterile}}$ neutrino oscillations with the atmospheric neutrino data in Super-Kamiokande, Ph.D. thesis, University of Tokyo (1999).
69. K. Okumura, Observation of atmospheric neutrinos in Super-Kamiokande and a neutrino oscillation analysis, Ph.D. thesis, University of Tokyo (1999).
70. W. Flanagan, A study of atmospheric neutrinos at Super-Kamiokande, Ph.D. thesis, University of Hawaii (1997).
71. V. Barger, D. Marfatia, and B. Wood, Phys. Lett. B **498**, 53 (2001); arXiv:hep-ph/0011251.
72. A. De Rujula, M.B. Gavela, and P. Hernandez, Nucl. Phys. B **547**, 21 (1999); arXiv:hep-ph/9811390.
73. V. Barger, S. Geer, and K. Whisnant, Phys. Rev. D **61**, 053004 (2000); arXiv:hep-ph/9906487.
74. M. Freund, M. Lindner, S.T. Petcov, and A. Romanino, Nucl. Phys. B **578**, 27 (2000); arXiv:hep-ph/9912457.
75. I. Mocioiu and R. Shrock, Phys. Rev. D **62**, 053017 (2000); arXiv:hep-ph/0002149.
76. V. Barger, S. Geer, R. Raja, and K. Whisnant, Phys. Rev. D **62**, 073002 (2000); arXiv:hep-ph/0003184.

77. M. Freund, P. Huber, and M. Lindner, Nucl. Phys. B **585**, 105 (2000); arXiv:hep-ph/0004085.
78. A. Bueno, M. Campanelli, and A. Rubbia, Nucl. Phys. B **589**, 577 (2000); arXiv:hep-ph/0005007.
79. A. Cervera et al., Nucl. Instrum. Meth. A **472**, 403 (2000); arXiv:hep-ph/0007281.
80. V. Barger, S. Geer, R. Raja, and K. Whisnant, Phys. Rev. D **63**, 113011 (2001); arXiv:hep-ph/0012017.
81. M. Campanelli, A. Bueno, and A. Rubbia, Nucl. Instrum. Meth. A **451**, 207 (2000).
82. M. Freund, M. Lindner, S.T. Petcov, and A. Romanino, Nucl. Instrum. Meth. A **451**, 18 (2000).
83. I. Mocioiu and R. Shrock, J. High Energy Phys. **0111**, 050 (2001); arXiv:hep-ph/0106139.
84. O. Yasuda, arXiv:hep-ph/0111172.
85. V. Barger et al., Phys. Rev. D **65**, 053016 (2002); arXiv:hep-ph/0110393.
86. A. Bueno, M. Campanelli, S. Navas-Concha, and A. Rubbia, Nucl. Phys. B **631**, 239 (2002); arXiv:hep-ph/0112297.
87. O. Yasuda, arXiv:hep-ph/0203273, and references therein.
88. G. Barenboim, A. De Gouvea, M. Szleper, and M. Velasco, arXiv:hep-ph/0204208.
89. M.M. Alsharoa et al., arXiv:hep-ex/0207031.
90. P. Huber, talk presented at the Neutrino Factory Working Group Meeting, CERN, 11 June 2002.
91. K. Eguchi et al., Phys. Rev. Lett. **90**, 021802 (2003); arXiv:hep-ex/0212021.

Index

Springer Tracts in Modern Physics

Springer Tracts in Modern Physics

Printing: Saladruck Berlin
Binding: Stürtz AG, Würzburg